AF577064

PLUMBING FUNDAMENTALS

CONTEMPORARY CONSTRUCTION SERIES

CARPENTRY FUNDAMENTALS
BY GLENN BAKER AND REX MILLER

ELECTRICAL WIRING FUNDAMENTALS
BY VOLT INFORMATION SCIENCES, INC.

HEATING, VENTILATING, AND AIR CONDITIONING FUNDAMENTALS
BY RAYMOND HAVRELLA

PLUMBING FUNDAMENTALS
BY JAMES L. THIESSE

READING CONSTRUCTION DRAWINGS
BY PAUL I. WALLACH AND DONALD E. HEPLER

PLUMBING FUNDAMENTALS

CONTEMPORARY CONSTRUCTION SERIES

JAMES L. THIESSE

GREGG DIVISION
McGRAW-HILL BOOK COMPANY

New York / Atlanta / Dallas / St. Louis / San Francisco
Auckland / Bogotá / Guatemala / Hamburg / Johannesburg / Lisbon
London / Madrid / Mexico / Montreal / New Delhi / Panama / Paris
San Juan / São Paulo / Singapore / Sydney / Tokyo / Toronto

SPONSORING EDITOR: CARY BAKER
EDITING SUPERVISOR: KAREN SEKIGUCHI
DESIGN SUPERVISOR: NANCY AXELROD
PRODUCTION SUPERVISOR: S. STEVEN CANARIS
ART SUPERVISOR: GEORGE T. RESCH

TECHNICAL ILLUSTRATOR: VANTAGE ART, INC.
COVER DESIGNER: AMPERSAND STUDIO
PHOTO CREDITS: CHAPTER 1 OPENER, U.S.D.A. SOIL CONSERVATION SERVICE
CHAPTER 3 OPENER, THE BETTMAN ARCHIVE

Library of Congress Cataloging in Publication Data
Thiesse, James L
Plumbing fundamentals.

(Contemporary construction series)
Includes index.
1. Plumbing. I. Title. II. Series.
TH6123.T44 696'.1 80–19905
ISBN 0–07–064191–9

In memory of Leonard A. Thiesse and Fred Cohrs

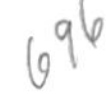

PLUMBING FUNDAMENTALS

1 2 3 4 5 6 7 8 9 0 SMBP 8 9 8 7 6 5 4 3 2 1

ISBN 0-07-064191-9

· CONTENTS ·

· PREFACE ·

Plumbing Fundamentals was designed and written to provide instructional material for secondary-level vocational programs. Some vocational programs are based solely on the plumbing trade, while others deal with plumbing as a portion of a construction cluster. This book was written to serve both of these educational concepts.

The language, figures, and design of the book are based on the skills and needs of students in the tenth, eleventh, and twelfth grades. The tools, fixtures, and other photographic subject matter are taken from real life. They are not new, and most show signs of wear and heavy use. It has been my experience that there are no new, shiny tools on the job.

The text organization guides the reader from basic information to increasingly complex applications. The first three chapters provide a look into some of the history of plumbing, a review of the mathematics of measuring, and a discussion of the need for and the processes of water purification and waste disposal.

The next group of chapters (4 through 9) deals with the tools, fixtures, materials, and common procedures that form a part of every plumber's day. These are the basics of the trade, and while they appear simple enough to master, errors made in these areas are costly both in time and material. Chapter 8 contains sections on oxyacetylene and arc welding—two processes that may or may not be used in the plumbing trade, depending on the local code. This chapter may be included in the vocational plumbing course, or it may be excluded when plumbing is being taught as a portion of the construction cluster. Chapter 9, which covers plumbing accessories and supplemental plumbing systems, explains many of the fixtures and systems that have become part of our daily lives. The knowledge to be gained here is a necessity for today's plumber.

Chapter 10, Drafting Fundamentals, and Chapter 11, Blueprints and Blueprint Reading, provide basic skills in the areas of mechanical drawing interpretation and blueprint reading. These skills are all too often developed without a basic understanding of what formation these two "tools on paper" can supply.

Use of drawings of component parts and of the parts list of complex equipment makes the servicing and repair of that equipment less expensive and time-consuming. The ability to identify a needed part from a drawing is a valuable addition to the plumber's tool box. The misreading and the misinterpretation of a blueprint are two errors that are very costly when corrections must be made. Many of these blueprint errors can be prevented with instruction in the basics of blueprint reading.

Chapter 12 provides the student with some basic information on the various heat distribution systems commonly used in residences today. The information in this chapter may not be valid instructional material for all areas of the country, or for the construction cluster course; and can be used as elective instructional material at the discretion of the instructor.

After studying the material given in this book, the student should have a firm knowledge of those basic skills required of a second-year apprentice. The student will be able to continue study under a journeyman plumber.

Acknowledgments

I would like to acknowledge the help and other assistance received from the following sources in the writing of this book: City of Savannah, Water Pollution Control Plant, Savannah, Georgia; Thiesse Plumbing, Forest Park, Illinois; Sears, Inc., Statesboro, Georgia; Water Department, Village of Forest Park, Illinois; I would also like to express my appreciation to my wife, Jane, for keeping me active on the project.

James L. Thiesse

PLUMBING FUNDAMENTALS

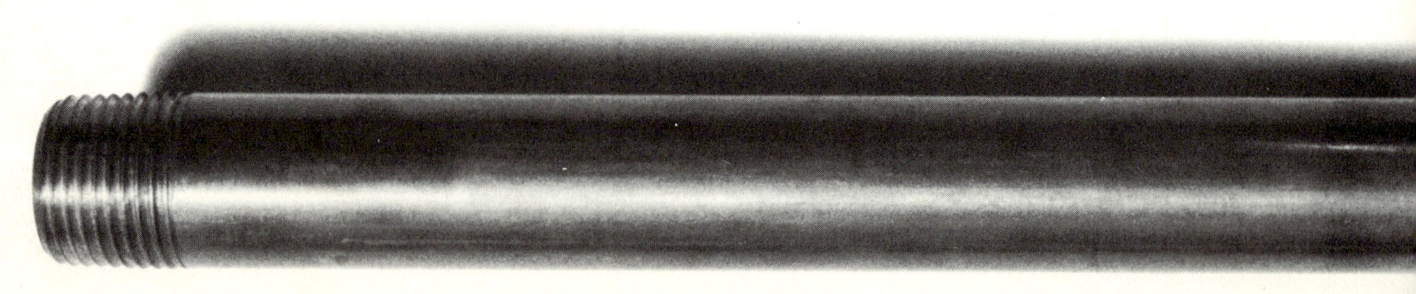

1
PLUMBING AND PLUMBERS

The plumbing system as we know it today, with a pure water supply and provisions for sanitary waste removal in most houses and buildings, is a recent development. Not too many years ago, water was supplied from a well and carried into the house in a bucket or other container. Human waste was disposed of in a pit, and the outhouse was a common sight.

At the completion of this chapter, you will be able to:

- Identify the origin of the word "plumber."
- Discuss early water transportation systems.
- Discuss early waste-disposal systems.
- Identify steps in the development of today's modern plumbing.
- Identify the levels of a plumber's training.
- Identify typical types of work and work settings of the plumbing trade.

• TRANSPORTING WATER •

Most history books have pictures of the ancient Roman aqueducts. An aqueduct is a man-made channel for carrying water. The water supply of ancient Rome was brought into the city through different aqueducts by the year 100 A.D. Over 1800 years ago, the Romans had more than 270 miles of water transportation systems which supplied the entire city with water.

These aqueducts moved the water through tunnels in mountains, and huge stone arches supported the aqueducts over low-lying valleys. One of the later Roman aqueducts (built about 700 A.D.) is about 300 feet high. The Romans also built aqueducts in the countries they conquered. One of these, built in Segovia, Spain, in 109 A.D., is still supplying water to that city today.

After the water was brought to the city through the aqueduct, there had to be a storage area and a distribution system. The storage areas were called reservoirs, and Rome had more than 240 water-storage reservoirs. The water was distributed to the public through water-supply basins, where water was sold as it ran from a lead waterspout. People purchased the water and carried it home for use.

The lead pipes through which the water ran gave plumbing its name. The Roman word for lead is *plumbum,* and a Roman who worked on the lead water-supply pipes was called a *plumbarius.* According to history, these ancient plumbers made their own pipe. There was no manufacturing process to supply the pipe, so each plumber formed a sheet of lead over a round wooden pole or stock and then soldered the edges together.

The Romans were not the only "old culture" to have an aqueduct system, although theirs was by far the largest and best organized. The ancient Greeks built aqueducts in Athens, Samos, and Syracuse as early as 600 B.C. (over 2500 years ago). The system at Syracuse is still in use today. Jerusalem was supplied with water through an aqueduct as early as 700 B.C.

Since those days, aqueducts have been built in almost all countries. The United States has many which furnish water to large cities, such as New York, San Francisco, and Los Angeles, and others which furnish fresh water to the Florida Keys or some desert areas.

• WASTE DISPOSAL •

Historically, waste has been disposed of either by using a waste pit or by pouring it directly into a river, lake, or ocean. Either of these methods is unsanitary and can cause sickness throughout an entire community or city.

The waste pit is generally used with an outhouse, and the waste remains in the bottom of the pit. This waste accumulates, and in time the hole must be filled and the outhouse moved to a new location. The waste in the pit is treated by spreading quicklime (calcium oxide) over it. Quicklime combines with the waste material and speeds up the decaying process as well as killing many germs and bacteria.

Disposal of human waste directly into a river, lake, or ocean pollutes the water with germs, bacteria, and solids, making the water unfit for use by humans or animals. Many of the plagues and sicknesses which killed thousands of people were directly linked to a waste-polluted water supply.

About 4000 years ago, the Minoan civilization was ruled by King Minos from his palace at Knossos in Crete. Recent findings have shown that this palace had a waste-removal system much like the ones we have today. Flowing water removed the waste, and the fixtures were trapped to prevent sewer gases from entering the building and vented so that high pressures would not occur in the sewer. But, after the fall of this civilization, these advances were largely forgotten until Sir John Harrington introduced the water closet (flush toilet) in the sixteenth century.

• PROGRESS •

Throughout history, as cities and towns have grown larger, there has been more demand for water. The ancient civilizations built aqueducts to furnish water for the citizens. The fact that this water came from areas far away kept it from being polluted with the waste of the city. Transportation and treatment of today's water will be covered in Chapter 3.

Waste removal, if done at all, was completed by transportation in large tank vehicles; the untreated waste was then dumped into either a large pit or a natural water source. The severe plagues of history, in which thousands of people died, were partly due to the lack of quick removal and treatment of human waste. Rodents and insects carried germs from the waste and transmitted them to humans. Development of the water closet, waste-removal systems (sewage systems), and waste treatment has done much to remove deadly plagues from our civilization.

If it were not for the modern water-supply and waste-disposal systems, people in today's large cities would fall victim to one of the vicious plagues of historical times.

• THE PLUMBER'S WORK •

As we mentioned above, the "plumbarians" in ancient Rome made their own pipes and installed them so that people could have a water supply close to their homes. Today's plumbers do not have to make their own pipe, but they do many more things than just put pipes together. Plumbers work both inside and outside the building; they work in new buildings and in old buildings installing pipe roads for hot water, cold water, and waste removal. They hook up or install sinks, bathtubs, toilets, water heaters, dishwashers, and a variety of other water-handling and labor-saving devices.

In many communities, plumbers install heating and air-conditioning systems. They may work with steam, hot-water, or hot-air heating systems one day, and the next day they may install or repair air-conditioning systems.

Outside the building, plumbers may work in ditches, connecting large-diameter pipes for a city water or sewer system. They may also work high above ground on a water-storage tank (Figure 1-1). They may work with large water pumps in a pumping station or the city's waterworks. In other words, plumbers do many different things in their day-to-day jobs. (See page 4.)

With the many different jobs plumbers may have to do, you can see that they must be well trained. They must be able to read, understand, and follow a blueprint when working on a new building. They will have to use some mathematics to determine water flow in a pipe or the number of gallons of water it takes to fill a tank or swimming pool. They have to be able to make accurate measurements, and then be just as accurate in cutting and installing the pipe.

Plumbers may be called upon to weld pipe together, so they will have to have some knowledge about gas or electric welding. When plumbers work with heating or air conditioning systems, they will also have to have an understanding of pressure and temperature.

• TRAINING THE PLUMBER •

There are three grades or levels recognized in the plumber's trade: the apprentice plumber, the journeyman plumber, and the master plumber. Each of these grades is based on experience in and knowledge of the trade.

Apprentice Plumber

The apprentice plumber is the would-be plumber who has had little or no experience in the trade and is in training to learn. The apprentice will work daily with a journeyman plumber learning the tasks of the trade in an on-the-job-training situation. The apprentice may also be required to attend classes at the local trade or technical school to receive instruction on blueprint reading, mathematics, welding, and local regulations or "codes."

This apprenticeship period lasts 5 years, and the apprentice is paid a percentage (usually 50 percent to start) of the journeyman's hourly rate. Apprentice pay increases every 6 months throughout the 5 years, so that at the end of the 5-year period, an apprentice is earning nearly the hourly rate of the journeyman. At the end of the 5-year period, the apprentice must take a test which covers the skills and knowledge required to become a journeyman plumber.

The road to becoming a journeyman plumber is most commonly completed by an apprenticeship program, either the Plumbers and Pipefitters Union apprenticeship program or an industry-sponsored apprenticeship program. Some vocational or technical schools also offer courses to prepare the student for a job in the plumbing trades. This may open the door to positions in fields which require a high level of technical education, such as building or plant engineering.

Journeyman Plumber

The journeyman plumber is the second grade level in the plumbing trade. Journeymen are the every-

Figure 1-1. Modern-day water-storage tank

day, working plumbers. They may work for plumbing contractors, industry, cities, states, or the federal government. Journeyman plumbers can remain at this grade for the rest of their lives, and many of them do.

The hourly rate which journeymen earn varies with the part of the country in which they live. In large cities, such as Chicago, they may make as much as $12.00 per hour, while in other areas, their pay may be as low as $7.00 per hour.

A journeyman plumber who wishes to advance to the highest grade in the plumbing trade, the master plumber will have to spend about 6 years as a journeyman and complete another, much harder, skill and knowledge test.

Master Plumber

The master plumber has many years of experience in the trade and is ready to become a supervisor of a plumbing construction crew or to start a business as a plumbing contractor. Many localities require a plumbing contractor to be licensed as a master plumber.

• THE PLUMBERS AND PIPEFITTERS UNION •

The United Association of Journeymen and Apprentices of the Plumbing and Pipefitting Industry, better known as the Plumbers and Pipefitters Union of the United States and Canada, consists of 725 (1968) local unions in the United States and Canada. The organization controls most plumbing construction in the two countries.

The Plumbers and Pipefitters Union was begun to make sure that working plumbers received a fair wage and good working conditions. A person who wishes to work as a plumber is required to become a member of this union. There are three ways to become a member: (1) come in as an apprentice and spend 5 years learning the trade; (2) come in as a journeyman, which requires proof of 5 years' experience in the trade; or (3) be working in a nonunion shop when that shop is organized by the union.

To join the union as an apprentice, you must apply to the local union. Selection of persons for apprenticeships is done at the local union level, and competition for the few openings is keen. Each union has an apprenticeship committee which is made up of local employers and union members. This committee screens the apprentice applicants and selects a certain number for entry into the program. There are few openings and there are many applicants, so some must be disappointed.

The union is the bargaining agent for the plumber. One plumber asking an employer for better or safer working conditions or for better wages might be ignored. But the union, which represents all the plumbers, can ask for the same things, and the employer will listen. This is the principle of unionization for worker protection.

• SUMMARY •

This chapter was an introduction to the plumbing system and the plumbers of today. It covered efforts to provide pure water and sewage disposal from ancient times to the present. It also provided a sketch of the present-day plumber's duties, training, and organizational structure.

• WORDS PLUMBERS USE •

aqueduct
reservoir
unsanitary
sanitary
decaying process
decayed
pollute
diameter
weld
apprentice plumber
journeyman plumber
master plumber
Plumbers and Pipefitters Union

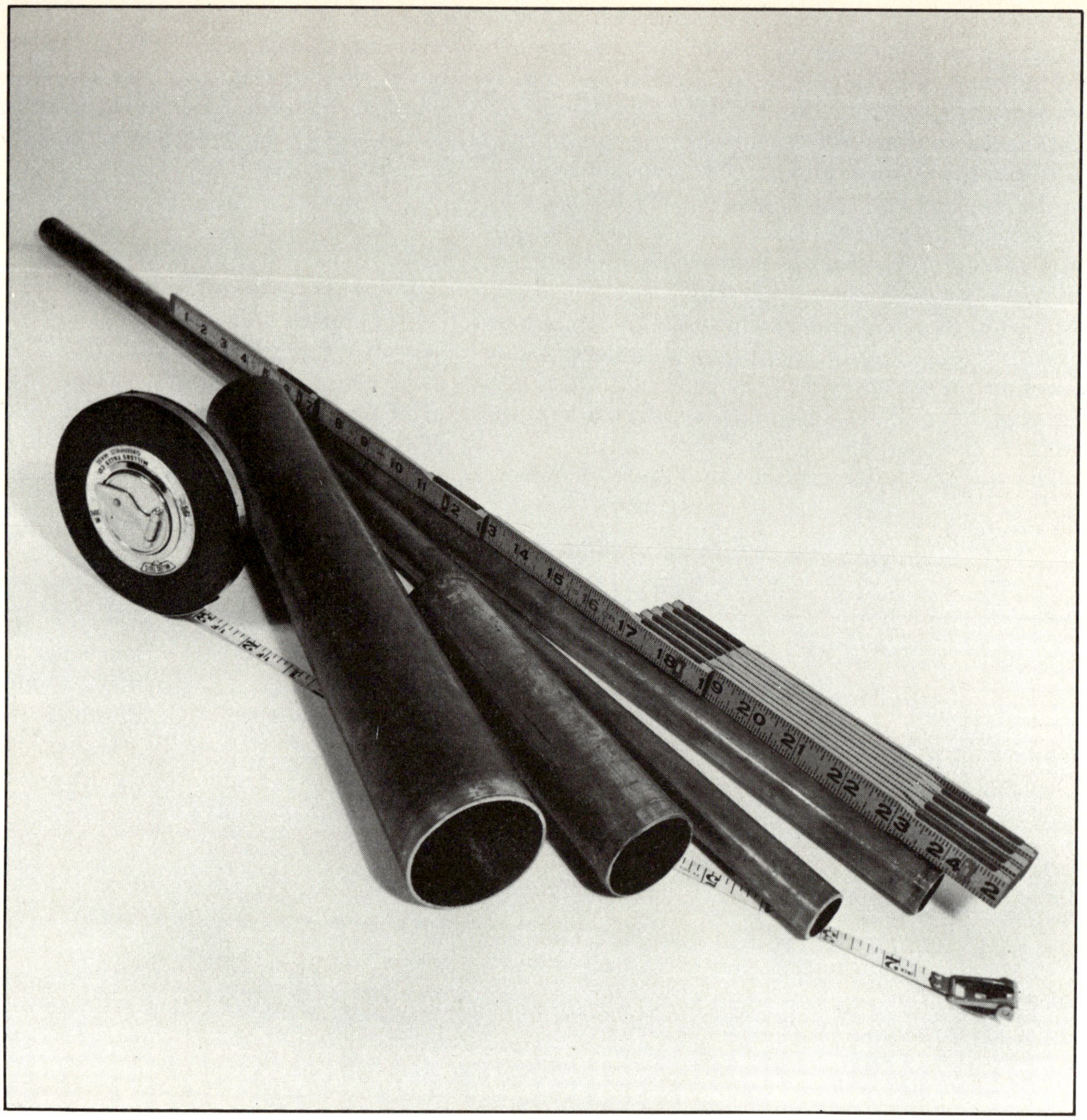

2 PLUMBING MATHEMATICS

There will be many times during your career as a plumber when you will have to use mathematics. You will be required to compute such things as the flow of water through different-sized pipes, the number of gallons of water in a pool or reservoir, the water pressure required for a special project, or the size of a sewer for a large building. You will also have to be able to measure and cut pipe accurately, adding and subtracting distance measurements. This chapter will identify the methods of mathematics used in both the U.S. Customary measuring system and the metric measuring system.

Upon completion of this chapter you will be able to:

- Identify the differences between the U.S. Customary and the metric measuring systems.
- Add, subtract, multiply, and divide measurements accurately.
- Apply formulas and compute answers for the following:
  Area Volume Circumference

• U.S. CUSTOMARY MEASURING SYSTEM MATHEMATICS •

The most common problems in mathematics a plumber faces are adding, subtracting, multiplying, and dividing using the U.S. Customary measuring systems. This section of the chapter will deal with four computations in measurements. For many years, the standard units for length measurements in the United States have been the inch, the foot, the yard, and the mile. The plumber has generally used only the inch and the foot to describe short distances from one point to another. In recent years, many other countries have converted to the metric measurement system, and the United States will convert to the metric system in the future. This chapter will deal with the length problems facing the plumber today and will identify the units in the metric system.

Using the correct procedures when figuring distance measurements will save you work and save the boss money and material. If you make an error in your mathematics, you will have to do extra work to correct it, and the pipe or other material which was wrong will be wasted.

Let us look at Table 2-1 and identify the units of measurement commonly used by a plumber today in the United States. The difficulty in using these units in mathematics is that there are 12 inches in a foot. The following examples show the most common mistake made when adding inches and feet, as well as the correct answer.

TABLE 2-1
MEASUREMENTS

1 foot = 12 inches
1 yard = 3 feet
1 mile = 5280 feet

Examples

Wrong		Right	
2′	8″	2′	8″
+1′	9″	+1′	9″
4′	7″	3′	17″
			−12″
		+1′	
		4′	5″

Wrong		Right	
1′	6″	1′	6″
3′	9″	3′	9″
2′	9″	2′	9″
8′	4″	6′	24″
			−12″
		2′	
		8′	0″

The only difference between the right and the wrong methods shown in the examples is that the right way takes 12 inches from the inch column before carrying over the extra foot. The second example has the same error; the right way carries over 2 feet for 24 inches instead of for 20 inches.

The same error can be made when subtracting distance measurements. The following example identifies the proper way to subtract inches and feet. You must remember that each foot contains 12 inches and that if you "borrow" from the foot column, you must carry 12 to the inches column.

Example

Wrong			Right		
5′	2″		5′	2″	
−2′	4″		−2′	4″	
is changed to			is changed to		
4′			4′		
~~5~~′	12″	(2 + 10)	~~5~~′	14″	(2 + 12)
−2′	4″		−2′	4″	
2′	8″		2′	10″	

The same type of error is easily made when multiplying or dividing distance measurements. The following examples show the proper method of multiplying and dividing inches and feet by a whole number.

Example. Multiply 3′4″ times 4.

Wrong		**Right**	
1		1	
3′	4″	3′	4″
×	4	×	4
12′	16″	12′	16″
or		or	
13′	6″	13′	4″

When you "carry" one foot over to the foot column, you must remember to subtract 12 from the inch column.

Example. Divide 13′4″ by 4.

Wrong		**Right**	
3′	3″ + ½	3′	4″
4)13′	4″	4)13′	4″
	+10		+12
	14″		16″

When you "borrow" one foot from the foot column, you must remember to add 12 to the inch column.

The following exercises are to help you practice the right way of figuring distance measurements. Work each problem and check your answers with your instructor.

• EXERCISES •

Addition

1. 2′ 5″ + 4′ 11″

2. 3′ 11″ + 1′ 6″

3. 10″ + 2′ 2″

4. 4′ 7″ + 9′ 8″ + 1′ 11″

5. 1′ 1″ + 2′ 3″ + 1′ 5″

6. 17′ 2″ + 1′ 10″ + 2′ 11″

Subtraction

7. 9′ 9″ − 2′ 10″

8. 4′ 1″ − 1′ 6″

9. 21′ 10″ − 2′ 9″

10. 6′ 11″ − 5′

11. 8′ 3″ − 11″

12. 2′ 1″ − 10″

Multiplication

13. 2′ 5″ × 6

14. 11′ 7″ × 2

15. 2′ 9″ × 5

16. 1′ 9″ × 4

17. 4′ 11″ × 2

18. 3′ 3″ × 3

Division

19. 21′ 4″ ÷ 4

20. 9′ 6″ ÷ 2

21. 5′ 3″ ÷ 3

22. 19′ 6″ ÷ 6

23. 11′ 8″ ÷ 5

24. 5′ 3″ ÷ 2

Before we begin working some of the mathematics problems a plumber may face on the job, it is important that we all know some of the symbols we will be using. The letters you will be using in some of the problems are symbols; that means they "stand for" or "mean" something.

The first three symbols we will look at use the small letters *l*, *w*, and *h*. *l* stands for length, *w* for width, and *h* for height. By using these three symbols, we can describe the size of almost anything.

The first letter we will use is the small letter *l*. *l* is the symbol for length, or how long something is. A 50-foot mobile home would be shown as $l =$ 50 feet, or the length of a piece of pipe 12′6″ long would be shown as $l =$ 12′6″.

The second letter, *w*, is the symbol for width. This tells us how wide something is. That 50-foot mobile home is 12 feet wide, and this would be shown as $w =$ 12 feet.

The third letter we will use is the small letter *h*. This letter is the symbol for height, or how high something is. If that 50-foot-long, 12-foot-wide mobile home were 11 feet high, this would be shown as $w =$ 11 feet.

Now that we have seen what the symbols for length, width, and height are, and how they are used, look at Figure 2-1 and identify the *l*, *w*, and *h* of each one.

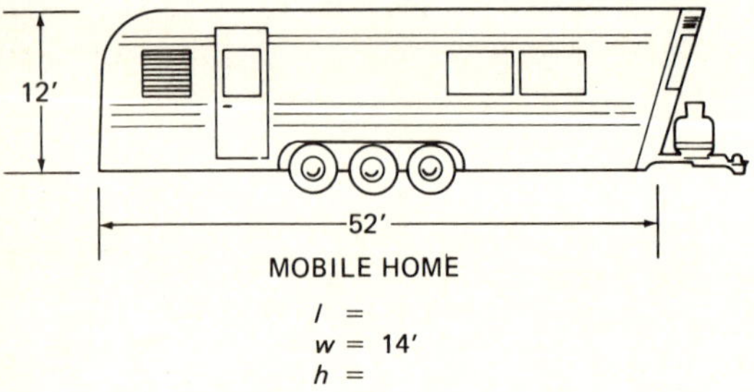

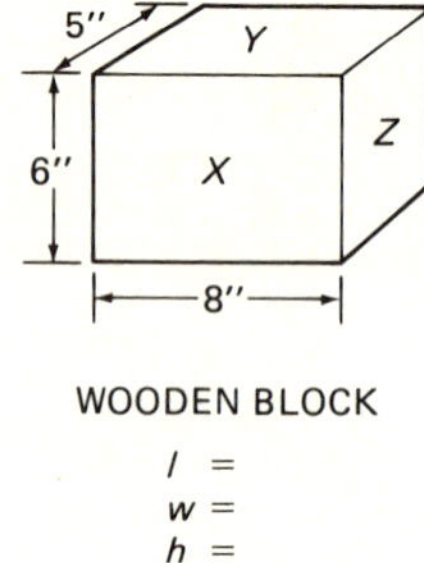

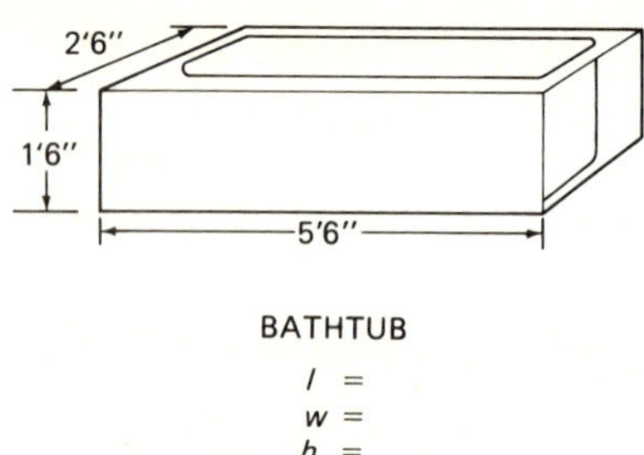

Figure 2-1. Length, width, and height problems

• AREA *(A)* •

The three symbols you have just used are sometimes used to find other values which are important to a plumber. These other values are the area and the volume. The area is the number of square inches or square feet that the side, the bottom, or the front of an object would take up. The symbol for area is a capital *A*. To figure the area of an object, we multiply the size of two edges of the object. For instance, if we wanted to know the area of side *X* of the wooden block in Figure 2-1, we would multiply the length by the height, and the formula would be

$$A = l \times h$$

If we wanted to find the area of side *Y*, the formula would be

$$A = l \times w$$

and the formula for the area of side *Z* would be

$$A = h \times w$$

Remember, the area uses only two sides of the object, the $l \times h$, the $l \times w$, or the $h \times w$, and the answer is always in square inches. Look at Figure 2-2 and follow through the steps as we work some area problems.

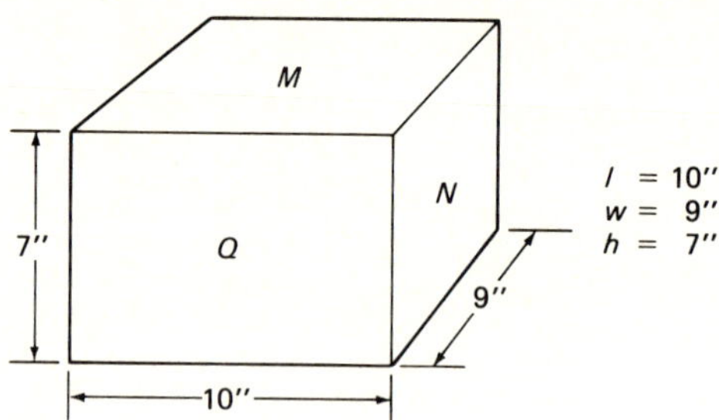

Figure 2-2. Figuring the area of an object

Side *Q*: $A = l \times h$
$A = 10 \times h$
$A = 10 \times 7$
$A = 70$ square inches

Side *M*: $A = l \times w$
$A = 10 \times w$
$A = 10 \times 9$
$A = 90$ square inches

Side *N*: $A = h \times w$
$A = 7 \times w$
$A = 7 \times 9$
$A = 63$ square inches

Which of these three sides has the largest area?

Now return to Figure 2-1 and figure the areas for the mobile home and the bathtub.

• VOLUME *(V)* •

The next very important value is the volume; this is the amount of space inside an object. This object could be a water tank, a swimming pool, a water reservoir, a septic tank, or any one of the many items a plumber works with. Knowing the volume of these objects is important so that the plumber will know how much water or sewage each object can hold.

The symbol for volume is a capital *V*, and the answer is always in cubic feet or cubic inches. The formula to figure it uses all the symbols *l*, *w*, and *h* and is much simpler to use than the area formula. The volume formula is

$$V = l \times w \times h$$

Look at Figure 2-3 and follow through as we work a volume problem.

To find the volume in the tank, use the formula $V = l \times w \times h$:

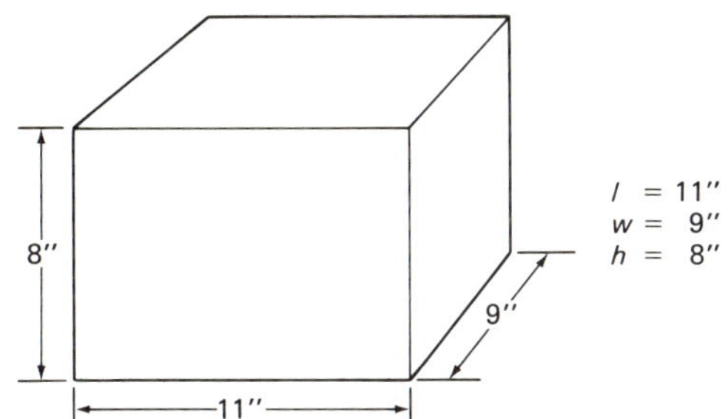

Figure 2-3. Figuring volume $l \times w \times h$

$V = l \times w \times h$
$V = 11 \times w \times h$
$V = 11 \times 9 \times h$
$V = 11 \times 9 \times 8$
$V = 99 \times 8$
$V = 792$ cubic inches

Many of the products with which a plumber works are round: pipes, tanks, and so forth. In many cases, the plumber must decide how large a pipe or tank to use so that the water flow (through the pipe) or the amount of water being stored (in the tank) will be adequate. Just as length *(l)*, width *(w)*, and height *(h)* are used to describe the size of a box-shaped object, similar symbols describe a round or cylindrical-shaped object.

If we look at the end of a piece of pipe or the end of a round tank, we see a circle. The circle we see has two symbols which describe its size. The small letter *r* is the symbol for the radius of the circle. The radius *(r)* is the distance from the center to the edge of the circle. (See Figure 2-4.) If the distance from the center to the outside of the circle were 1 inch, the symbol would be shown as $r = 1$ inch.

Another symbol for a circle is the small letter *d*. This stands for the diameter of the circle. The diameter *(d)* of the circle is the distance covered by a straight line from one side of the circle to the other which passes through the center of the circle. (See Figure 2-5.) If the distance from one side of the circle to the other side through the center is 2 inches, then the symbol would be shown as $d = 2$ inches. Many people are confused by the difference between *r* and *d*, but it is really very simple, as you can see in Figure 2-6. Just remember:

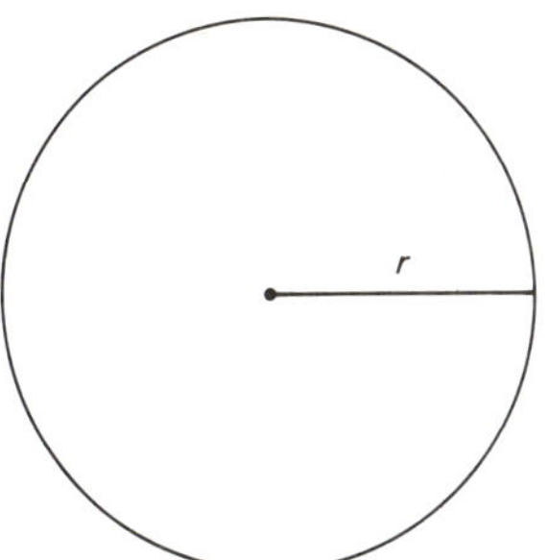

Figure 2-4. The radius of a circle

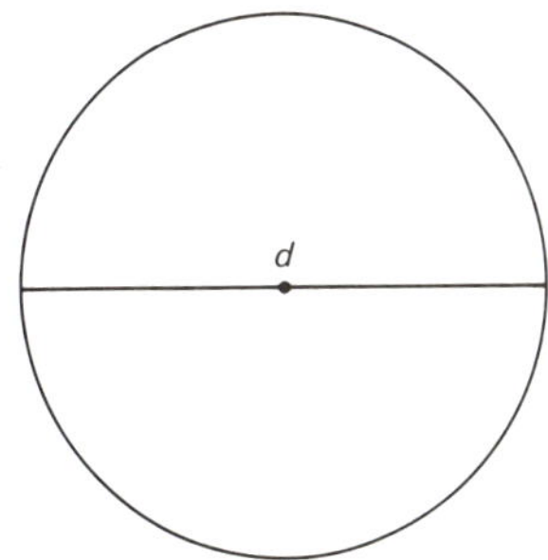

Figure 2-5. The diameter of a circle

$2 \times r = d$ and $d \div 2 = r$

There is one other symbol that is used when people work problems with circles. This symbol is called pi (π), and it is really a Greek letter. The symbol pi (π) is used to find two measurements of a circle, and it stands for the number 3.14.

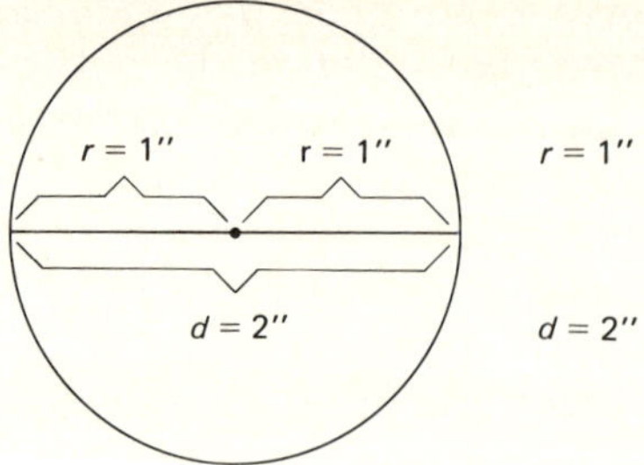

The radius (r) on one side of the circle is 1 inch. The radius (r) on the other side is 1 inch. The diameter (d) is 2 inches.

2 times the radius of any circle is equal to the diameter of that circle.

$$2 \times r = d$$

One-half the diameter of a circle equals the radius (r) of that circle.

$$\frac{d}{2} = r$$

Figure 2-6. Measuring a circle, radius, and diameter

• CIRCUMFERENCE •

The circumference of a circle (Figure 2-7) is the distance from any point on the circle all the way around the circle, and its symbol is the capital letter *C*. It is important for a plumber to know the circumference of a pipe or tank so that enough insulation can be ordered to cover the object.

The formula for finding the circumference *(C)* of a circle is

$$C = \pi \times d$$

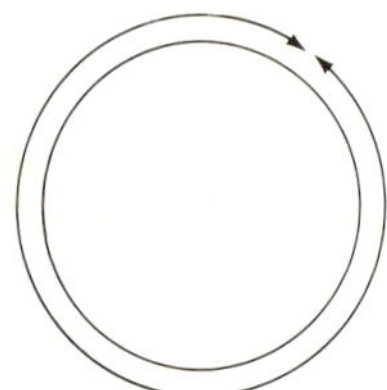

Figure 2-7. Circumference of a circle

(circumference equals pi times diameter). Figure 2-8 shows how the formula for the circumference of a circle is used.

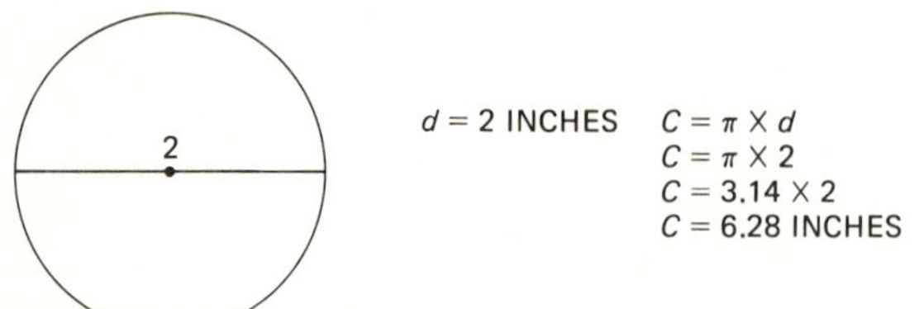

Figure 2-8. Finding the circumference of a circle

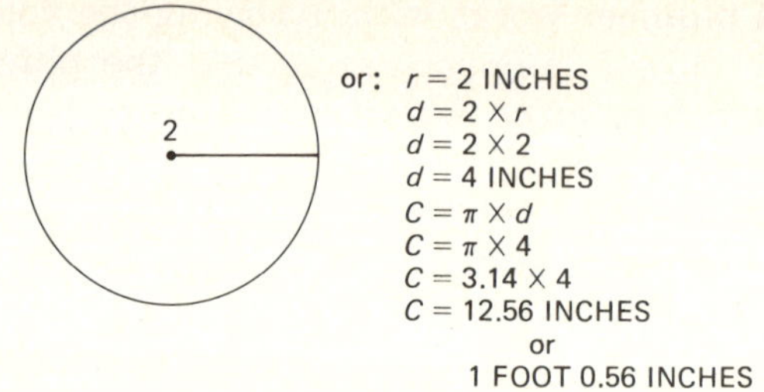

CIRCUMFERENCE IS ALWAYS IN LENGTH UNITS, SUCH AS INCHES OR FEET

Figure 2-8. (cont'd.)

• AREA •

Just as we had an area in square inches for each side of a box, we also have an area for a circle, and the symbol π helps us to find that area. The formula for the area of a circle is

$$A = \pi \times r^2$$

(area equals pi times r squared). The small 2 above the r means that r is multiplied by r, and the formula could be written as

$$A = \pi \times r \times r$$

We will use r^2 (r squared) because it is more generally used. Figure 2-9 shows how the formula for the area of a circle is used.

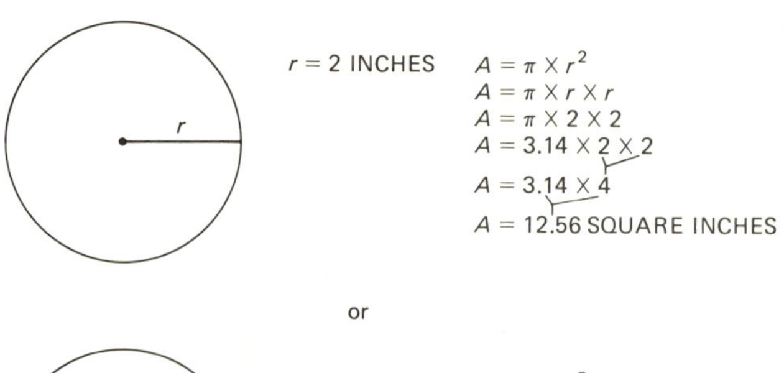

or

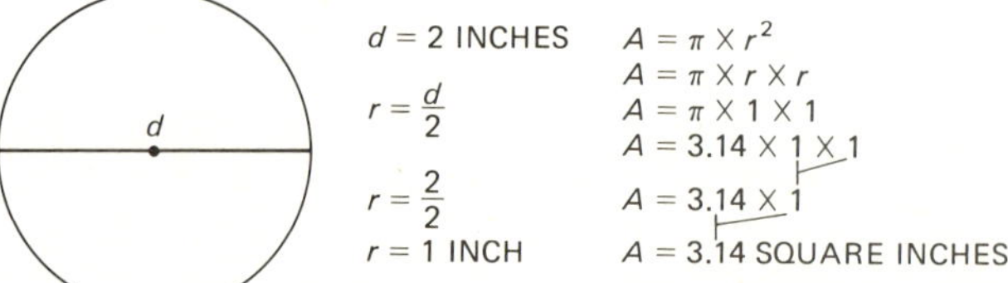

AREA IS ALWAYS IN SQUARE UNITS

Figure 2-9. Finding the area of a circle

• VOLUME OF A CYLINDER •

Earlier we worked on figuring the volume of a rectangle to find the number of cubic feet or inches. These figures were needed to determine how much room we would have for water or sewage storage. We are going to do the same thing for a cylindrical tank.

A cylindrical tank has a radius *(r)*, a diameter *(d)*, and a height *(h)*. You now know that the radius *(r)* and the diameter *(d)* are the measurements for the circular part of the tank. The height *(h)* describes how high the tank is. The formula for the volume of a cylinder is

$$V = \pi \times r^2 \times h$$

(volume equals pi times radius squared times height). You should recognize the first part of this formula; it is the same as the formula for the area of a circle. Figure 2-10 will show you how the formula for the volume of a cylinder is used.

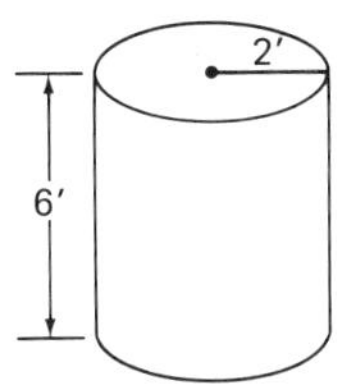

$r = 2$ FEET
$h = 6$ FEET

$V = \pi \times r^2 \times h$
$V = \pi \times r \times r \times h$
$V = \pi \times 2 \times 2 \times 6$
$V = 3.14 \times 2 \times 2 \times 6$
$V = 3.14 \times 2 \times 12$
$V = 3.14 \times 24$
$V = 75.36$ CUBIC FEET

or

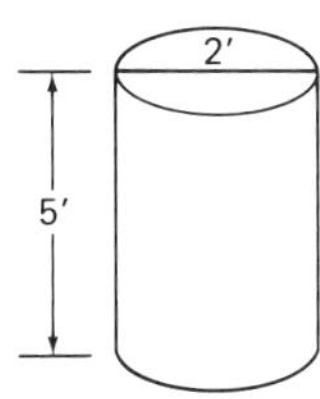

$d = 2$ FEET
$r = \frac{d}{2}$
$r = \frac{2}{2}$
$r = 1$ FOOT
$h = 5$ FEET

$V = \pi \times r^2 \times h$
$V = \pi \times r \times r \times h$
$V = \pi \times 1 \times 1 \times 5$
$V = 3.14 \times 1 \times 1 \times 5$
$V = 3.14 \times 1 \times 5$
$V = 3.14 \times 5$
$V = 15.70$ CUBIC FEET

VOLUME IS ALWAYS IN CUBIC UNITS

Figure 2-10. Finding the volume of a cylinder

• EXERCISES •

1. Find the area of side x and the volume of the rectangular tank shown.

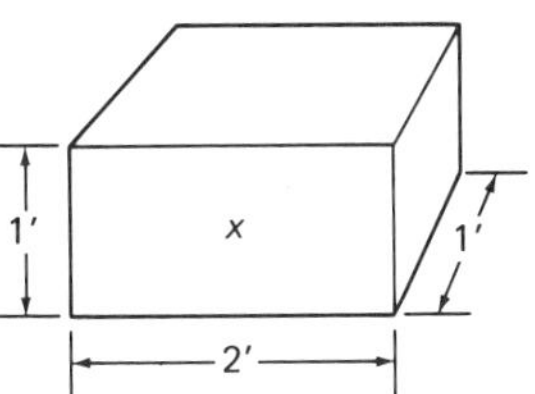

2. Find the area of side z and the volume of the square shown.

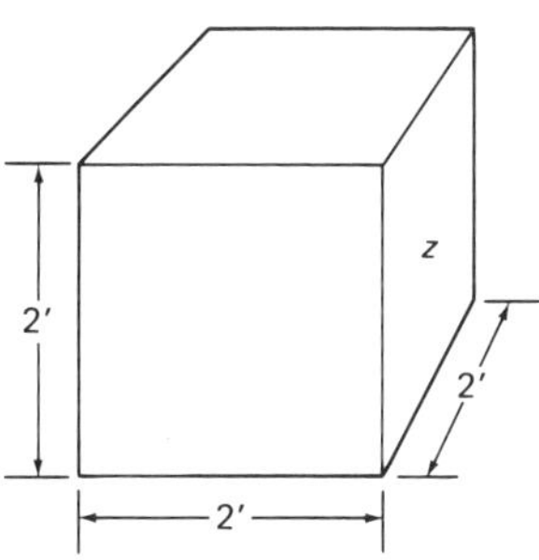

3. Find the circumference of the circle.

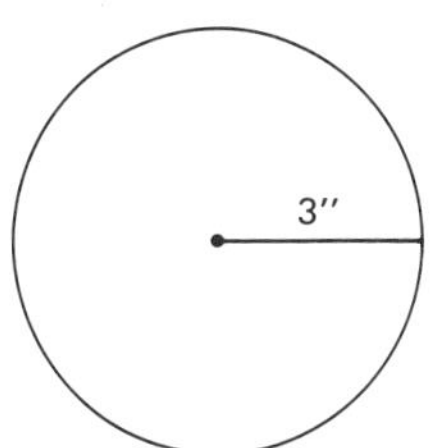

4. Find the volume of the cylinder shown.

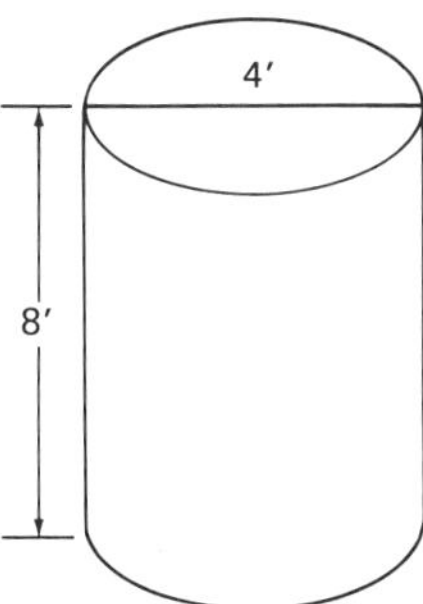

You probably have never thought of it, but water is quite heavy. A square box, 1 foot on each side, filled with water would weigh over 60 pounds. A hot-water heater with a tank 2 feet in diameter (d = 2 feet or r = 1 foot) with a height of 5 feet (h = 5 feet) would weigh about 1000 pounds. Each gallon of water weighs about 7 pounds, so 100 gallons of water in a tank would weigh 700 pounds (7 × 100).

This becomes important to a plumber because sometimes a customer wants a large water tank or heater located on a weak floor area. Although most floors are built to carry a 30- or 40-gallon hot-water heater or tank almost anywhere, some could not hold up under the load of a 100-gallon tank. The customer would be very upset if a misplaced 100-gallon tank broke through the floor.

• THE METRIC MEASURING SYSTEM •

The metric measuring system has been used for many years in other countries of the world. Length measurements in this system are based on the meter (m), and a meter is about 39.5 inches.

Lengths based on the meter (m) are much easier to use than those which are based on the foot. The metric system was designed so that each unit is ten times or one-tenth the size of the next. The problem is that we must learn different words to describe the length measurements.

The largest length measurement normally used in the metric system is the kilometer (km). Anytime the word kilo is used it means one thousand (1000). Therefore, a kilometer (km) means one thousand (1000) meters (m).

The next length measurement normally used is the meter. As you read earlier, this is the basic length measurement of this system. All the other length measurements are based on it. The next measurement is called the decimeter (dm). The word deci means one-tenth (1/10) of something, so a decimeter would be one-tenth of a meter. Ten decimeters (dm) equals one meter (m).

The next smaller measurement is the centimeter (cm). The word centi means one one-hundredth (1/100) of something; therefore, a centimeter (cm) would be one one-hundredth (1/100) of a meter (m). The two parts of the metric system which a plumber would normally use are the centimeter (cm) and the meter (m).

Let us go over these two again. The meter (m) is the basic measurement of the system and is about 39.5 inches long. A centimeter (cm) is one one-hundredth (1/100) of a meter; one hundred centimeters (cm) equals one meter (m). Look at Figure 2-11. It shows the relationship between these measurements.

Metric system mathematics is easier because all the measurements are based on sets of tens, hundreds, or thousands. Look at the following examples and check the addition and subtraction of metric system measurements. Notice that there is no need to worry about an odd number like 12 inches.

1 KILOMETER (km) = 1000 METERS (m)
1 METER (m) = 10 DECIMETERS (dm)
1 DECIMETER (dm) = 10 CENTIMETERS (cm)
1 METER (m) = 100 CENTIMETERS (cm)

Figure 2-11. Metric-system length measurements

Examples

Addition

5 m	65 cm
+ 3 m	40 cm
9 m	05 cm
	52 cm
+ 1 m	83 cm
2 m	35 cm

Subtraction

3 m	122 cm
− 1 m	63 cm
2 m	59 cm
1 m	83 cm
−	47 cm
1 m	36 cm

Another way of doing the same problem is to change all the measurements into centimeters and add or subtract them. Check the following example for a simple way of combining these metric measurements using three or more columns.

Example

5 m 25 cm = 525 cm
1 m 76 cm = 176 cm

(m)	(cm)
5	25
+ 1	76
7	01

or 7 m 01 cm

(m)	(cm)
5	25
1	76

(m)	(cm)
4 ~~5~~	25
− 1	76
3	49

or 3 m 49 cm

Multiplication and division in the metric system are done in the same way. See the example.

Example

Division

(m) (cm)
2 m 44 cm ÷ 2 = 2 44 ÷ 2
or 2)2 44 or 1 m 22 cm
5 m 36 cm ÷ 4 = 5 36 ÷ 4
or 4)5 36 or 1 m 34 cm
1 m 15 cm ÷ 5 = 1 15 ÷ 5
or 5)1 15 or 23 cm

Multiplication

```
                  (m) (cm)
2 m  44 cm × 2 or  2   44  × 2
                 or 2   44
                      × 2
                    4   88 or 4 m 88 cm
5 m  36 cm × 4 or  5   36  × 4
                 or  5  36
                      × 4
                   21   44 or 21 m 44 cm
```

• SUMMARY •

In this chapter, you were introduced to the two measuring systems used today. One system uses feet and inches to describe the distance from one point to another, while the other uses meters and centimeters. Using either of these systems may require addition, subtraction, multiplication, or division of the measurements. This chapter identified the procedures to be used when using mathematics within either of the measuring systems.

• WORDS PLUMBERS USE •

U.S. Customary measuring system	area	diameter	cylinder
metric measuring system	volume	pi (π)	meter
symbol	radius	circumference	centimeter

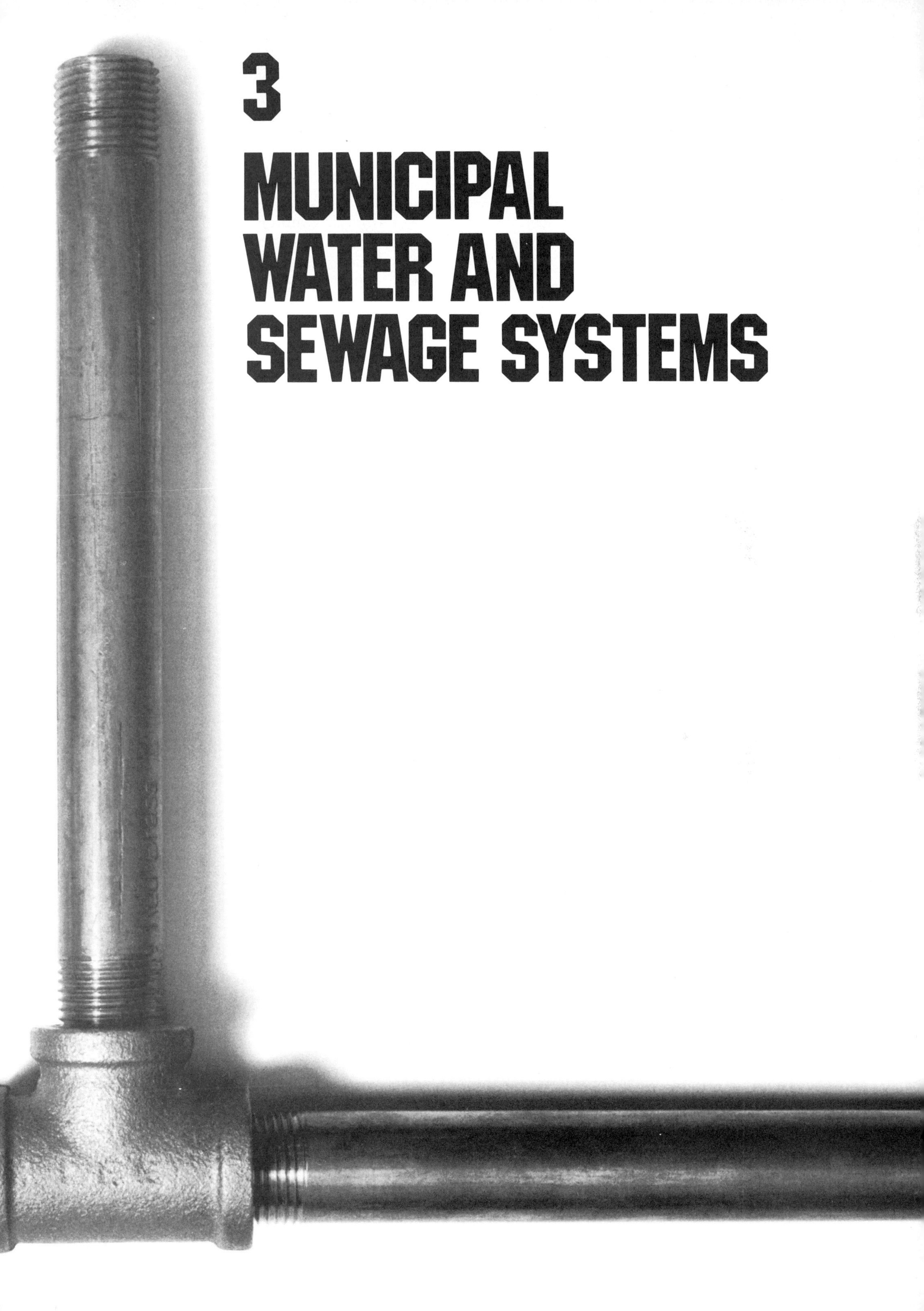

3

MUNICIPAL WATER AND SEWAGE SYSTEMS

There are different ways of supplying water to a building or removing sewage (waste). Not too many years ago, the main supply points were a lake, a river, or a well, and there was no sewage or human-waste removal. None of the supply points actually brought water into the building through pipes. The water was carried from the supply point to the building in containers, and the sewage was disposed of in pits.

Today, most cities and towns have a system of underground pipes which supply water to each of the buildings. This system is normally owned and operated by the city, and each building occupant is charged for the amount of water used. A similar system of pipes, called sewerage, removes the waste materials from the building.

At the completion of this chapter, you will be able to:

- Identify the parts of a basic municipal water-supply and sewage-treatment system.
- Identify the different types of water supplies.
- Name the different methods of water and sewage treatment.
- Identify the pumps used in water-supply and sewerage systems.

• PURPOSE •

The purpose of this chapter is to introduce you to the different types of municipal water and sewage-treatment systems. We will investigate the different types of water supplies, the treatment of water and sewage, the types of storage facilities used, the system control devices, and the underground water-distribution and sewage-collection system.

• SUPPLIES •

Most municipal water systems furnish water to many thousands of people, maybe millions of people if the system is for a city as large as New York or Chicago. For this reason, the system needs a large water supply to meet the needs of the people.

Have you ever noticed that most of the large cities in the United States are near a big river or lake? This happened because of water transportation, not because of the water supply.

You can easily see that a city located next to a big river or lake would not have much trouble getting a large amount of water for its people. It would be a matter of running some large pipes out into the water and then pumping the water into the city system. This is the method many cities use as their primary method of getting fresh water. Figure 3-1 shows a typical fresh-water supply point.

There are other large cities, however, that do not have this ready supply of fresh water. They have a different supply point. Some, like the ancient Romans, pipe their water in from reservoirs many miles away. Others drill many large, deep wells and pump out the water they need. The city of Miami in Florida is surrounded by salt water and supplies its people with fresh water from large wells like this. Figure 3-2 shows a picture of a typical water-system well.

Very few municipal systems rely on a single supply source; most use a combination of different sources.

Figure 3-2. Water-system well

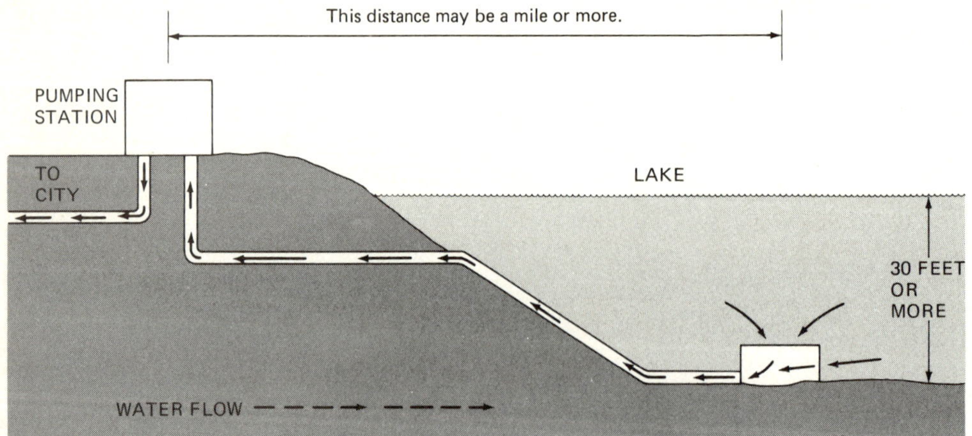

Figure 3-1. Lake-fed fresh-water supply point

• PUMPS •

Since water is a liquid and the law of gravity makes it impossible for a liquid to flow uphill without some help, water needs some mechanical help to move from a lower area to a higher area. In most cases, the help is supplied by a pump. A centrifugal pump is shown in Figure 3-3.

Figure 3-3. Large water pump used in a municipal pumping station

Suction Pump

The suction pump is generally used for shallow wells in places where large amounts of water are not needed. It is the type of pump which would be seen outside an old farmhouse or used with a windmill. Figure 3-4 is a cutaway drawing of a hand-operated suction pump.

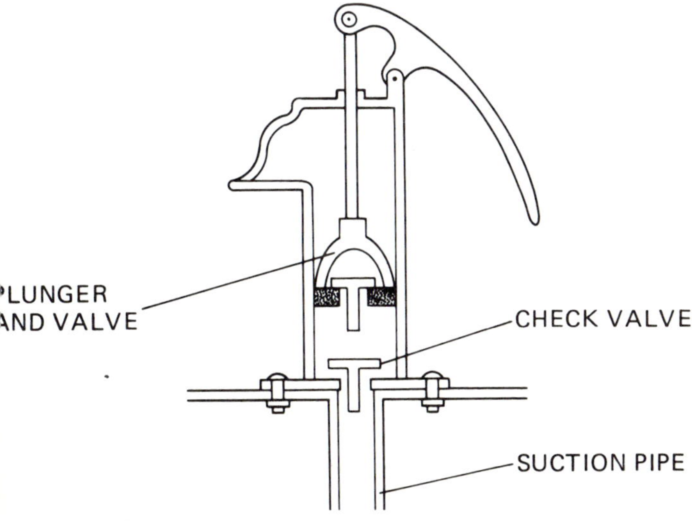

Figure 3-4. Hand-operated suction pump

This pump can only bring up about 5 gallons of water per minute. You can see that a large municipal system which used millions of gallons of water per day could not use this type. However, many of these pumps are in use on farms and ranches, since they can supply all the water needed there.

Centrifugal Pump

The centrifugal pump is a relatively low-cost pump that can deliver large amounts of water efficiently. The centrifugal pump is so named because the main or pumping section (called the impeller) spins around very rapidly and forces the water or other liquid out by centrifugal force. The impeller is connected directly to a motor and is mounted in a housing. Figure 3-5 is a drawing of an impeller and its housing.

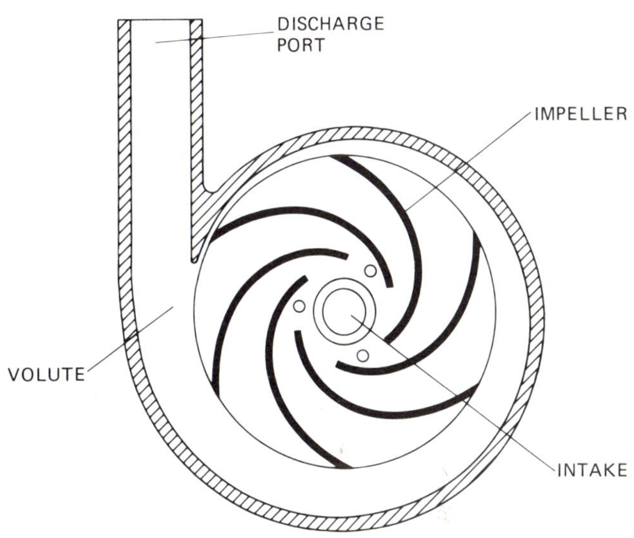

Figure 3-5. Main operating parts of a centrifugal pump

You can see that the impeller is not mounted in the center of the housing. The channel formed by the offset mounting is called the *volute,* and it allows the water to flow around the impeller and out the discharge point. The intake for the pump is at the center of the impeller. As the water or other fluid comes through the intake, the impeller spins it out to the volute and the discharge point.

The drawing in Figure 3-5 is a very simple one, but the same principle applies to the large centrifugal pumps used in municipal water systems, like the one shown in Figure 3-3.

Jet Well Pump

The jet well pump is used with a centrifugal pump. It brings the water from the well up to the inlet

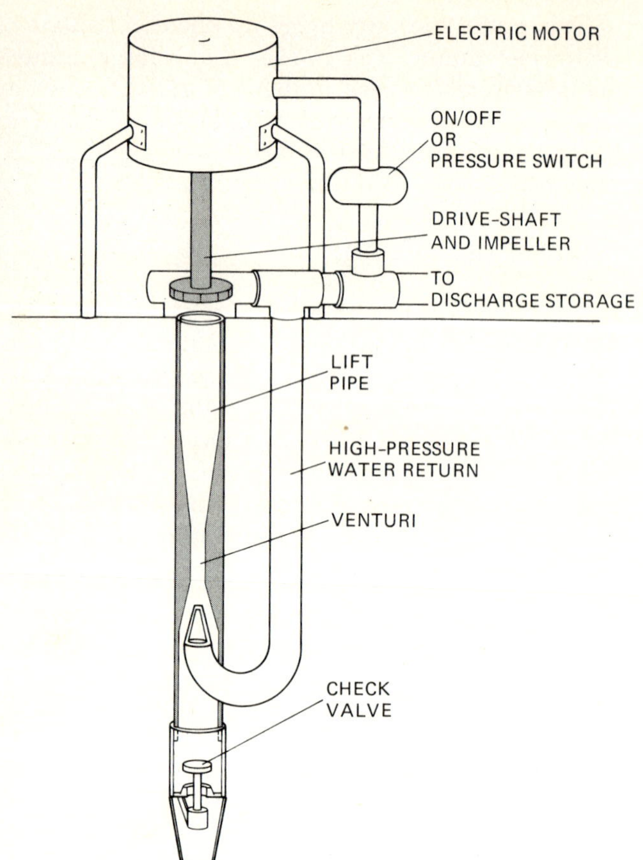

Figure 3-6. Jet well pump

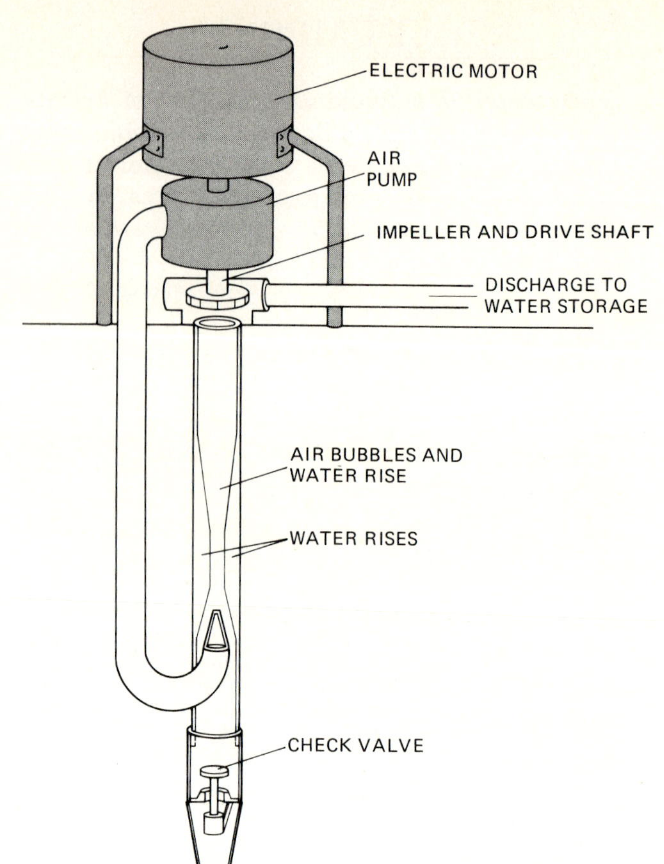

Figure 3-7. Air-lift pump

of the centrifugal pump. Figure 3-6 is a drawing of a jet pump.

The discharge port of a centrifugal pump at the top of the well has two discharge pipes. One goes to the storage tank or water-treatment area and supplies water under pressure. The other goes to the bottom of the well, and the stream of water from this pipe is directed back up toward the top of the well. As this water flows up, it brings some of the surrounding water up to the inlet of the centrifugal pump, and the cycle begins again.

Air-Lift Pump

An air-lift pump works in much the same way as a jet pump. A stream of air is blown into the water at the bottom of the well, and the water is carried upward with the air bubbles. There is a slight difference in the construction of the well, though. Two pipes, one inside of the other, are used. The water with the air bubbles rises in the center pipe, while the outside pipe only has water. A drawing of an air-lift pump is shown in Figure 3-7.

The air-lift pump just brings the water up to the intake of a second pump. Most of the time, this second pump is a centrifugal pump which moves the water from the top of the well to a storage tank or into the water-treatment area.

• WATER STORAGE •

After the water is brought from the source area, some systems must have a place to store it. Most smaller municipal systems use a large tank on the top of a high tower. Water can be stored here before or after treatment. This method of storage not only provides a certain amount of water in case of an emergency, it also provides a source of pressure. Figure 3-8 shows a water-storage tank. Water pressure is required to move the water through the pipes and out of the faucet or other control device at the end. There are many different ways of getting water pressure in a municipal system, but the easiest is to raise the storage tank above the ground.

Other systems use a storage area which looks like a big swimming pool. This type of storage is generally used when the water supply is very reliable, that is, when there is little or no chance of pump or other failure in the supply of water. Figure 3-9 shows a picture of this type of water-storage area.

Another method of water storage is the use of a reservoir. A reservoir is an artificial lake; it can be made by building a dam across a river or stream. This method is normally used in an area of the country which has hills or mountains. Figure 3-10 is a picture of a water-system reservoir.

Figure 3-8. Water-storage tank

Figure 3-9. Indoor, ground-level water-storage

Figure 3-10. Outdoor water reservoir

• WATER TREATMENT •

You have probably visited a lake or a stream and noticed that the water looked muddy or dirty. This is because some of the surrounding dirt had been carried into the lake while it was mixed with the water. This material mixed with the water is called sediment. There are many other materials which the water also carried but which you could not see. These other materials were dissolved in the water the same way sugar is dissolved in coffee or tea. Iron, calcium, lead, and magnesium are just a few materials which may be dissolved in water.

Germs and bacteria which are dangerous to our health can also be carried by water. The water which is supplied through the municipal system must be free of these germs and bacteria to prevent sickness in the community. It should also be cleaned of the sediment and the dissolved materials so it will look and taste good. The reasons for the treatment of water in the municipal system are to supply clean and healthful water to the community.

• CLEANING THE WATER •

There are many methods of cleaning the water of the sediment and dissolved materials. Two which we will discuss are sedimentation and filtration. Sedimentation is allowing the water to remain in a pool so that the sediment settles to the bottom. Your instructor can give you a demonstration of this. Filtration is passing the water through a series of materials which will catch and hold the mixed solids, the dissolved solids, and many germs and bacteria.

A drawing of a filter is shown in Figure 3-11.

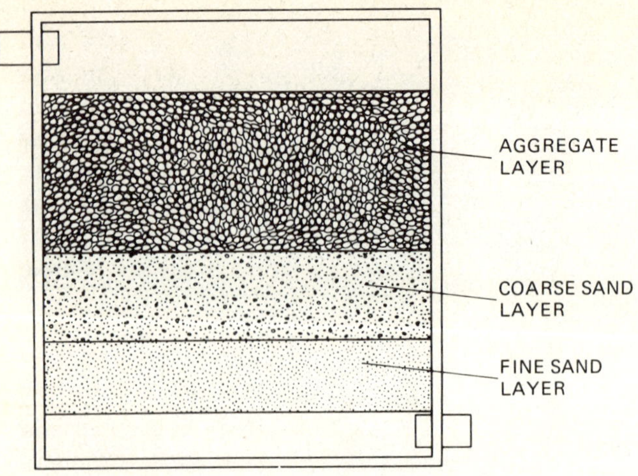

Figure 3-11. Simple water filter

• WATER PURIFICATION •

Many of the germs or bacteria will be removed through the filtration process, but in order to be sure that they are all removed, the water is treated again. This final purification treatment is the addition of a chemical which will eliminate the remaining germs and bacteria so that the water can be safely used.

Two of the chemicals used for this final purification step are chlorine and ozone. The one used most often is chlorine. Chlorine is a very dangerous chemical, and it must be handled with care. It is normally supplied to the water-treatment plant as a gas or a solid (powder). The chlorine gas is very poisonous and is connected directly to a special machine that regulates the flow of the gas and supplies the amount needed to the water supply. Figure 3-12 is a picture of an automatic chlorine-gas regulator.

Figure 3-12. Automatic chlorine-gas regulator

The solid chlorine or powder must be handled with care, for it can be dangerous if it gets into the eyes or mouth, even though it forms poisonous gas only when it is mixed with water and not controlled. The water-treatment plant in a municipal system which uses this chlorine will also use a special machine to mix the chlorine and supply it to the water.

• WATER DISTRIBUTION •

Treatment of the water is not the last responsibility of the municipal water system. The water still must be transported to the houses and businesses of the community. Remember, we heard earlier that water must be pumped upward, but it will flow downward all by itself.

Many towns and cities use this downward flow to supply water to the community. After the water is treated, it is pumped up and stored in a water tower. The level of the water inside the tower is controlled automatically so that as water is used and the level goes down, the pumps send more water up and refill the tower.

The water in the tower, high above the houses and buildings of the town, provides enough pressure to supply water to the community. Figure 3-13 shows a picture of a water tower. Other cities and towns use pumps to supply water under pressure to their houses and businesses.

Figure 3-13. Water-storage tank

The last part of the municipal water system is the series of large underground pipes, called water mains, which bring the water to each house or building. These water mains are normally underneath the streets and have connections to each of the houses or buildings of the town. Figure 3-14 shows a diagram of an entire municipal water system.

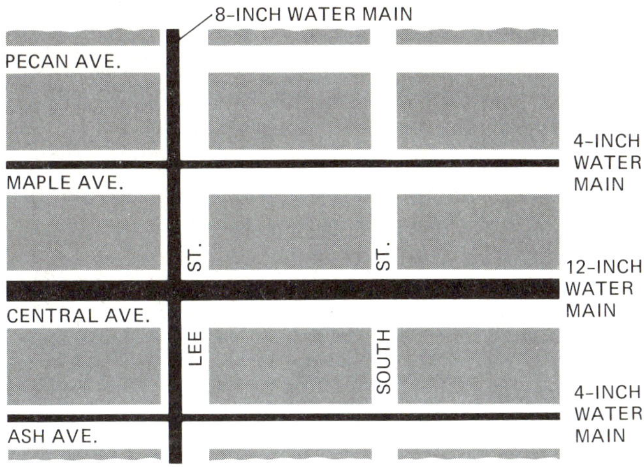

Figure 3-14. Municipal water-distribution network

• QUESTIONS •

1. Lakes, rivers, reservoirs, and wells are the principal water-supply points for a municipal system. (T or F)
2. A reservoir is a natural water suppy. (T or F)
3. What are the four types of pumps used to bring water out of wells?
4. What kind of pump has an impeller?
5. What are the two steps in water treatment?
6. Sedimentation and filtration are two methods of cleaning water. (T or F)
7. What are the two main methods of purifying water?
8. Filtration removes dirt, dissolved materials, and germs and bacteria. (T or F)
9. Purification removes dirt from the water. (T or F)

• MUNICIPAL SEWAGE-TREATMENT SYSTEM •

You can look at the municipal sewerage system as the same type of service the town or city provides when it supplies water to all the community. This system removes the sewage (waste water) from the buildings and helps to prevent disease and sickness.

The water-supply system removed water from some natural source, cleaned and purified it, and provided it to the community. The sewage system removes the sewage from the community, removes the waste material from the water and sanitizes it, and returns the water to nature. When waste water has been treated properly in a sewage-treatment plant, it is cleaner than the water in a natural lake or river.

Collection System

The municipal sewage-collection system is made up the same way as the water-distribution system. It is a series of pipes which are normally buried under the streets of the community. Each house, building, or lot has an outlet to these pipes, called sewer mains. The basic difference from the water-distribution system is that the pipes in the sewer (sewage) system are larger. Another difference is that the lowest point in the sewage-collection system is at the treatment area. A well-designed sewer system will have the waste (sewage) running downhill to the treatment area. Remember, this waste is carried by water, and the water will run or flow downhill.

Treatment Plant

The treatment area, the place where all the waste arrives, is called a sewage or waste-water treatment plant. There are three different types of treatment plants, a primary (first stage) plant, a secondary (second-stage) plant, and a tertiary (third-stage) plant. We will discuss each of these plants and discover the difference between them.

The primary treatment plant uses one stage to treat sewage. It receives the community's waste water and then removes the solid particles by sedimentation. Remember, this same type of treatment was used to remove dirt and other mixed particles at the beginning of the water treatment.

Sedimentation of waste water can be completed in different ways. The most common way is to allow the waste water to flow into a large settling or sedimentation basin. This basin looks like a shallow swimming pool with a deep end and a shallow end (Figure 3-15).

The operation of the sedimentation basin is simple. As the waste water flows into the deep end of the basin, the sediment mixed with the water slowly sinks or settles to the bottom of the basin. The basin is made long enough so that by the time the water reaches the shallow end, all the sediment (solid waste) has settled to the bottom of the basin.

If a sedimentation basin were operated for a long time, the waste which settled to the bottom would

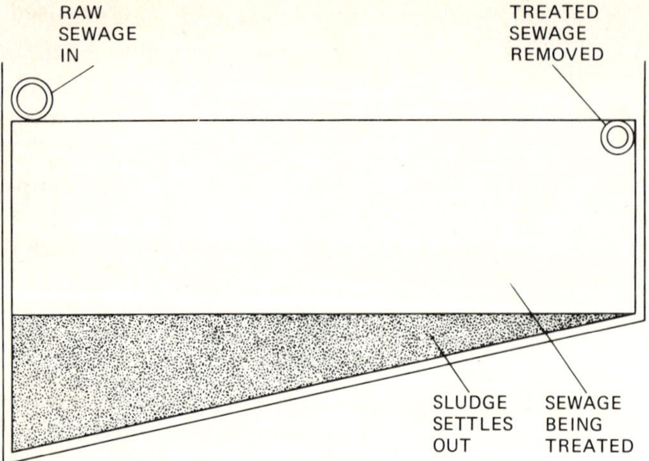

Figure 3-15. Settling or sedimentation basin

fill the basin. So, after a basin has been used for a length of time, it must be cleaned out. The flow of waste water is stopped and the settled waste, called sludge, is pumped out. Some sedimentation basins have an automatic sludge-removal system. These automatic systems use large paddles or chains to move the sludge to a collection point while the basin is operating. Pumps then remove the sludge from the collection point.

This process removes only the solid waste, not the dissolved waste or the germs and other bacteria which are in the sewage. Since this primary treatment of sewage is incomplete, most municipal treatment plants are classed as secondary plants.

Secondary Treatment Plants

The secondary treatment plant uses two stages to treat sewage. The first stage is sedimentation, the same process used in the primary treatment plant. This removes all the sediment from the sewage.

The second stage of sewage treatment at this plant is the use of a biological process to remove the dissolved materials from the sewage. The process most commonly used is the trickling filter. This uses a filter bed of gravel covered by a slime. As the water flows down through the gravel, the dissolved materials are removed. These processes are difficult to understand and are best left to the engineers who design and operate them.

The secondary sewage-treatment plant uses two stages of waste-water treatment, sedimentation to remove the mixed sediment and a biological process to remove the dissolved material from the water. This leaves only some germs and other bacteria which pollute the water.

Tertiary Treatment Plant

The tertiary treatment plant uses many different methods of waste-water treatment. The sedimentation stage may be used to remove the sediment, a biological process may be used to remove the dissolved materials, and a third stage is added to remove the remaining pollution by a physical-chemical process. Again, these processes are best left to the engineers.

The waste-water treatment plant can be one of three different types. The primary treatment plant uses one stage of treatment, sedimentation, and removes about 40 percent of the sewage. The secondary treatment plant uses two stages of treatment, sedimentation and biological treatment, and removes about 80 percent of the sewage. The tertiary treatment plant uses three stages of treatment and removes 99 percent of the sewage. The secondary and tertiary treatment plants return the treated water, now as pure as lake or river water, to a natural setting.

Sludge Disposal

The sludge gathered from the sedimentation basins is usually treated with some chemicals, dried, and used as fertilizer. The chemical treatment and the drying sanitize, or purify, the sludge and remove the odor, making it useful as a garden, lawn, or farm fertilizer.

Figure 3-16 is a picture of a typical secondary sewage-treatment system.

Figure 3-16. Municipal secondary sewage-treatment system

• QUESTIONS •

1. Name the pipes which bring the sewage to the treatment plant.

2. What are the three types of sewage treatment plants?
3. Which type of treatment plant removes about 40 percent of the sewage?
4. Which type of treatment plant removes about 80 percent of the sewage?
5. Which type of treatment plant uses a physical-chemical process?
6. The sludge is sanitized and dried and then used for fertilizer. (T or F)

• SUMMARY •

This chapter discussed the municipal water-supply and sewerage systems. This should promote a general understanding of the where, how, and why of the water-supply and sewerage systems.

• WORDS PLUMBERS USE •

gravity
centrifugal force
impeller
volute
discharge
sediment
bacteria
filtration
purification
sanitize
primary sewage treatment
secondary sewage treatment
tertiary sewage treatment
biological process
sludge

4
HAND TOOLS AND SAFETY

This chapter will acquaint you with the many common hand tools used by the plumber and make you aware of potential safety hazards.

At the end of this chapter, you will:

- Be able to identify the various common hand tools used by the plumber.
- Be able to identify potential safety hazards in a working area.
- Be aware of the Occupational Safety and Health Act.

• COMMON HAND TOOLS •

Many of the tools used by a plumber are common to all the building trades, and many are used around the house. The major part of a plumber's toolbox consists of common tools such as pliers, screwdrivers, hammers, chisels, saws, wrenches, and many others. This section will identify the different types of common hand tools.

Pliers

You are probably familiar with the three types of pliers, the straight pliers and two types of offset pliers. These three types of pliers are pictured in Figure 4-1.

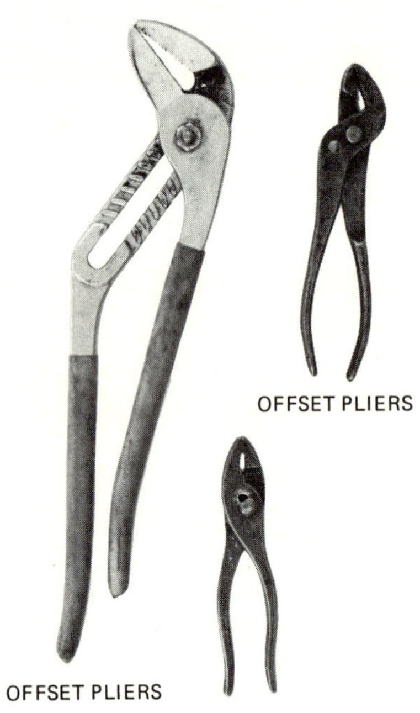

Figure 4-1. Common pliers

In addition to the pliers pictured above, there are some special-use pliers which are used by plumbers; these are needle-nose pliers and side-cutting pliers. The needle-nose pliers have a long thin blade for reaching into a close area. The side-cutting pliers,

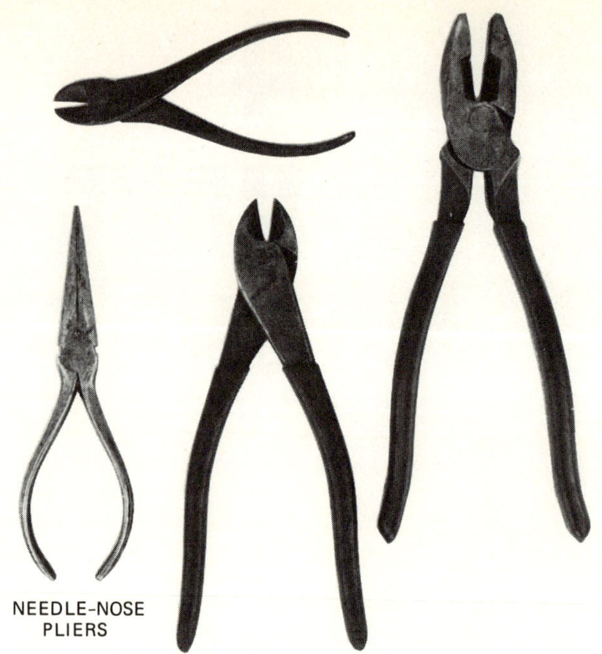

Figure 4-2. Special pliers

also known as electrician's pliers, are used for cutting wire, pipe straps, and other thin metals. Figure 4-2 pictures these different pliers.

Screwdrivers

Two basic types of screwdrivers used by a plumber are the blade screwdriver and the Phillips screwdriver. The blade screwdriver is the most common

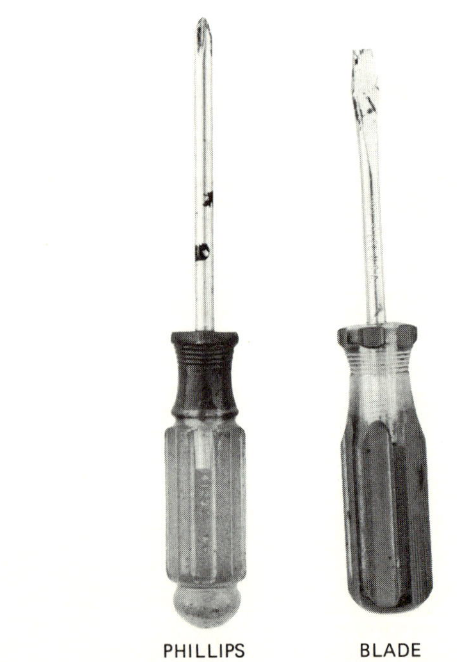

Figure 4-3. Common screwdrivers

type. The tip is a single flat blade, while the Phillips screwdriver has a pointed tip with four short blades. Figure 4-3 shows pictures of these two types of screwdrivers.

Each of the different types of screwdrivers is available in different sizes. A large flat screwdriver would not work with a small screw, nor would a small Phillips screwdriver work with a large Phillips screw. It is important for the plumber's toolbox to contain some different sizes of each type of screwdriver. The right tool for the job saves money, time, and mistakes. Figure 4-4 shows a set of different-sized blades and Phillips screwdrivers.

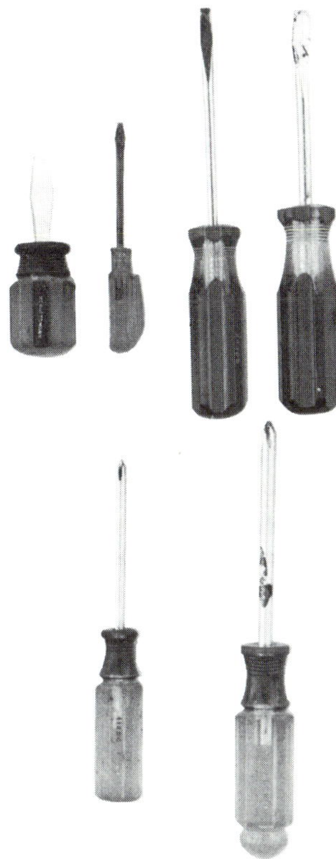

Figure 4-4. Sized screwdrivers

Chisels

The plumber's toolbox should also contain two different types of chisels, the wood chisel and the cold chisel. The wood chisel is made for chipping or cutting wood or other soft material and has a long sharp cutting edge, while the cold chisel, made for chipping concrete, metal, or other hard material, has a short, more blunt cutting edge. These different types of chisels are shown in Figure 4-5.

Like the screwdriver, both wood and cold chisels are made in different sizes. It would be difficult to cut a small hole with a large chisel. Again, the plumber's toolbox should contain a few different sizes of each chisel.

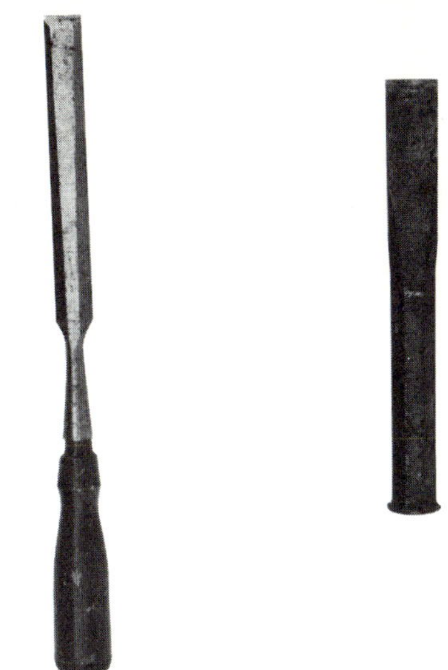

Figure 4-5. Common chisels

Another item important to a plumber is a star drill. A star drill is a special type of chisel which is used for making round holes through concrete, brick, or tile walls. The tip of the star drill has four cutting edges shaped like a star. Figure 4-6 shows several star drills.

Star drills are also manufactured in different sizes, each size made to drill a hole of a certain diameter. Since plumbers work with different-sized pipes, it is important that they have a few different-sized star drills.

Figure 4-6. Star drills

Hammers

The plumber will normally carry two different hammers, a claw hammer for driving and removing nails, and a heavier hammer (ball peen) for use with the chisels or star drills. The claw hammer will normally have a 16- to 20-ounce head, while the heavier hammer will have a head that weighs 32 to 48 ounces. Pictures of these hammers are shown in Figure 4-7.

Figure 4-7. Common types of hammers

Wrenches

The wrenches that will be covered here are not used for pipe or plumbing fittings. They are the wrenches a plumber would use to tighten nuts and bolts used with the plumbing fixtures. Three types of wrenches are used, the adjustable open-end wrench, the box-end wrench, and the Allen wrench or hexagon key set. Figure 4-8 shows these three different types of wrenches.

The adjustable open-end wrench has a fixed jaw and an adjustable jaw, and it can be used on various sizes of nuts and bolts. These wrenches are manufactured in different lengths. A 6- to 8-inch length will normally be large enough for most plumbing tasks.

Box-end wrenches are manufactured in many different sizes, each one referring to the size of the nut or bolt with which it can be used. The most common sizes used by a plumber are $\frac{3}{8}$, $\frac{7}{16}$, $\frac{1}{2}$, $\frac{9}{16}$, and $\frac{5}{8}$. The metric sizes would be 9, 10, 11, 12, 13, 14, and 15. Figure 4-9 is a picture of a set of box-end wrenches.

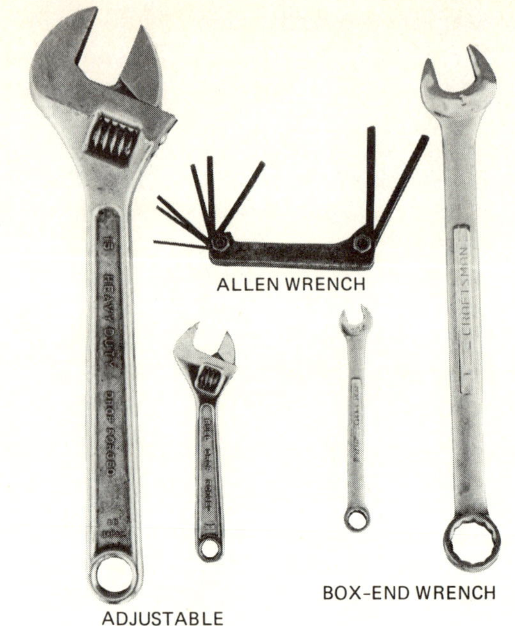

Figure 4-8. Types of wrenches

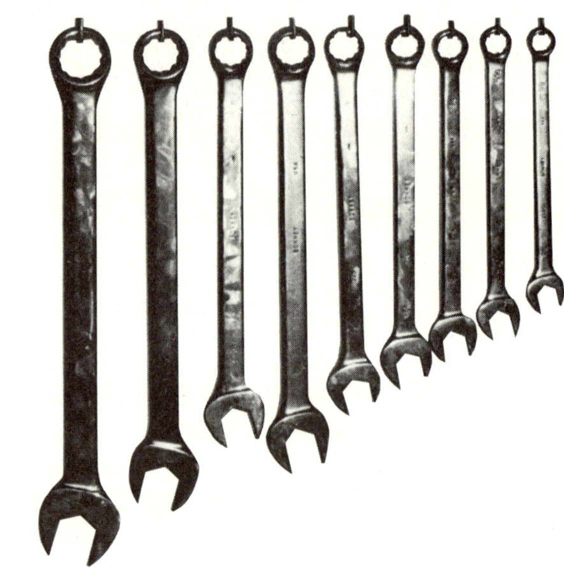

Figure 4-9. Box-end wrenches

The Allen wrench is a different type of wrench. It is used to tighten setscrews or flush-head bolts. These wrenches are six-sided pieces of strong steel and come in sets. Figure 4-10 is a picture of an Allen wrench set.

Figure 4-10. Allen-wrench set

Hand Snips

The plumber often uses hand snips to cut sheet metal or chimney flue pipe. There are many different types of hand snips, but the type most often used by the plumber is aviation snips. Figure 4-11 shows a pair of aviation snips.

Figure 4-11. Metal hand (aviation) snips

Saws

Three different types of saws are commonly used by the plumber. They are the handsaw, the keyhole saw, and the hacksaw. These three different saws are shown in Figure 4-12.

The handsaw is used to cut boards, flooring, or wall sections, while the keyhole saw is used to begin a cut in the floor or the wall. The hacksaw is used to cut through steel pipes or other hard materials.

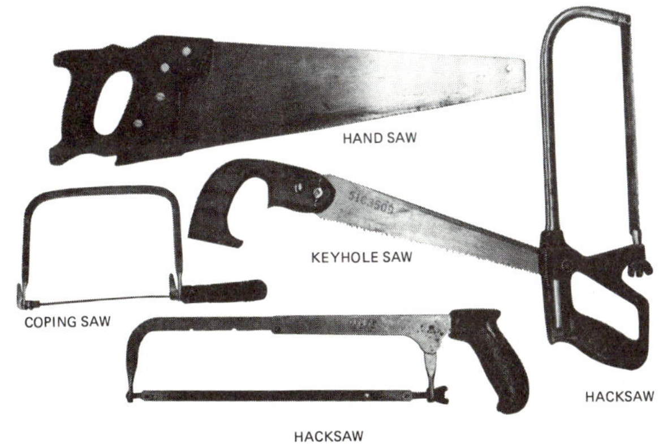

Figure 4-12. Common handsaws

Hand-Held Power Tools

Though plumbers will select these power tools themselves, the most common ones used are the electric drill, the saber saw, and the skill saw. Figure 4-13 shows each of these three power tools.

Figure 4-13. Hand-held power tools

• SPECIAL PLUMBING TOOLS •

Many special tools, such as pipe wrenches, monkey wrenches, stocks and dies, pipe cutters, pipe reamers, and others, are also part of the plumber's toolbox. These tools will be introduced and discussed in later chapters.

• SAFETY •

We have all become familiar with the word safety, but what it really means is preventing accidents. We all want to keep our fingers, no one wants a broken arm or leg, and very few people want to die. These are just a few of the things an accident can cause, even an accident on a plumbing job.

Tools left lying around can cause someone to trip and fall, or they might be knocked off a ledge and hurt someone down below. One of the major causes of minor injury is a cluttered or messy work area.

Another cause of injury is the improper use of a tool. A screwdriver is not a chisel. If it is used as one, the handle can shatter and cause, at the least, cuts or, at the worst, blindness. Each tool is designed to be used for a certain task and should be used in a certain way. If tools are used incorrectly, injury can result.

Tools which are improperly cared for are another source of injury. Tools which are dull or not correctly sharpened can slip off the work surface and cause injury. If chisels and caulking irons have a "mushroom" head (Figure 4-14), small pieces of steel may

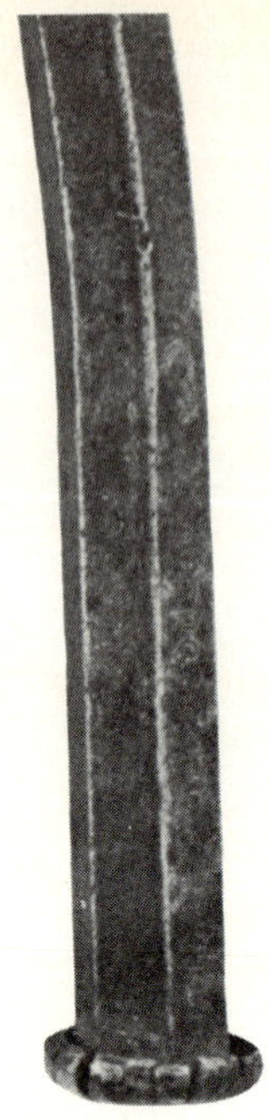

Figure 4-14. Mushroom head of chisel

fly off and injure or blind someone. If hammers, axes, and sledges have cracked handles or improperly seated heads, the head may come off. If a hammer head that weighs 1 pound or more flies through the air, it can seriously injure someone.

A shovel with a cracked handle is very dangerous. If the handle breaks, the splintered shovel is just like a spear pointed at the chest of the person using it. All tools should be kept sharp and in good repair. Sharp tools work better, and there is less chance of injury.

The electrical power tools on the market today have a built-in safety feature. The entire tool is grounded by a wire so that if the tool develops a short circuit (short), the person using the tool will not be shocked. The ground wire from the tool is connected to the third (round) prong on the electrical plug. If this prong is cut off or is not connected, a short circuit could result and the person using the tool could be electrocuted. A drawing of safe and unsafe plugs and connection practices is shown in Figure 4-15.

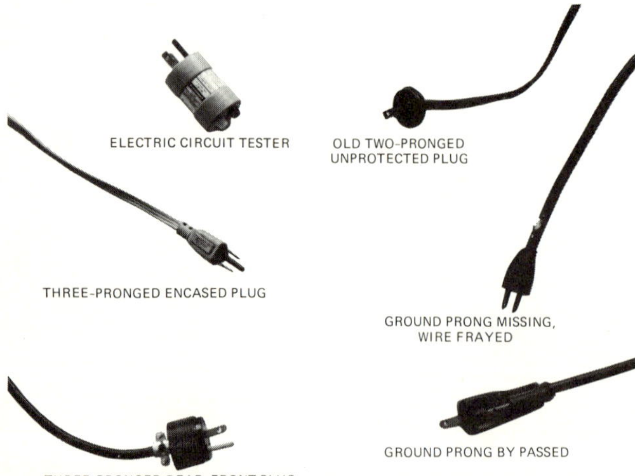

Figure 4-15. Safe and unsafe electric plugs and cords

These are just a few of the safety practices which could make your job as a plumber safer. The biggest safety factor on the job is you. If *you* use the correct tools for the job, if *you* keep your tools sharp and in good repair, if *you* keep the work area clean, and if *you* use common sense and good work habits, then *you* and the people around you will be safe from injury.

• EXERCISE •

Identify the different types of common hand tools used by a plumber that are available in your classroom shop.

• OSHA •

OSHA is a term you will be hearing for many years. It stands for the Occupational Safety and Health Act of 1970, an act passed by the U.S. Congress to "assure safe and healthful working conditions for working men and women" (Public Law 91–596, 1970). This act sets up many safety standards and practices for almost all the occupations in the United States, and it applies to any business that employs people.

One of the reasons for this act was the many deaths and injuries which were caused by accidents while people were working. In 1970, more than 14,000 people were *killed* in on-the-job accidents, and over 2,000,000 were injured badly enough that they had to take time off from their jobs. Those figures are large, but the amount of money lost because of these injuries was even larger: over 9 billion dollars.

OSHA sets up some rules for the employer and the employee, and you should be aware of these rules. The rules for employers are:

1. They must provide a safe and healthful place of employment.
2. They must buy, maintain, and require employees to use safety equipment and devices necessary to reasonably protect the employees. This would be things like safety glasses, earplugs, and hard hats.
3. They must keep records of all on-the-job accidents and injuries.
4. They must also make reports to the Department of Labor at certain times.

5. If they do not do these things, they can be fined.

The rules for the employee are that:

1. They should report any violations of the act by the employer.
2. They can aid an OSHA inspector on the inspection.

Though employees are not required to use all the safety equipment and devices supplied by the employer, they may be fired if they do not.

There are many different rules and regulations in this act. The main ones will be discussed during the following chapters as they apply.

• SUMMARY •

This chapter described those tools used by a plumber which are common to many different building trades. In addition to these tools, there are many special tools common only to the plumbing trade. These special tools will be introduced throughout this book when their use is discussed.

Certain general safety rules and regulations were also discussed. In addition, specific safety rules and procedures will be discussed in later chapters as the need arises.

• WORDS PLUMBERS USE •

offset pliers
needle-nose pliers
side-cutting pliers
blade screwdriver
Phillips screwdriver
wood chisel
cold chisel
star drill
claw hammer
ball peen hammer
adjustable open-end wrench
box-end wrench
Allen wrench
aviation snips
hand saw
keyhole saw
hacksaw
safety hazards
short circuit (short)
OSHA

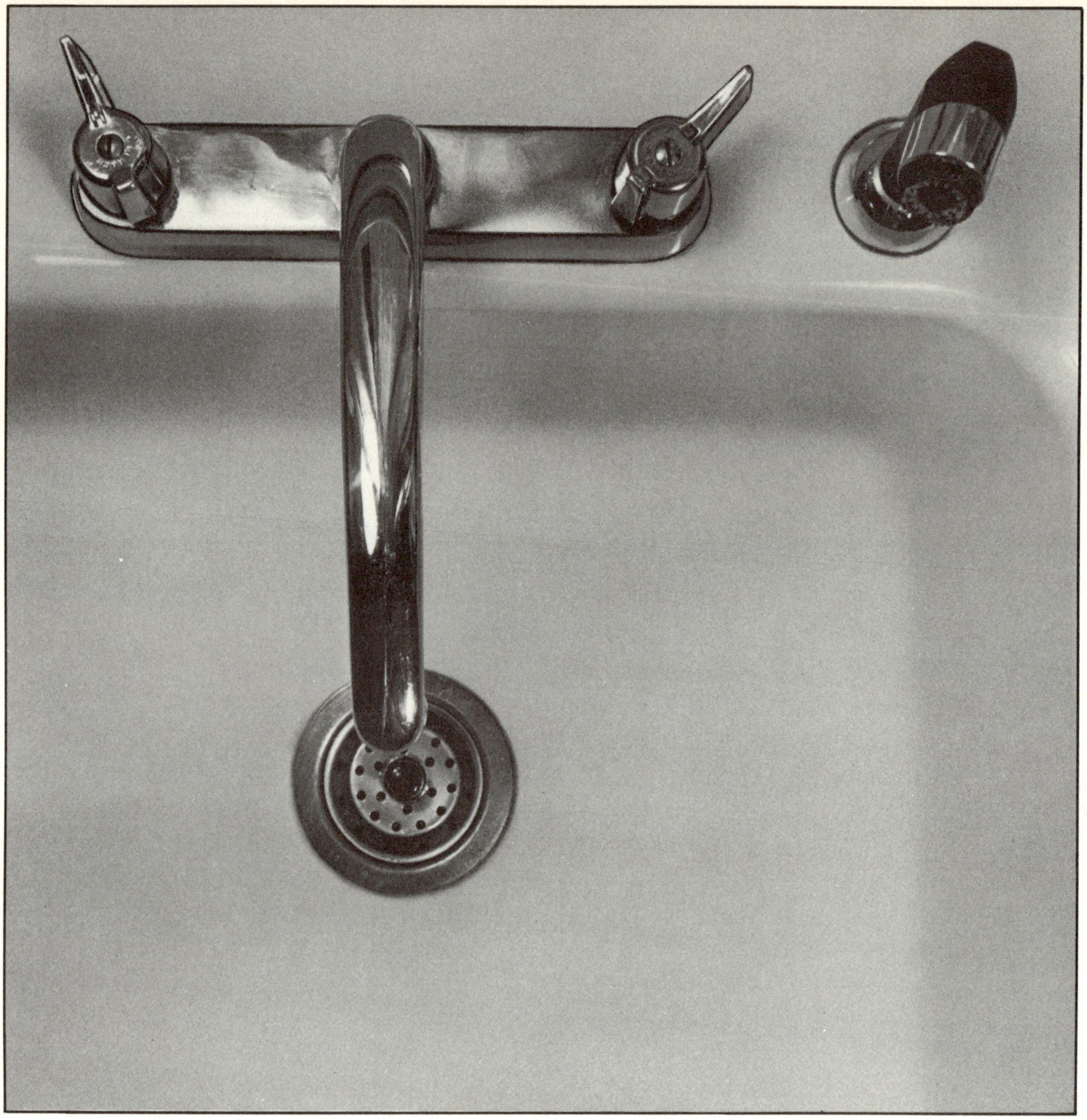

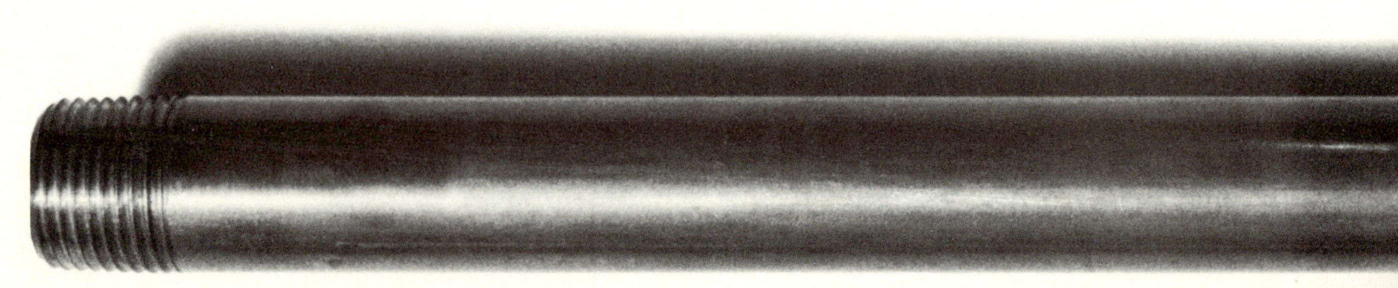

5

PLUMBING FIXTURES

If you have never worked on a plumbing system before, the components with which you are most familiar are those terminal fixtures in your home: the sink, lavatory, bathtub or shower, and commode. Though these are basic fixtures a plumber works with, there are many more used for special purposes which the plumber will install.

The primary purpose of this chapter is to acquaint you with the special tools and the different designs and uses of many of the fixtures you will be working with in the years to come. The secondary purpose of this chapter is to make you familiar with the many different types of faucets, valves, traps, wastes, and other equipment (commonly called brass or brassware) used with those fixtures.

At the completion of this chapter, you will be able to:

- Describe the component parts of the plumbing fixtures used in a typical dwelling.
- Identify the tools, materials, and procedures used in assembling the component parts of the plumbing fixtures used in a typical dwelling.
- Assemble the component parts and install the plumbing fixtures using appropriate tools, materials, and procedures.

• THE PLUMBER'S TOOLS •

In addition to the common hand tools described in Chapter 3, the tools described below are special tools used when working on the fixtures and their components.

Monkey Wrenches

Much of the plumber's trade consists of working with chrome or other metal-plated fixtures. Use of a pipe wrench on the smooth, bright surface of these fixtures would mar the finish. The monkey wrench, with its smooth, long, nonmarring jaws, is ideally suited for use with these plated finishes.

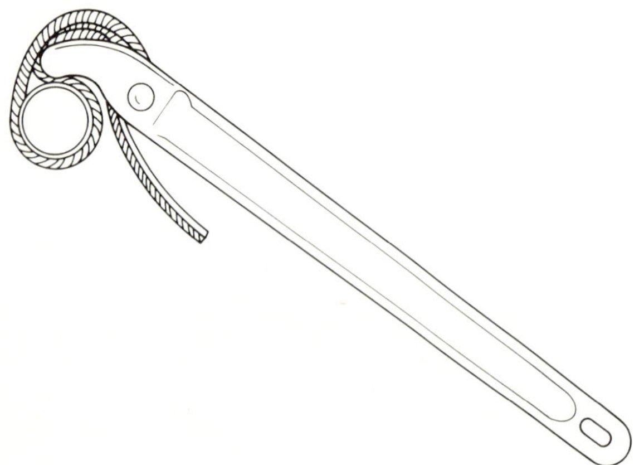

Figure 5-1. Strap wrench

Strap Wrenches

Many times the plumber is also called upon to work with chrome or other plated pipe. As with the fixtures, the use of a pipe wrench would mar the finish. The strap wrench is the tool used to apply enough pressure to tighten the pipe without marring the finish (Figure 5-1).

• KITCHEN SINKS •

The first fixture we will discuss is the kitchen sink. There are many different models available on the market today made from a variety of materials. Today's most popular kitchen sink is a single- or double-bowl, stainless-steel, countertop sink used with a cabinet or other base. Stainless steel is a tough, corrosion-resistant metal which is particularly suited to heavily used fixtures such as kitchen sinks. It resists damage by hot or cold liquids, can absorb heavy blows from pots or pans without denting or marring, and is relatively easy to keep clean. Figure 5-2 shows two types of stainless-steel sinks.

As you can see, these stainless-steel sinks come in a variety of sizes (length and width) as well as

Figure 5-2. Stainless-steel sink

TABLE 5-1 COMMON SINK SIZES

Type A* Single Bowl		Type B* Double Bowl		Type C* with Disposal Bowl	
y	*x*	*y*	*x*	*y*	*x*
24	21	32	21	33	22
25	22	33	22		
31	21				

* Many other styles and sizes are available.

different bowl depths. The depth of the bowl on most sinks will be proportionate to the size—the larger the size, the deeper the bowl—but normally the bowl will be between 6½ and 7½ inches deep. Some sinks are manufactured with extra-deep bowls for convenience. Table 5-1 lists some of the more common sink sizes.

Kitchen sinks are also made of porcelainized cast iron. This type of sink was popular with home builders in the first half of the twentieth century. These sinks are first cast out of molten iron, then a glazed porcelain (glasslike) finish is applied and baked onto the cast sink. Most of the older porcelainized sinks were white, and they had other built-in features,

Figure 5-3. Porcelainized cast-iron sink

such as a single or double drainboard and a splash guard at the rear or side. Figure 5-3 is a drawing of a type of porcelainized cast-iron sink.

Kitchen Faucets

If you have taken a close look at both the stainless-steel sink and the porcelainized cast-iron sink, you have noticed small holes in the top or back of the sink and larger holes in the bottom of the bowl. The small holes are for the faucets, and the larger holes are for the waste connection. All sinks will have the waste-connection hole, but some may not have the faucet holes. Faucets used with those sinks are mounted on the wall or in the wooden countertop. This practice has been the exception rather than the rule, since water tends to leak around the base of the faucet and rot out the wooden countertop or damage the wall.

Most kitchen-sink faucets today are of one of two types, the single-control faucet and the dual-control faucet. Pictures of both types of faucets are provided in Figure 5-4.

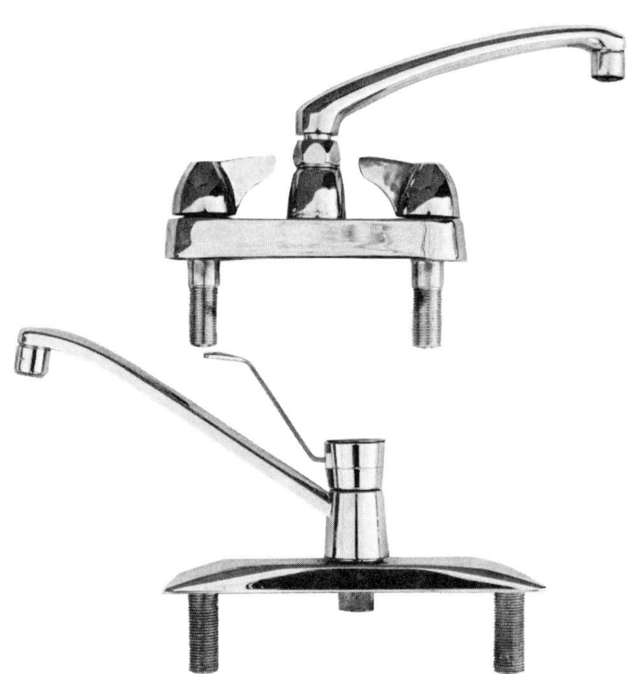

Figure 5-4. Types of sink faucets

THE DUAL-CONTROL FAUCET • As you can see from the pictures in Figure 5-4, the dual-control faucet has two handles, one for the cold water and one for the hot. An expanded view of the dual-control faucet is shown in Figure 5-5. Look it over carefully.

This faucet uses a rubber (or other soft compound) washer and a valve seat to control the flow of water. This faucet is a dual-control compression type. Refer back to the diagram of the faucet as we discuss the parts and their function. The valve stem holds the washer, and the valve seat is directly below the stem in the body of the faucet. The middle part of the valve stem has a series of threads which allow the entire stem to be raised and lowered within the body of the faucet; as the stem is raised, the washer is moved away from the seat and water is allowed to flow through the opening and out the spigot. As the valve stem is lowered, the washer comes into contact with the seat and the flow of water is stopped. There are two other parts which are important, since they keep the water from leaking up around the stem of the faucet. These are the valve-stem compression nut and the valve-stem packing washer. As you can see, the valve-stem packing fits around the upper part of the valve stem; as the valve-stem nut is tightened on the body of the faucet, it compresses the packing and keeps the water from leaking around the stem. The remaining two parts are the handle and the screw which holds it in place. Most compression faucets use a very similar arrangement.

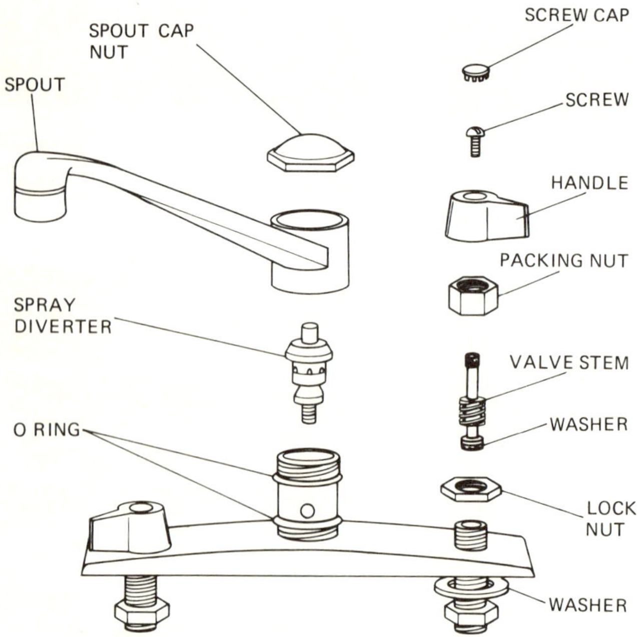

Figure 5-5. Parts of a dual-control faucet

THE SINGLE-CONTROL FAUCET • The single-control faucet does not have a washer. It uses a smooth ball-and-socket arrangement to regulate the flow of water. An expanded view of the single-control faucet is shown in Figure 5-6. Compare this faucet with the one in Figure 5-5 and note the differences.

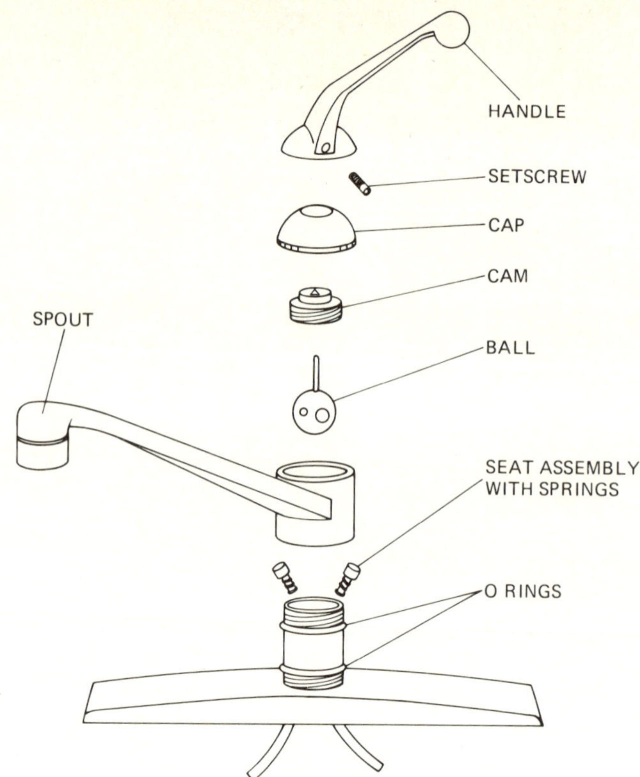

Figure 5-6. Parts of a single-control faucet

The Waste and Trap

The last two pieces of equipment needed to complete the kitchen sink are the waste and the trap. These two parts connect the sink to the sewer of the building and allow all the waste water to drain from the sink into the sewer. The waste connection fits into the large hole in the bottom of the sink and is normally made up of about five pieces. Figure 5-7 is a cutaway drawing of the waste connection.

Refer to Figure 5-7 as we discuss the various parts of the waste connection. First, take a close look at the part labeled the plug top. This part fits inside the drain hole in the bottom of the sink with a rubber or neoprene washer between the sink and the plug top. The large locknut is then tightened up from the bottom of the sink and compresses the washer between the plug top and the sink to stop leaks. The next part, the tailpiece, slides up into the plug top and is held in place by a washer and compression nut. As the nut is tightened, the washer is compressed and holds the tailpiece tightly in position and seals the joint.

There are many different types of waste connections. Some have large strainers and some connect directly to a disposal, but all basically connect to the sink in the same manner as the plug-top type. Some pictures of the different waste connections are shown in Figure 5-8.

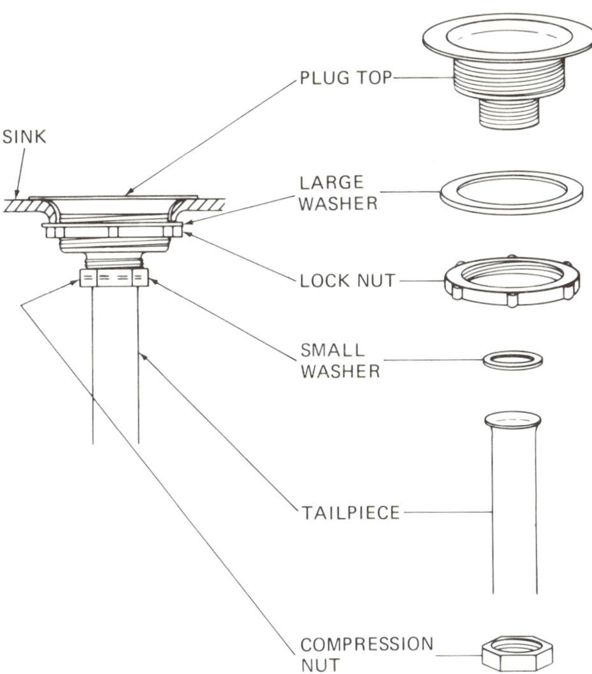

Figure 5-7. Waste connection

Figure 5-8. Types of waste connections

The Trap

Before we discuss the trap as it applies to the kitchen sink, it is important that you know why a trap is used in the drainage system of a house or building. A trap is used to seal off the sewer from the inside of the building. This solves several different problems. You can imagine the problems that might occur if this seal were not used. The sewer gas (a very flammable gas) could enter the building and cause an explosion, or at the very least could fill the building with a noxious (bad) odor. There are various disease germs that could be carried into the living or working space and cause sickness, or certain vermin or insects would be able to come into the building from the sewer. For these and other sanitary reasons, the trap is a very important part of the plumbing system. Two types of trap are illustrated in Figure 5-9.

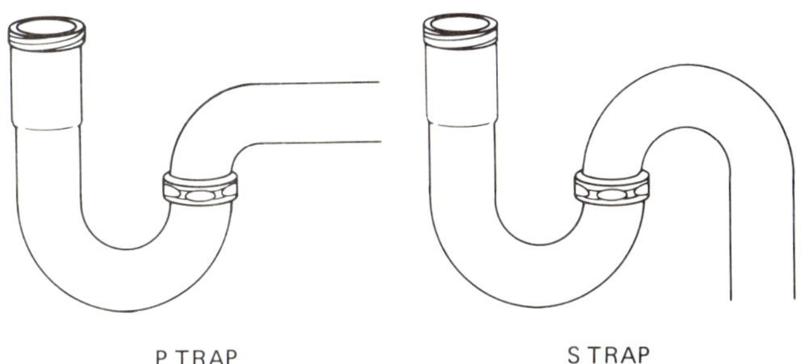

Figure 5-9. Typical waste trap

There are a few things that should be made clear about a trap. First of all, the seal of the trap is made with water that has been used and is destined for the sewer. When the water drains down the waste connection, it moves through the trap until the water in the sink is all gone. Since the trap is U shaped, the water never really drains all the way out of the trap (see Figure 5-10).

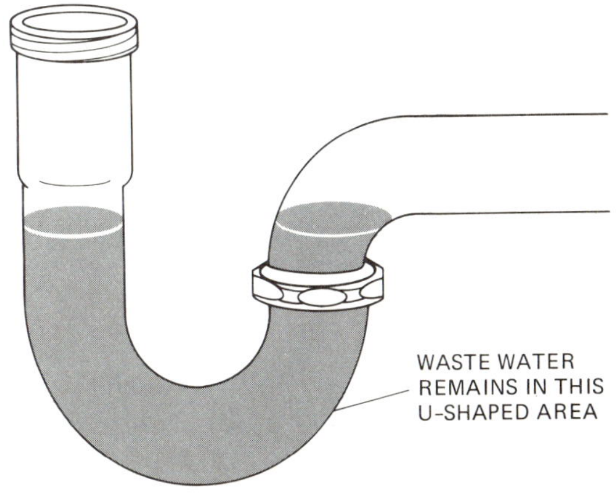

Figure 5-10. P-trap showing water seal

One other important point about the trap seal is that the sewer must be vented, or have an air passage to the outside air. This vent stops the water in the trap from siphoning out and breaking the seal. You have all heard about siphoning a liquid from one place to another, and this is the same principle which removes the water seal from the trap if the sewer is not vented. Figure 5-11 is a diagram of an experiment which will demonstrate the trap and vent principle.

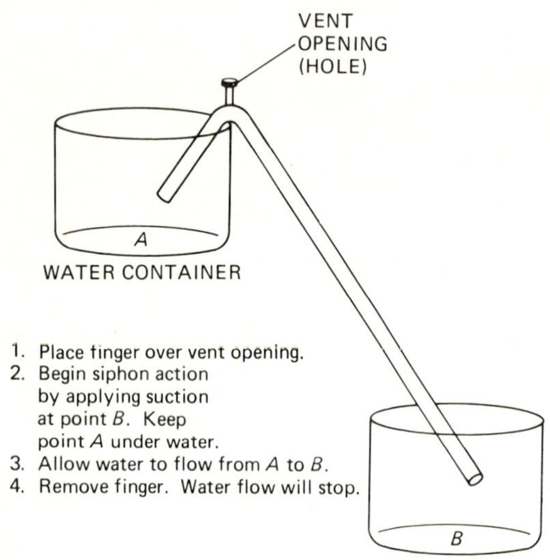

Figure 5-11. Trap and vent experiment

Many different types of traps are made, but the P trap has been used over the last 30 years or so and has almost become the standard for the plumbing industry. An expanded view of the P trap is shown in Figure 5-12.

Since the P trap is a basic design and most other traps have similar parts, learning the components of the P trap will help in installing many of the other traps. Follow through on Figure 5-12 as we discuss the P trap and its parts:

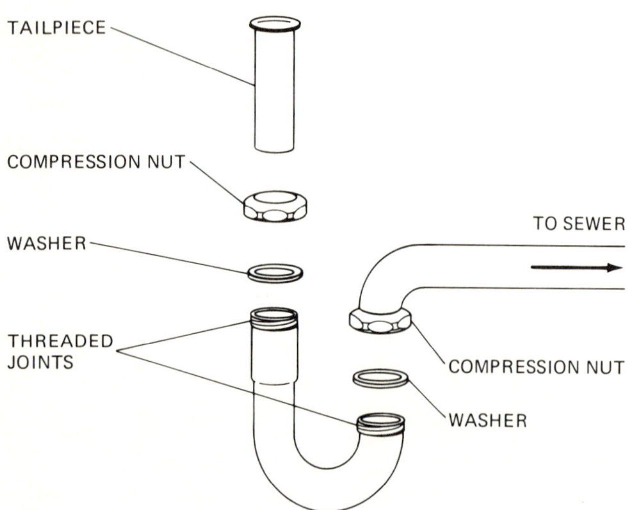

Figure 5-12. Parts of a P-trap

1. The inlet opening of the trap connects to the tail pipe of the waste connection and is held in position by the same type of compression joint used with the tail pipe, the washer and compression nut joint.
2. The outlet end of the trap slides into the sewer connection and is tightened into place with the same type of compression joint.
3. The middle joint also uses a compression joint, but it is of a slightly different type, since it uses a small lip on the outlet pipe as a compression flange.

Setting the Sink

Setting the countertop sink is accomplished by sawing out a section of the counter and using mounting brackets and screws to fasten the sink firmly to the countertop (Figure 5-13). The height of the top of the sink will be the same as that of the counter, and the carpenter will normally install the cabinets so that the counter is 36 inches [92.5 cm] above the floor.

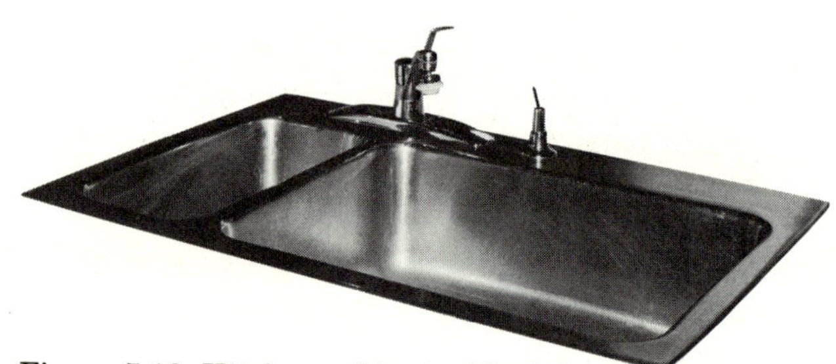

Figure 5-13. Kitchen cabinet with sink

Porcelain kitchen sinks are set using different systems. Some are hung from the wall using brackets with legs to support part of the weight, while others use a cabinet as the main support. The use of wall brackets and legs or cabinets will be discussed in the next section of this chapter when we cover lavatory installation.

• LAVATORIES •

There are two basic kinds of bathroom sinks, normally referred to as lavatories in the plumbing trade. They are the vitreous-china lavatory and the porcelainized cast-iron lavatory. The vitreous china lavatory, simply called the china lavatory, is formed from wet clay, fired (baked) in an oven (kiln) to make it hard, coated with a glazing material, and refired to change the glazing material to a hard, glassy, waterproof finish. The china lavatory is rather brittle and must be handled with care. It cannot hold up under rough use or impact from sharp or heavy tools, since it is easily cracked or chipped. Installation of the china lavatory requires special care in handling.

The second kind of lavatory is the porcelainized cast-iron lavatory. This kind is manufactured by the same process as the cast-iron kitchen sink. It is more durable that the china lavatory, since it is not as easily cracked, but the porcelainized finish can be chipped by a hard blow from a tool or other object.

It is generally believed that the china lavatory has cleaner lines and a finer finish than the porcelainized cast-iron lavatory. For this reason, the china lavatory is often used in private homes while porcelainized ones may be used in either private, commercial, or public buildings. Figure 5-14 is a picture of a lavatory.

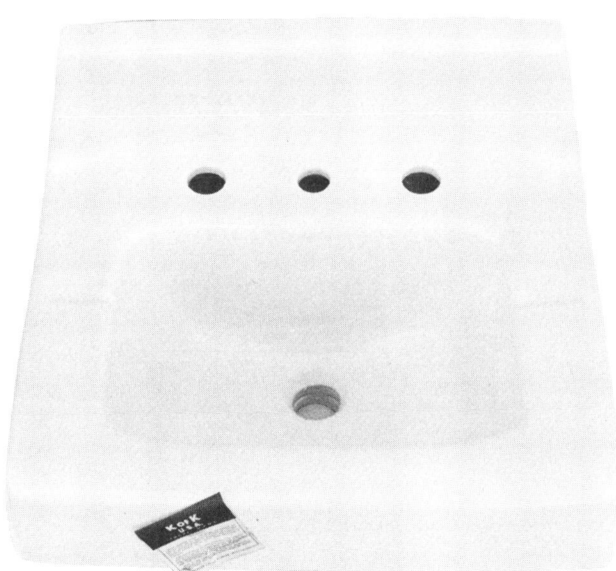

Figure 5-14. Typical modern wall-hung lavatory

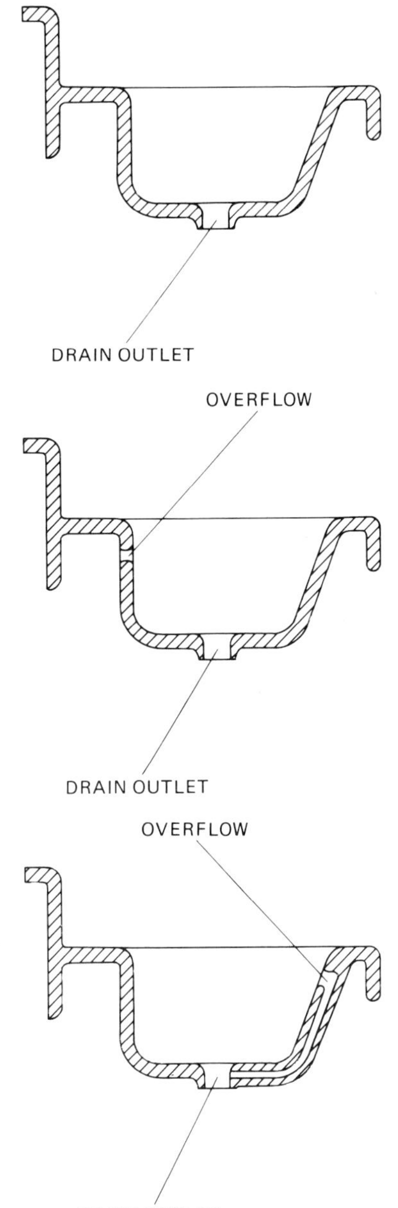

Figure 5-15. Views showing lavatory construction

Both kinds of lavatories are available in different styles, types, and sizes. The two basic styles of lavatories are the same as the styles of kitchen sinks, the countertop and the wall-mounted style.

Within each lavatory style group, there are three different types or configurations, the drain outlet only, the drain outlet and overflow type, and the integrated drain outlet and overflow type. Cutaway drawings of these different styles and types are shown in Figure 5-15.

Lavatory Faucets

Each lavatory has the same openings that the kitchen sink has, small openings or holes for the faucets and larger openings or holes for the drain and overflow. The smaller openings for the faucets will accommodate the same types of faucets which were used with the kitchen sink, a single-control faucet or a dual-control faucet. Many lavatory faucets have an extra lever or plunger which is used to raise or lower the plug in the waste of the bowl. Three different types of lavatory faucets are shown in Figure 5-16.

Waste Connections

Since there are three different types of waste or waste and overflow combinations, there are also three different ways of connecting the waste or waste and overflow connections. The first is the waste-only type. This is almost the same as the waste connection on the kitchen sink, a single plug top and tailpiece which is attached to the drain hole in the bottom of the bowl. The cutaway drawing of this type in Figure 5-17 will refresh your memory.

The waste connection for the drain and overflow type of lavatory begins with the same type of plug top but uses a T fitting to connect the overflow to the waste. This connection is made before the waste connects to the trap, as indicated in the cutaway view of Figure 5-18.

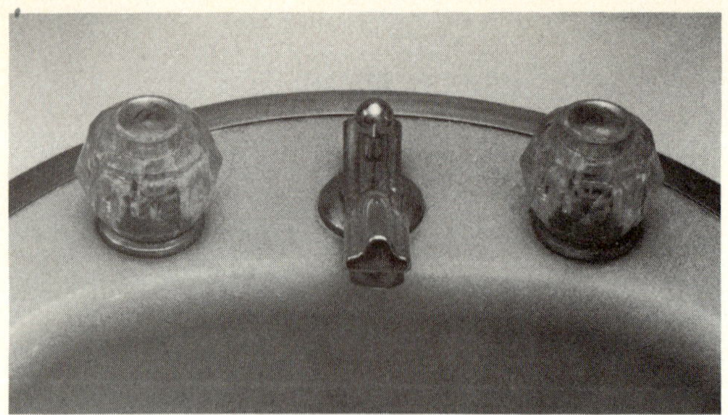

Figure 5-16. Lavatory faucets

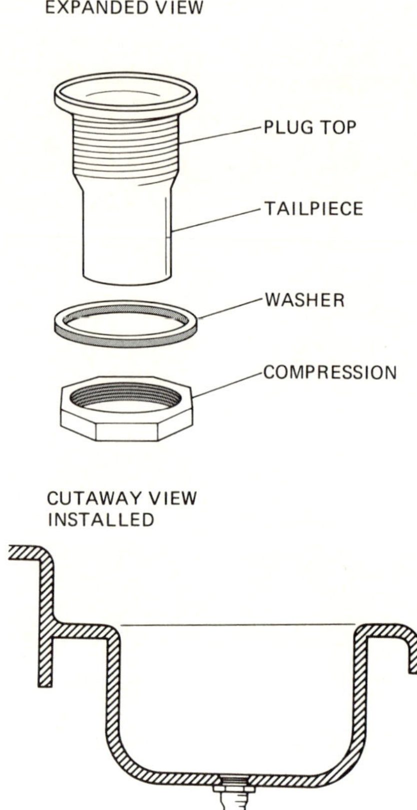

Figure 5-17. Waste connections, waste-only lavatory

The waste connection which is used with the integrated drain and overflow type of lavatory almost looks the same as the one used for the kitchen sink or the drain-only lavatory. There is one major difference. The plug top has holes or ports so that if the water does overflow, it can flow through these ports and out through the trap. Compare the cutaway in Figure 5-17 with the one in Figure 5-19 and note the difference in the plug tops.

You can see the difference between these two waste connections. The connection for the integrated drain and overflow is longer and has ports or holes in the plug top. These two connections are not interchangeable. If the one for the drain-only were used for the integrated type of lavatory, the overflow would not work. If the plug top designed for the

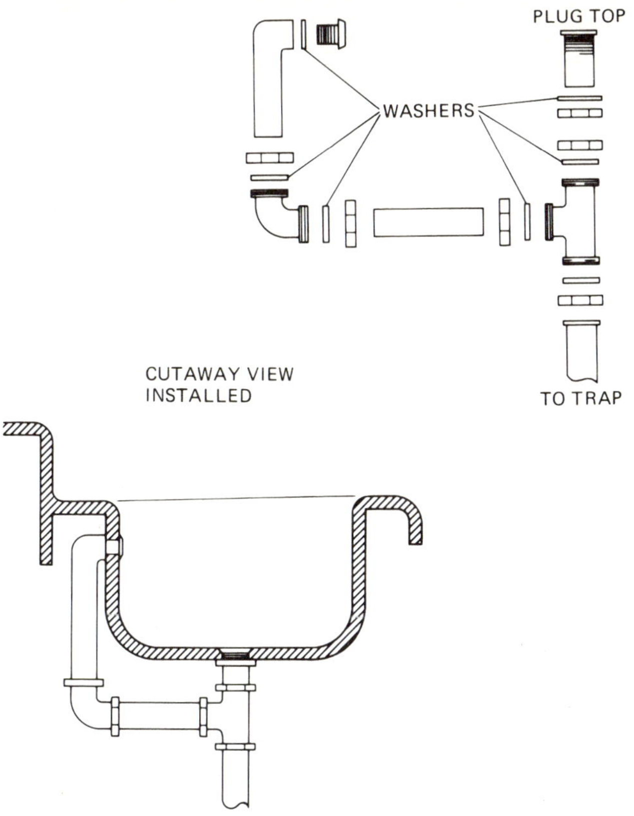

Figure 5-18. Waste connection, drain and overflow lavatory

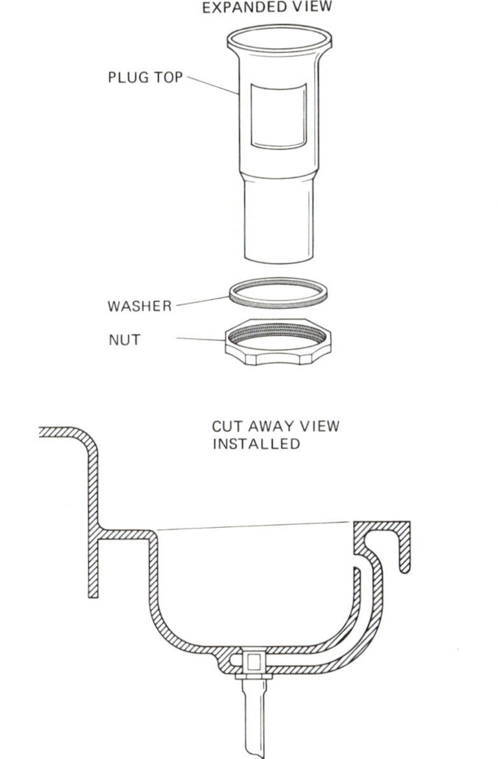

Figure 5-19. Waste connection, integrated drain and overflow lavatory

integrated type were used in the drain-only lavatory, the ports would allow the waste water to drain out of the lavatory onto the floor.

The waste connection in the lavatory is installed in the same manner as the waste connection in the kitchen sink. A minor difference is the connection required between the overflow and the waste of the second type of lavatory. Review Figure 5-18 to note this difference.

Lavatory Traps

Many different designs of lavatory traps have been used since the beginning of indoor plumbing facilities. As with the kitchen sink, the most commonly used trap is the P trap. This trap for the lavatory is slightly smaller in diameter than the kitchen-sink trap, but it operates on the same principle. The connections to the waste and to the sewer are the same. The cutaway view of this trap in Figure 5-20 will refresh your memory.

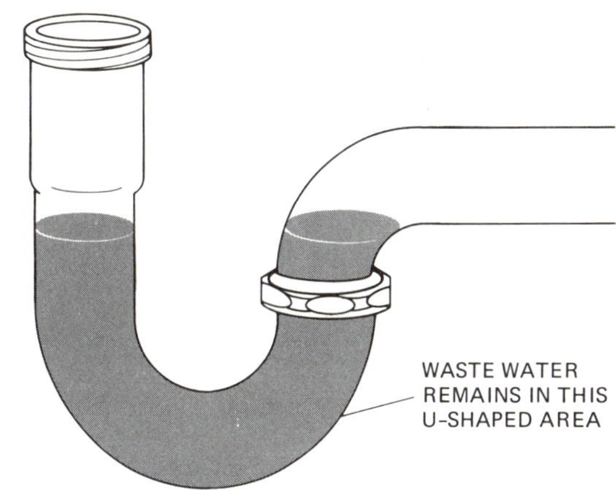

Figure 5-20. P-trap showing water seal

Setting the Lavatory

Lavatories are set the same way as kitchen sinks, either hung from the wall on a bracket or set into the top of a counter. The countertop installation is accomplished in the same way as it was for the kitchen sink. Simply cut out a hole of the correct size and then fasten the lavatory in position using the brackets and screws furnished. The lavatory is normally set at 32 inches [80 cm] off the floor.

The wall-hung lavatory (and the wall-hung kitchen sink) usually requires some preparation during the rough construction of the wall on which it is to be installed. The plaster, gypsum board, or tile found as a finished interior wall surface of a building does not have the strength to hold up a 20- to 30-pound [9- to 13-kilogram] lavatory. Some extra backing or bracing must be available behind the finished

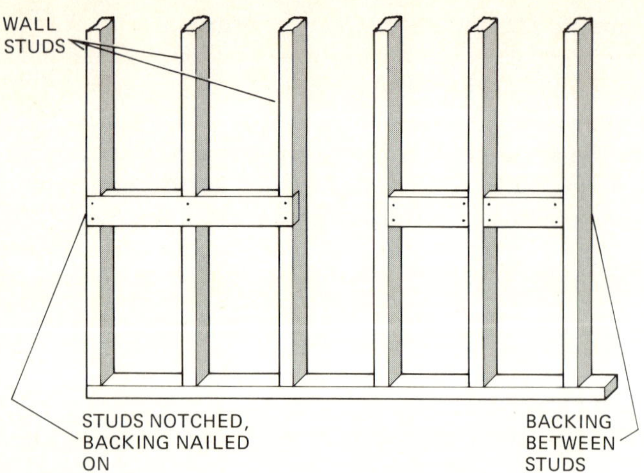

Figure 5-21. Installation of lavatory backing

wall to increase the holding power of the screws or other fastening devices. This backing is normally installed before the finished wall covering is applied and consists of a board (1 × 6) fastened between the wall studs. Two ways of installing this backing are shown in Figure 5-21.

The height at which this backing is installed is critical. To be sure that this measurement is correct, the lavatory and the type of bracket used must be taken into consideration. Various types of brackets are used in hanging either the sink or the lavatory. The most common types are the bar hanger and individual bracket hangers. The bar hanger is most often used with the porcelainized cast-iron sink or lavatory, while individual bracket hangers are used with vitreous-china lavatories. Figure 5-22 shows a bar hanger.

Figure 5-22. Bar lavatory hanger

To determine the height at which the backing should be installed, the following steps should be taken (see Figure 5-23):

1. Place the lavatory on the floor or table so that the wall side is easily accessible.
2. Place the bar hanger or the individual bracket

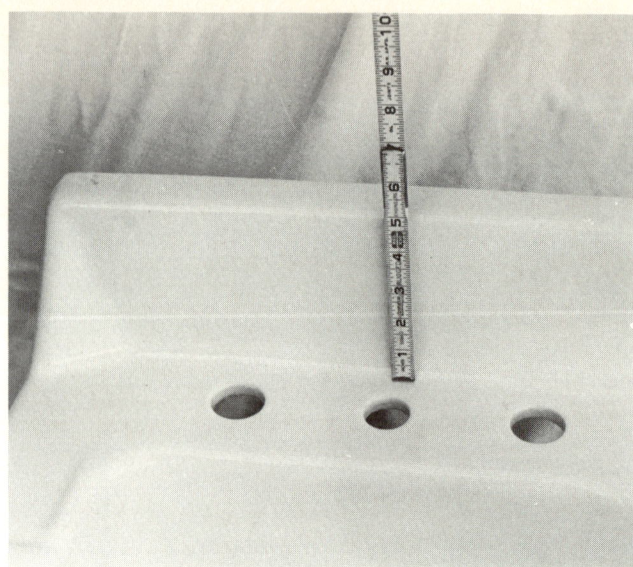

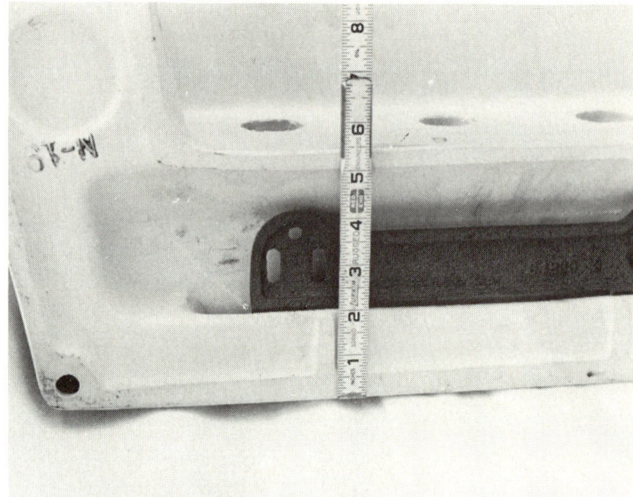

Figure 5-23. Installing the lavatory backing

hangers in position on the wall side of the lavatory.

3. Measure from the center of the hanger to the top of the lavatory bowl.
4. If the bracket is lower than the bowl, subtract that difference from 32 inches [80 cm]. If it is higher, add that difference to 32 inches.
5. Use the measurement obtained in step 4 as the center measurement of the backing board, and fasten the board into place.

After the backing is properly placed and fastened, the wall can be finished. Then hang the lavatory (or sink) by fastening the bar hanger or brackets into place at the correct height and placing the fixture into position.

Though a properly hung bracket will be sturdy enough under normal use, misuse may loosen the bracket or break the fixture from the wall. To provide extra support and decoration, adjustable legs may be installed under the front corners of the fixture. See Figure 5-24 for an example of these legs.

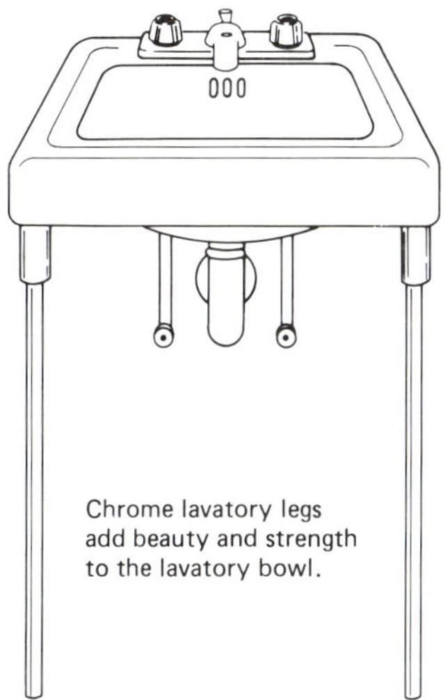

Figure 5-24. Lavatory legs

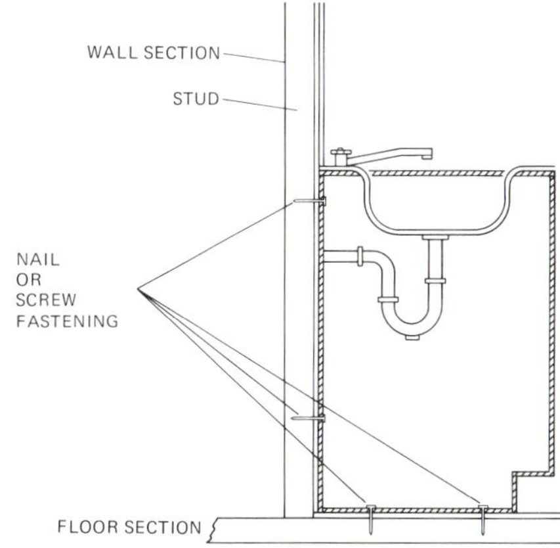

Figure 5-25. Cabinet-mounted lavatory

Cabinet-mounted lavatories (or kitchen sinks) are about the easiest to install. The fixture is placed on the cabinet and secured. The cabinet is then moved into position and fastened directly to the wall or the floor (or both). Backing is generally not required for this, since the cabinet can usually be fastened directly to the wall studs. Figure 5-25 shows cabinet mounting.

• QUESTIONS •

1. What are the two different materials used in the manufacture of lavatories?
2. The porcelainized cast iron lavatory is easily cracked or chipped by a blow. (T or F)
3. The china lavatory can be chipped, but it is very resistant to cracking. (T or F)
4. What are the three types of configurations of lavatories?
5. Which type of lavatory uses a plug top with holes or ports?
6. A lavatory faucet could differ from a sink faucet because it may have a lever to control the drain plug. (T or F)
7. The backing for a wall-hung lavatory is always placed 32 inches from the floor. (T or F)

• BATHTUBS •

Bathtubs and bathtub-shower combinations have changed dramatically over the past 30 to 40 years. The major changes have been in the materials used in their manufacture and in their sizes and shapes. The bathtub of the early 1900s is typified by the picture in Figure 5-26.

Figure 5-26. Early free-standing bathtub

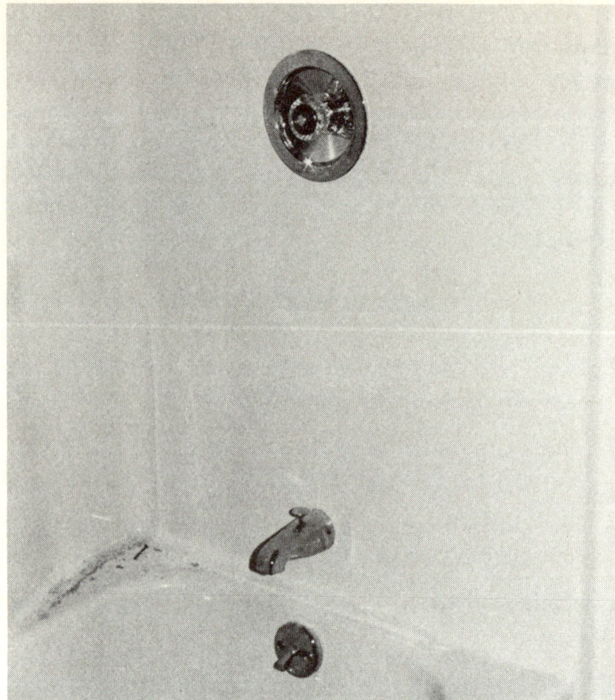

Figure 5-27. Left-hand Fiberglas tub

This type of bathtub was manufactured using the same process used for the porcelainized cast-iron fixtures discussed earlier in this chapter. This process is still used in the production of tubs, but two new materials are also widely used in bathtub manufacture: porcelainized pressed steel and Fiberglas. Figure 5-27 shows a Fiberglas tub.

There are two things which you should notice about the bathtub in Figure 5-27. The first is that there are holes for the faucet and for the waste connection and overflow, and the second is the relationship between the end with the holes and the finished side (apron). This designates the placement of the tub, with the apron on the left-hand side or the right-hand side. Figure 5-27 shows a left-hand tub. Since some tubs have no holes or openings for the faucets, the bathtub faucet will be mounted in the wall above the edge of the tub.

Bathtub Enclosures

Most of the bathtubs manufactured today are designed to be used as a bathtub-shower combination. This means that the tub must be built into the wall so that the shower water will run down the wall, return to the tub, and drain away. Porcelainized cast-iron or pressed-steel fixtures are normally set against the rough wall before the finished wall covering is applied. Plastic tubs can also be set in this manner, but they may also be installed using a prefabricated plastic enclosure. Cutaway examples of both installations are shown in Figure 5-28.

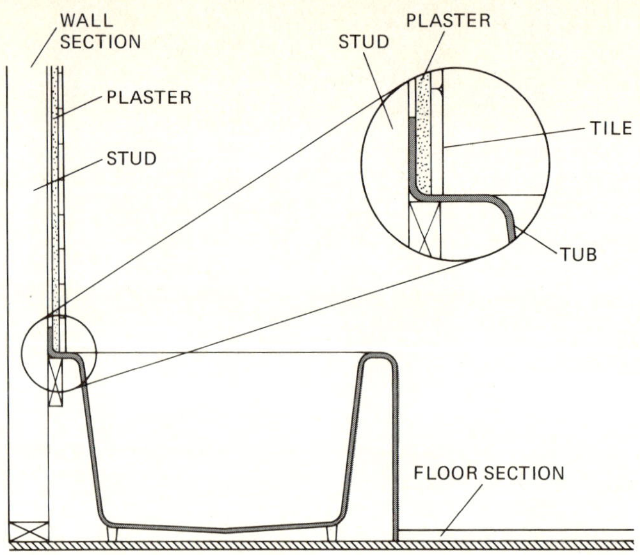

PRESSED STEEL OR CAST IRON TUB

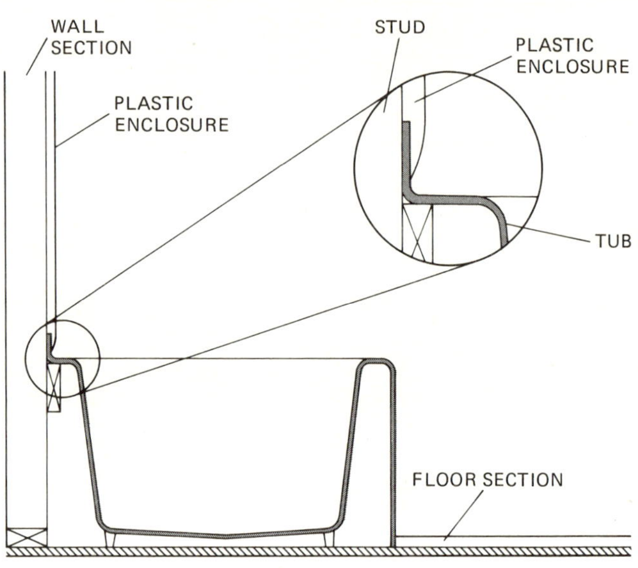

PLASTIC TUB

Figure 5-28. Bathtub installation

Setting the Bathtub

There are two methods of setting bathtubs, one for the cast-iron tub and one which should be used for the pressed-steel and plastic tubs. Setting the cast-iron tub is relatively simple if the floor is level. The tub is uncrated and moved into position and then checked with a level to make sure it will drain properly. The cast-iron tub is very strong, and the apron and the short legs under the body of the tub will support a person of almost any size with ease. If the floor is not level, or if a pressed-steel or plastic tub is used, a different method of installation will be required.

Both the pressed-steel and plastic tubs are strong, but each requires a little more support on the sides which are in contact with the wall. This support can be given by using a length of board (1 × 4) along the wall to support the edges of the tub in much the same manner as the bracing was used with the lavatory. The most important point in this method of setting or installing the tub is that the support boards should be placed so that the edges of the fixture will be level when the fixture is placed into position. A cross-section drawing of the different methods of setting a bathtub is shown in Figure 5-29.

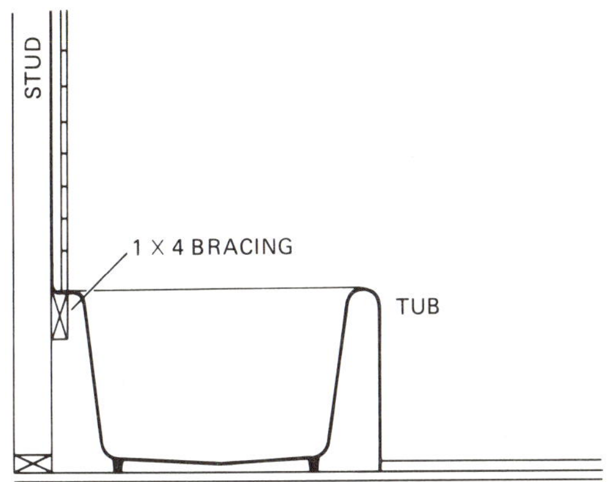

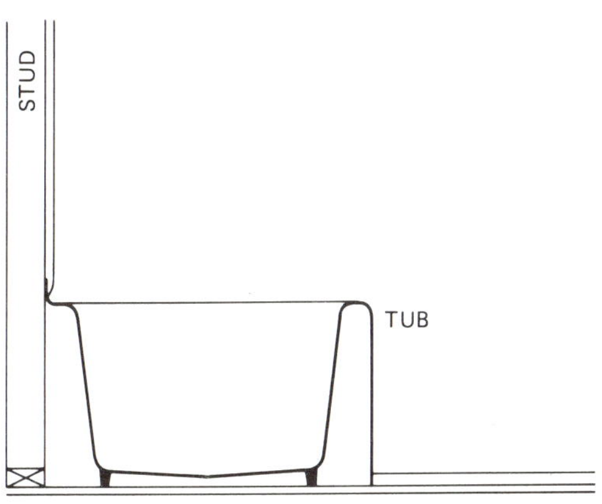

Figure 5-29. Setting the bathtub

Bathtub Faucets

Bathtub faucets, either with or without shower provisions, are available in either the single-control-lever or the dual-faucet model. Either will have provisions for a single spigot and a shower spigot. Pictures of each of the models are shown in Figure 5-30 above.

As mentioned earlier in this chapter, the bathtub faucet is installed in the wall above the edge of the tub. This can be done either before or after the tub is set into position. A typical faucet installation is shown in Figure 5-31.

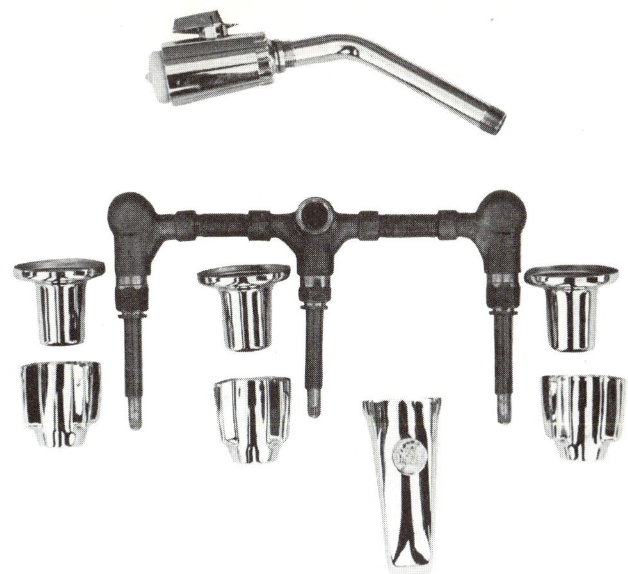

Figure 5-30. Bathtub faucets

Figure 5-31. Bathtub-faucet installation

Notice the use of bracing between the studs and behind the faucet and pipes. This bracing is important to keep the faucet and spigots steady.

Bathtub Waste Connections

Tubs have two holes for the waste connection. The bathtub has a separate drain and overflow arrangement like the drain and overflow lavatory. The waste and overflow connection consists of a waste elbow, an overflow elbow, a slip-joint T, and a tailpiece. The bathtub waste connection also has a lever and control rod to raise and lower the plug in the drain. Figure 5-32 is a cutaway drawing of a bathtub waste connection.

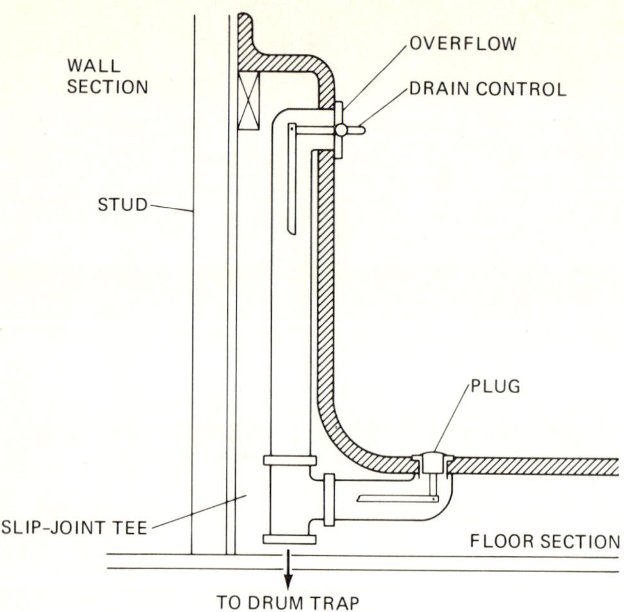

Figure 5-32. Bathtub waste installation

Bathtub Traps

Like any other fixture, the bathtub requires a trap to seal off the sewer from the house. In some areas the code allows a P trap to be used, while in other areas a "drum trap" is required. You should be familiar with the P trap from the discussion on kitchen sinks and lavatories. The drawings in Figure 5-33 will help you understand the operation of the drum trap.

The drum trap is normally made of cast iron and has threaded entry and exit holes. Connections from the waste to the drum trap and to the sewer are made using malleable-steel pipe (see Chapter 7). One advantage of the drum trap is that an opening is available for cleanout in case the drain line becomes plugged up or clogged.

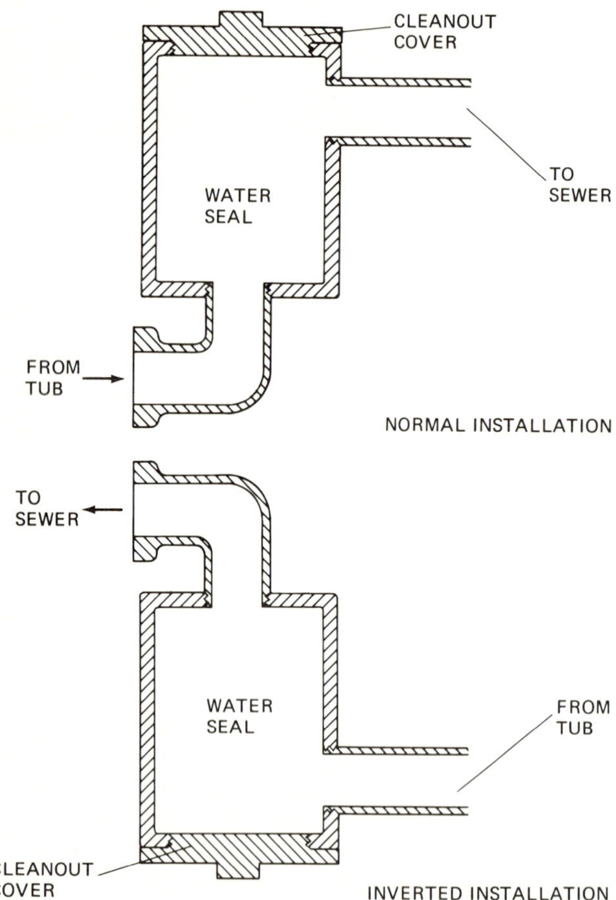

Figure 5-33. Drum-trap installation

• COMMODES •

The toilet, called the commode or water closet in the plumbing trade, is made by the same process and of the same material as the china lavatory. Like the china lavatory, the parts of the commode are

Figure 5-34. Types of commodes

easily chipped or cracked, and the fixture must be handled with care. There are generally two sections to the commode, the water tank and the bowl. The water tank contains the water-supply valve, the flushing mechanism, and a reserve of water for the flushing action. The bowl is the waste area and the trap combined.

There are many different types of commodes available today, although they will have the same basic components discussed above. Pictures of two different types of commodes are shown in Figure 5-34.

The Water Tank

As we mentioned earlier, the water tank is a storage receptacle as well as a container for the two water-control systems, the ball cock (water-supply valve) and the flushing valve. Figure 5-35 shows a cutaway view of the water tank and its components. Pay close attention to the two water-control systems.

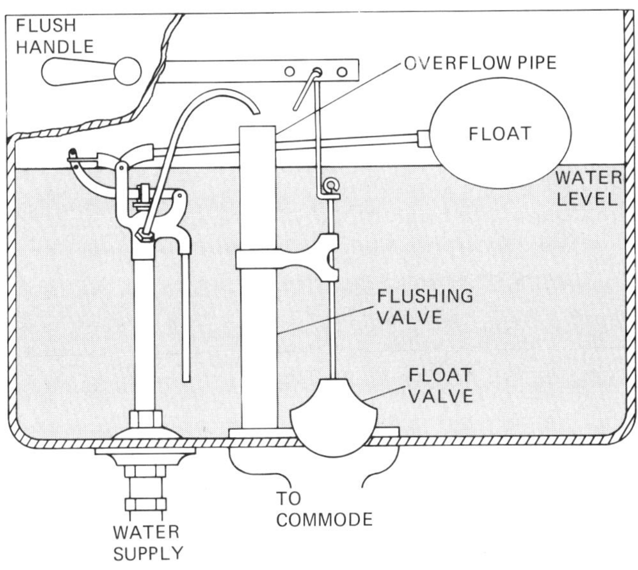

Figure 5-35. Commode water tank

The two water-control systems work independently. That is, they are not mechanically connected. They do depend on a certain level of water in the tank to operate properly. The operation of the two systems is shown in Figure 5-36.

In Figure 5-36*a*, the stop ball is raised from its seat, and as the water flows out of the tank and into the bowl, the ball-cock float and the stop ball begin to descend. In Figure 5-36*b*, the stop ball has reseated itself and the float ball is near the bottom of the tank. This position of the float ball opens the ball-cock valve and allows water to flow into the tank. In Figure 5-36*c*, the stop ball is in position, the tank is almost full, and the float ball is positioned near the top of the tank, which closes the ball-cock valve and stops the flow of water. To help you understand the operation of the ball cock more fully, Figure 5-37 is a cutaway view of a typical ball cock showing the open and closed positions.

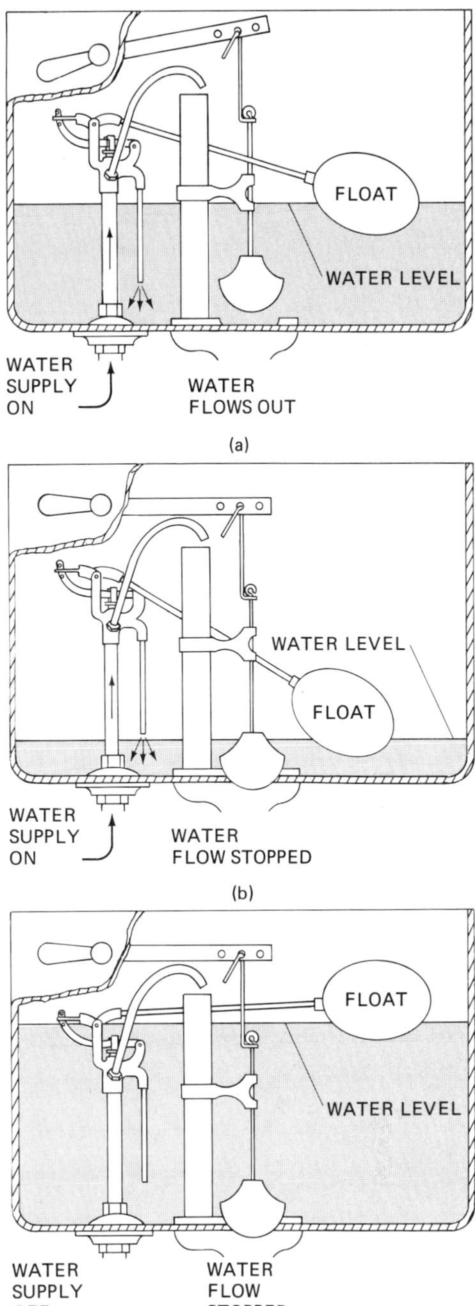

Figure 5-36. Commode water-tank system operation

The Commode Bowl

The two most commonly used types of commode bowls are the washdown and the siphon-jet types. These two types differ in the manner in which waste material is washed down the drain from the bowl.

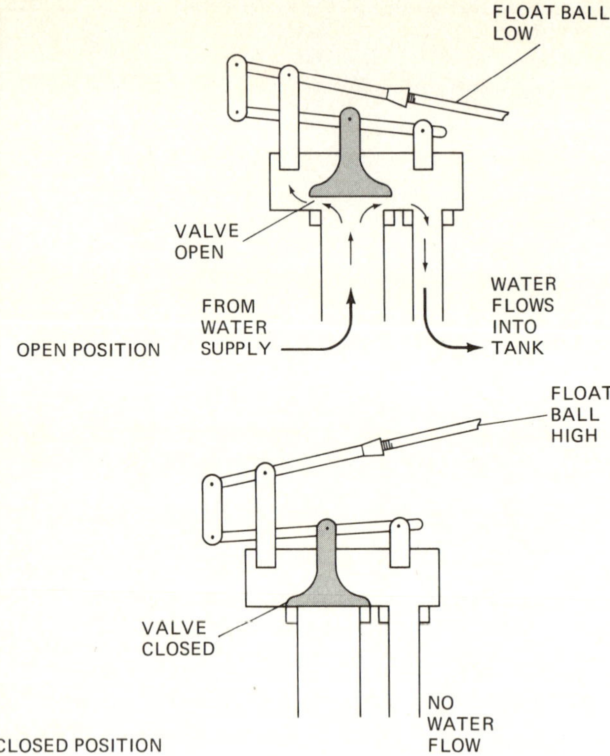

Figure 5-37. Water-tank ball-cock valve operation

Figure 5-38 shows cutaway views of these two types.

The washdown bowl uses the force of the water to carry away the waste, while the siphon-jet bowl applies the siphon principle. The small opening in the bottom of the bowl in the siphon-jet type directs the flow of water up through the trap and helps to begin the siphon action.

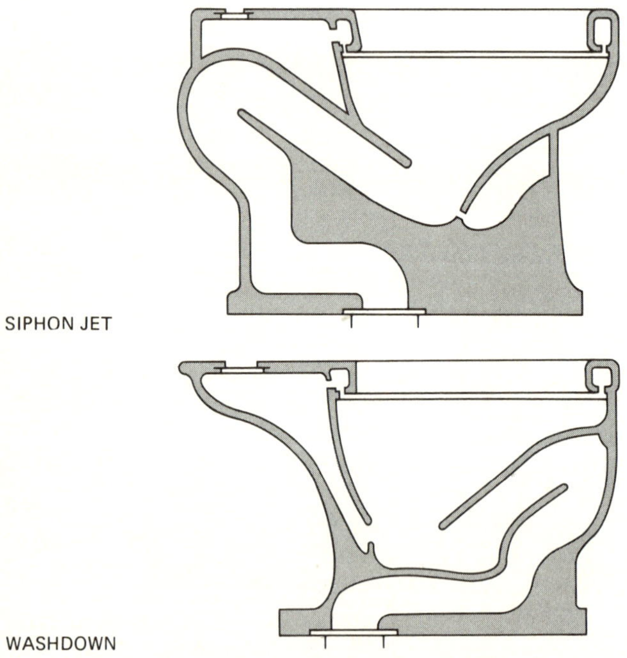

Figure 5-38. Washdown and siphon-jet commodes

Setting the Commode

The commode is probably the easiest plumbing fixture to set in position. The entire operation requires two floor bolts, two tank bolts, four nuts and washers, a wax ring (putty or a rubber gasket may also be used), and a rubber washer. Follow the steps in Figure 5-39.

1. Set the two floor bolts into the closet collar (floor flange).
2. Invert the bowl and slip the wax ring around the outflow horn of the bowl.
3. Turn the bowl upright and position it so the two floor bolts slide through the holes on each side of the base, and lower the bowl into position.
4. Using two nuts and washers, carefully tighten the bowl into position. Remember, this material is very brittle and breaks easily.
5. Place the rubber washer in position on the top of the bowl. Then place the tank bolts and washers into the small holes in the bottom of the water tank.
6. Position the water tank so that the tank bolts and the flushing hole are lined up with the appropriate holes on the bowl, then lower the tank into position.
7. Using the remaining nuts and washers, tighten the tank onto the bowl, being careful not to crack or break the components.

• LAUNDRY FACILITIES •

In contrast to the homes built in the first half of this century, very few present-day homes have a separate laundry tray. The modern home is equipped with an automatic washer, and though a laundry tray or tub may be installed, there is little need for it. This section will identify the laundry-area fixtures and the facilities used with the automatic washer.

Laundry Trays

Laundry trays or tubs are available in the same basic categories as kitchen sinks, with either a single or double bowl. The similarity ends there, for both the size and the material used in their manufacture differ greatly. Slate, plastic, and Fiberglas are the three main materials used in manufacturing the laundry tray, and the laundry tray is about three

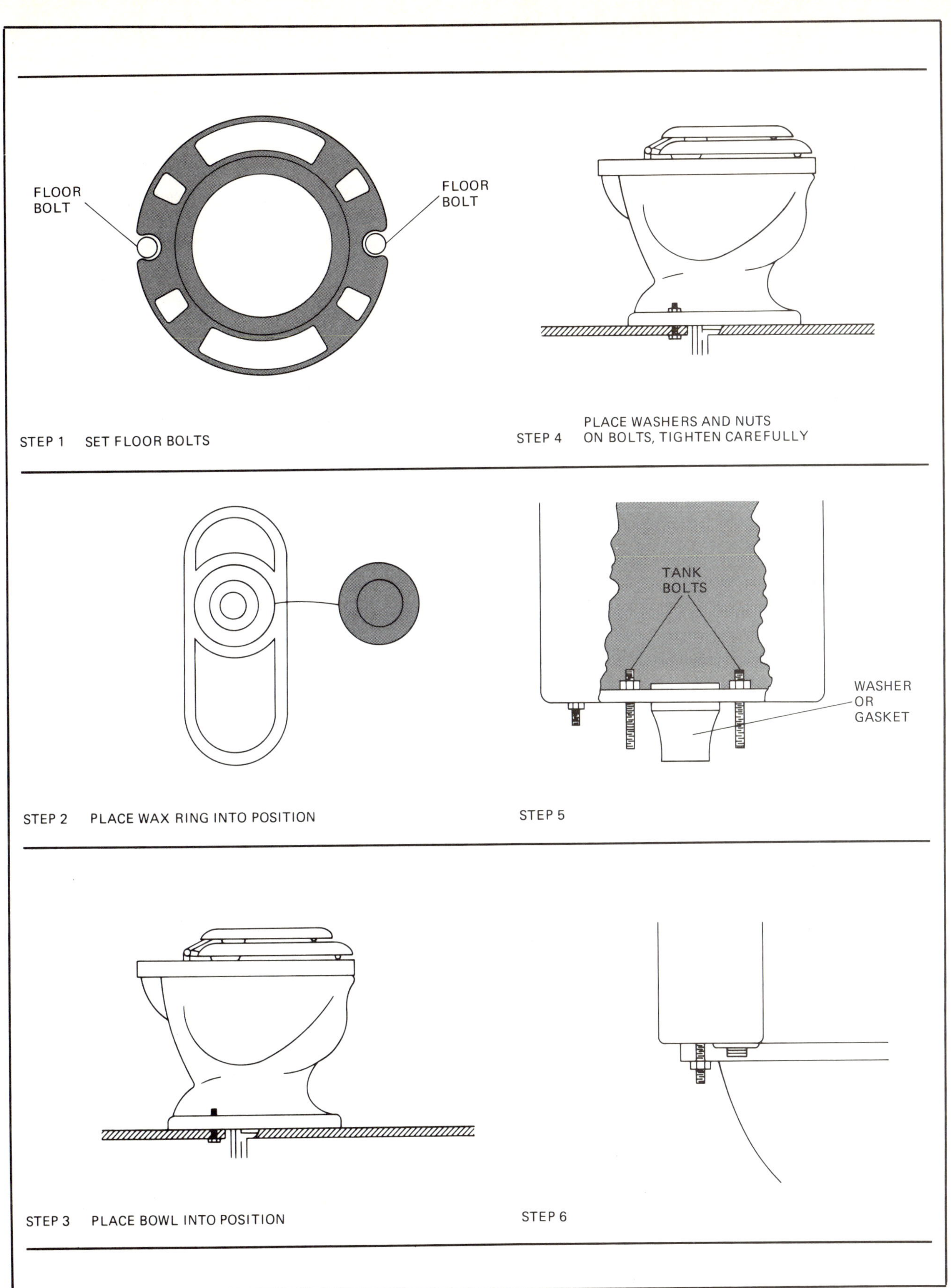

Figure 5-39. Setting the commode

times deeper than the sink. Compare the picture in Figure 5-3 with the pictures in Figure 5-40 and notice the difference in the depth of the bowl.

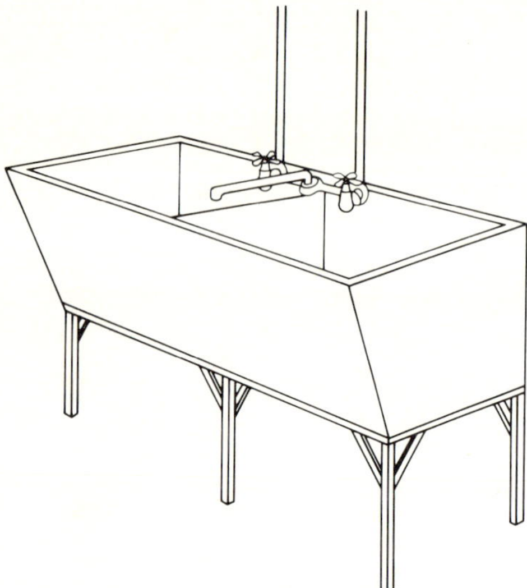

Figure 5-40. Laundry tray

There are also different ways used to mount the laundry tray. It may be on a stand, or it may be hung on the wall. The slate laundry tray is very heavy and is normally mounted using the stand, while the Fiberglas or plastic tray may be mounted using a wall bracket as was used with the lavatory. The laundry-tray waste is usually a simple plug top which uses a rubber stopper or plug, and the waste is installed in the same way as the sink or lavatory waste. A P trap is usually used to seal off the drain.

The faucets which supply water to the laundry trays are almost always of the compression type. They normally have separate controls for hot and cold water and use a single swing spout or spigot. A typical laundry faucet is shown in Figure 5-41.

Figure 5-41. Typical laundry-tray faucet

Automatic Washing Machines

Automatic washing machines require hot- and cold-water supplies and an adequately trapped drain. Each supply line is controlled by a compression valve with a hose connection, like the hose connection outside your home or school. Because the automatic washer controls the flow of water, these valves are left in the open position when the washer is connected. Figure 5-42 shows a typical automatic washer valve.

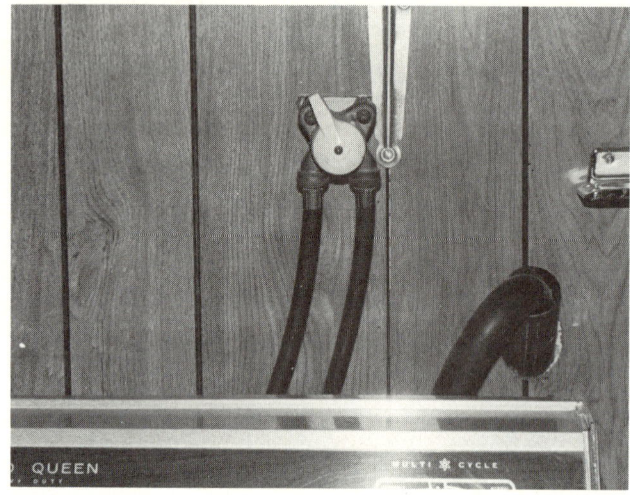

Figure 5-42. Automatic-washer valve

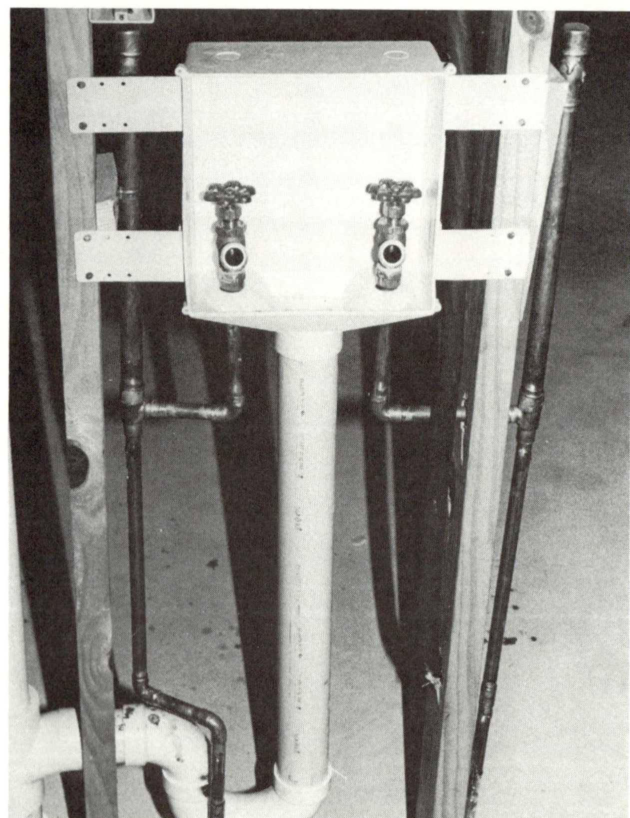

Figure 5-43. Automatic-washer drain

The drainage system normally installed with the automatic washer system consists of a P trap and a vented drain line. Figure 5-43 shows this typical installation.

• SUMMARY •

In this chapter we have identified and discussed the common plumbing fixtures used in a typical family dwelling. We have reviewed the procedures as they apply to the installation of the brassware (faucets, wastes, and traps) and the setting or hanging of the fixture. Later in this book, we will discuss the final connections of both the waste and supply to complete the building system.

• WORDS PLUMBERS USE •

monkey wrench
strap wrench
porcelainized cast iron
single-control faucet
dual-control faucet
compression
waste connection
plug top
tailpiece
trap (P trap)
vermin
vent
siphon (siphoning)
lavatory
vitreous china
drain and overflow type
integrated drain and overflow type
bar hanger
bracket hanger
drum trap
commode
ball cock
stop ball
float ball
washdown bowl
siphon-jet bowl
laundry tray

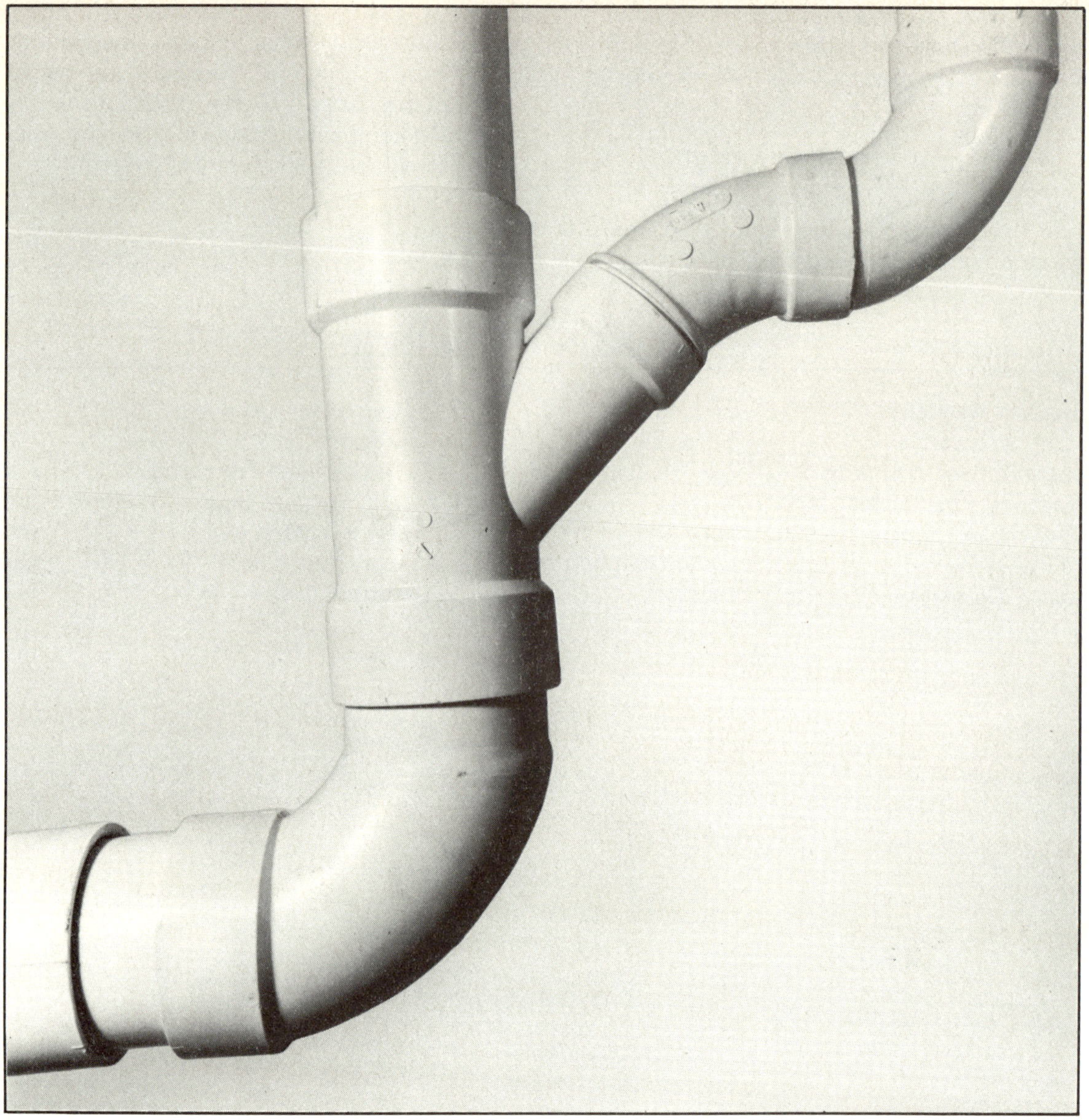

6

THE WATER-SUPPLY SYSTEM

In Chapter 5, we reviewed the different fixtures and components used in a typical dwelling. We also discussed the installation of some of the components. In this chapter, we will discuss the different types of materials and procedures used in the installation of a water-supply system for a new building.

The water-supply system of any building is there to route the water from a single supply point to the many different areas where it will be used. Until the 1930s, many water-supply systems were built using lead pipe as the carrier. Modern systems use a variety of materials. Steel, copper, and a plastic called polyvinyl chloride (PVC) are the main ones. This unit will introduce you to these three main types of pipe and discuss the characteristics of each.

At the completion of this chapter, you will be able to:

- Identify the tools, materials, and procedures used in the construction of a water-supply system for a typical dwelling.
- Construct a segment of a typical dwelling's water-supply system from a diagram using the appropriate tools, materials, and procedures.

• MALLEABLE-STEEL PIPE •

Most of you are familiar with steel pipe, but did you know that there are really three different ways of manufacturing it? The first and most common way is the butt-weld method. In butt-weld pipe, a flat sheet of steel with the edges squared is curved to form the proper size of pipe, and then the two square edges, almost touching, are welded together (Figure 6-1). Butt-weld pipe is not as strong or as resistant to rupture as either of the other two types, the lap-weld or seamless, because the weld is only as deep as the pipe wall.

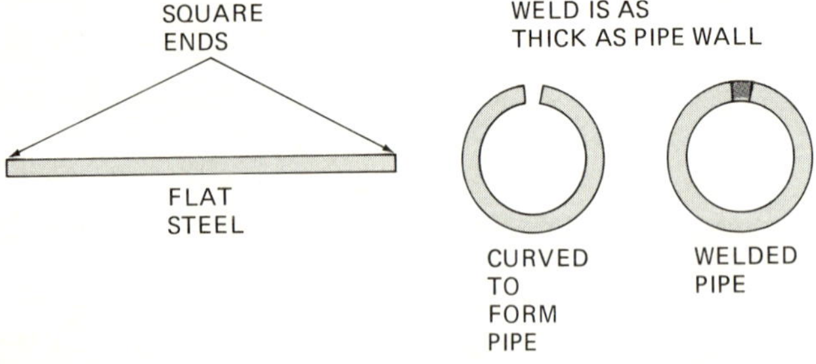

Figure 6-1. Making butt-weld pipe

The second type of manufactured pipe is the lap-weld pipe. This is made by much the same process as the butt-weld pipe, with the depth of the weld as the main difference between the two types (Figure 6-2). The angled edge depicted in the figure allows for a deeper weld, which increases the strength of the pipe and its resistance to rupture.

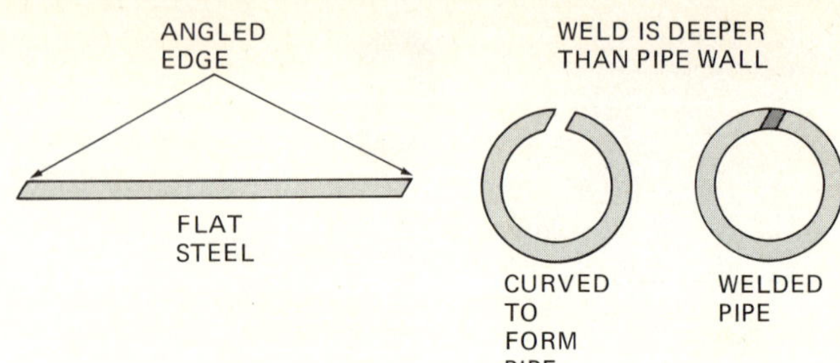

Figure 6-2. Making lap-weld pipe

The third process of pipe manufacturing is entirely different from the first two. The pipe is formed while the steel is still hot by a special forming machine. This results in a solid-wall pipe with no seams (Figure 6-3). Seamless pipe, having no weld at all, is the strongest of the three main types manufactured.

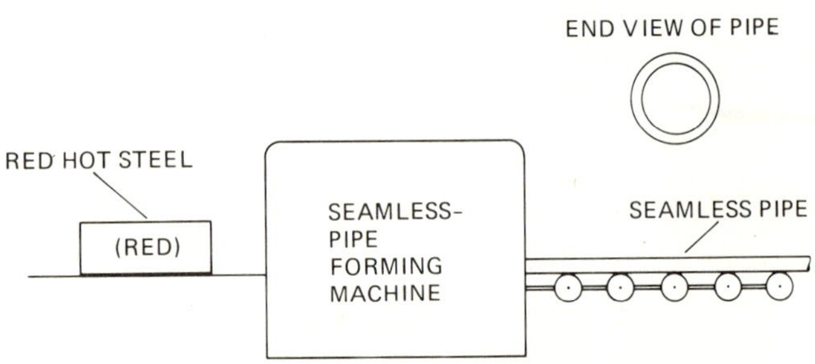

Figure 6-3. Making seamless pipe

Each of these manufacturing processes produces steel pipe which is not coated and is subject to rust and other deterioration. In order to make the pipe more resistant to rust, one of two common coatings is applied. The most rust-resistant pipe is pipe that has been coated with zinc. This pipe is normally referred to as galvanized pipe. The second type of coating is black paint, and it is applied through a dipping or spraying process. This pipe is referred to as black pipe.

• SIZES OF PIPE •

When a plan or a blueprint specifies ½-inch (½″) pipe for a certain water supply, it means that the pipe must have a ½-inch opening, or inside diameter (ID). Steel pipe, either galvanized or black, is made in all the sizes shown in Table 6-1, with each size referring to the size of the inside diameter of the pipe.

Although the table implies that pipe sizes stop at 12 inches, actually all pipe larger than 12 inches is measured by the outside diameter (Figure 6-4).

Up to this point, we have assumed that all pipe has the same wall thickness. Water-supply systems normally use the steel pipe designated as "standard"

TABLE 6-1 NOMINAL STEEL-PIPE SIZES

Nominal Size, Inches	ID, Inches	OD, Inches	Threads per Inch
¼	0.36	0.54	18
⅜	0.49	0.67	18
½	0.62	0.84	14
¾	0.82	1.05	14
1	1.04	1.31	11½
1¼	1.38	1.66	11½
1½	1.61	1.90	11½
2	2.06	2.37	11½
2½	2.46	2.87	8
3	3.06	3.50	8
3½	3.54	4.00	8
4	4.02	4.50	8
5	5.04	5.56	8

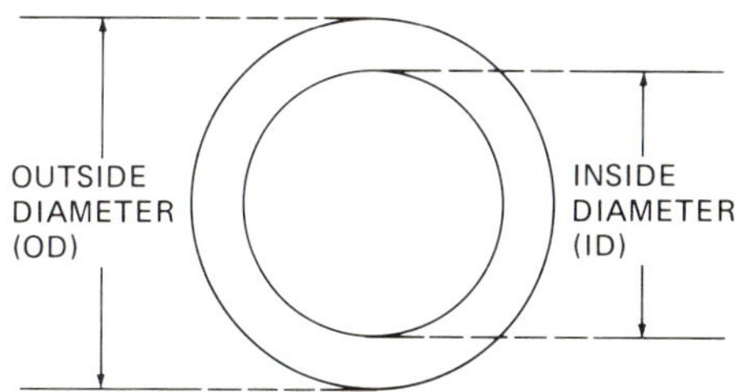

Figure 6-4. Identifying pipe diameters

pipe. Steel pipe is available in two other wall thicknesses for special needs, such as high-pressure steam lines or other high-pressure applications. These thicker-walled pipes are designated as extra heavy (extra strong) or double heavy (double extra strong). Figure 6-5 shows the differences between these three different pipe designations.

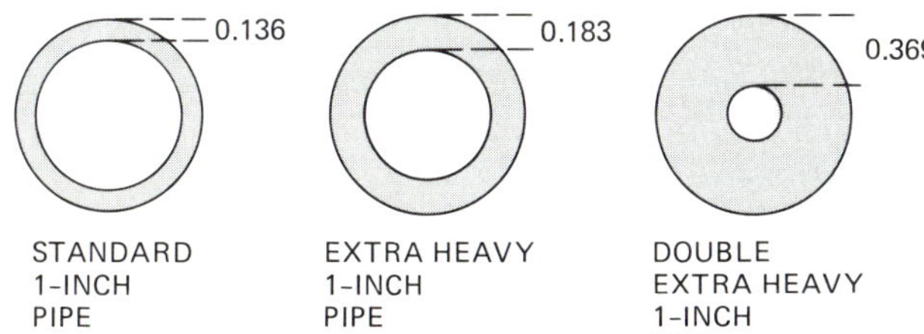

Figure 6-5. Pipe wall thickness

• QUESTIONS •

1. The butt-weld pipe is the strongest of the welded steel pipes. (T or F)
2. The steel pipe with no weld is called a lap-weld pipe. (T or F)
3. What is the ID of a ¾-inch pipe?
4. What is the ID of a 16-inch pipe?
5. A high-pressure steam line would use "standard" pipe. (T or F)
6. Black pipe is the most resistant to rust and corrosion. (T or F)

• JOINING STEEL PIPE •

There are two basic methods of joining steel pipe, either galvanized or black. The primary method is the thread joint, which ensures a watertight fit through pressure on the metal thread facings of the pipe and the receptacle fitting. The second basic method of joining steel pipe is by a welding process (either electric-arc or oxyacetylene) which produces a stronger and more durable joint for high-pressure applications.

The screw-on, or thread, joint is the most common joining process used in the plumbing industry. This joint makes use of tapered threads. The threads on the pipe are external, or male, and the threads in the fitting are internal, or female. The thread process can be used on all sizes of steel pipe (Figure 6-6). The primary method of cutting these threads

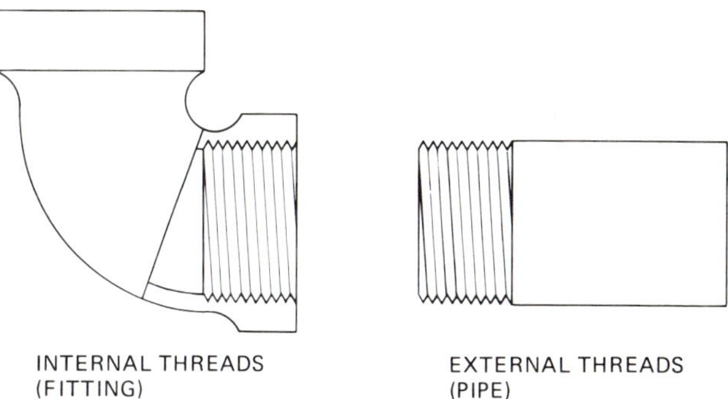

Figure 6-6. Pipe joint threads

is by using a stock equipped with a set of fixed or adjustable dies. Dies are manufactured for each pipe size to cut a specific number of threads per inch according to the American Standards for Pipe and Pipe Thread (Figure 6-7).

External threads should have a minimum of six to seven perfect threads. That is, the threads must be sharp at both the top and the bottom so they will match evenly with the internal threads of the fitting. Any imperfection in the threads such as nicks, cuts, broken threads, or not enough perfect threads will cause the joint to leak.

Another important tool in the cutting and threading of steel pipe is the pipe vise. This vise differs from others in that it has a latch or lock and the jaws are V-shaped and equipped with sharp teeth.

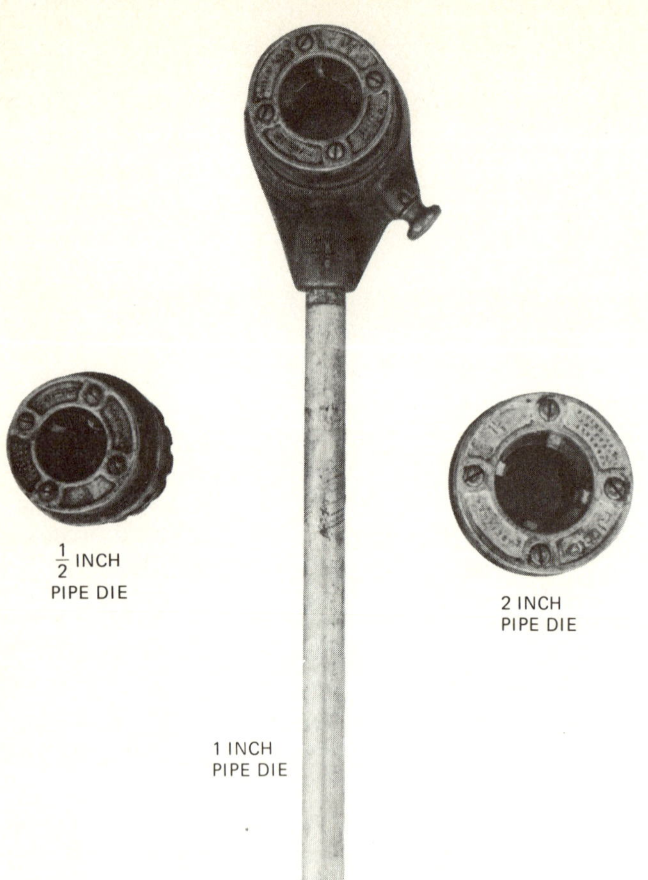

Figure 6-7. Pipe-thread dies

There are many different types of pipe vises. A ture of two of them is shown in Figure 6-8 (see figure photo at right).

• STEEL-PIPE FITTINGS AND NIPPLES •

Steel-pipe fittings are manufactured in a variety of shapes and in all sizes. Because of the rigidity of steel pipe, it is not practical to bend it to change directions. Most direction changes will therefore be made using manufactured fittings. These fittings are standardized in size, thread cut, and degree of bend.

Figure 6-9 identifies the most common fittings used with steel pipe. Each type of fitting can be purchased in either the galvanized or black finish. Care must be taken to use only galvanized fittings with galvanized pipe and black fittings with black pipe.

Steel-pipe nipples are short pieces of pipe threaded on each end. Hand-cutting threads on short (6 inches or less) pieces of pipe is difficult and requires special tooling or techniques. To save on-the-job time and special tool expense, nipples are available in precut and threaded form in lengths ranging from those designated "close" and "shoulder" up to 12 inches in ½-inch increments (see Table 6-2).

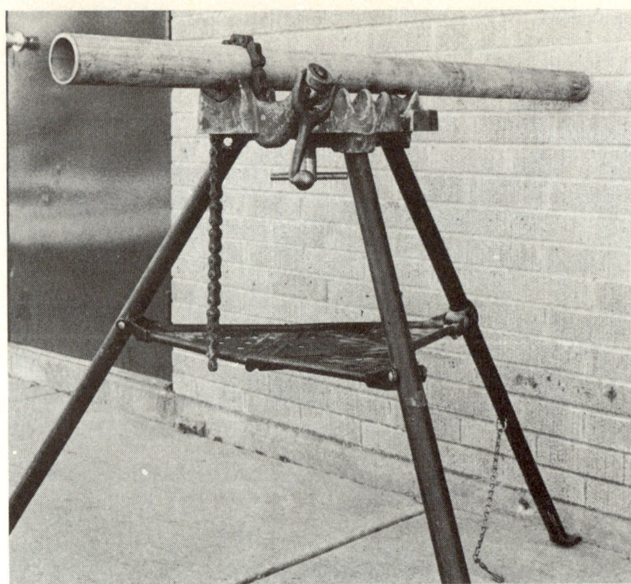

STANDARD VISE

ELECTRIC VISE

Figure 6-8. Standard and electric pipe vises

TABLE 6-2 PRECUT NIPPLE SIZES

Close	5	9
Shoulder	5½	9½
2	6	10
2½	6½	10½
3	7	11
3½	7½	11½
4	8	12
4½	8½	

Nipples are measured end to end.

• QUESTIONS •

1. A perfect thread must be sharp at both the top and bottom of the thread. (T or F)

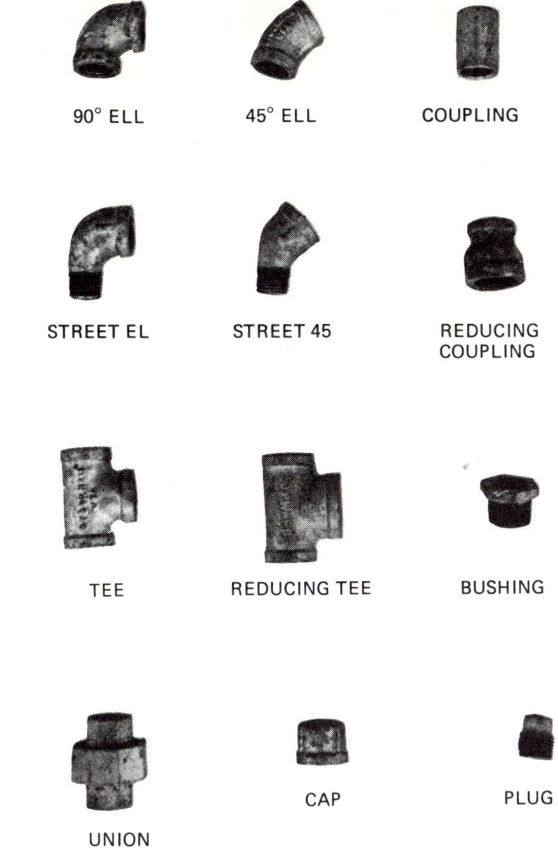

Figure 6-9. Common pipe fittings

2. Imperfections in a pipe thread probably will not cause leaks. (T or F)
3. Nipples have female threads on each end. (T or F)
4. What is a fitting with a 90-degree bend called?
5. What is the most common method of joining steel pipe?

• MEASUREMENTS •

Measurements for the water-supply system are made using the standard dimensions. At the present time in the United States, this standard is the foot and inch. Measurements in the future will be made using the metric system. It is necessary for every plumber to be able to use a ruler and to interpret those measurements properly down to ⅛ inch (⅛″). Measurements are generally specified as center-to-center, center-to-end, end-to-end, or end-to-back (Figure 6-10).

Allowances must be made for the amount of pipe which is threaded into the fitting (Figure 6-11). These allowances must be made when measuring, before cutting and threading the length of pipe. Normal allowances for different pipe sizes are given in Table 6-3.

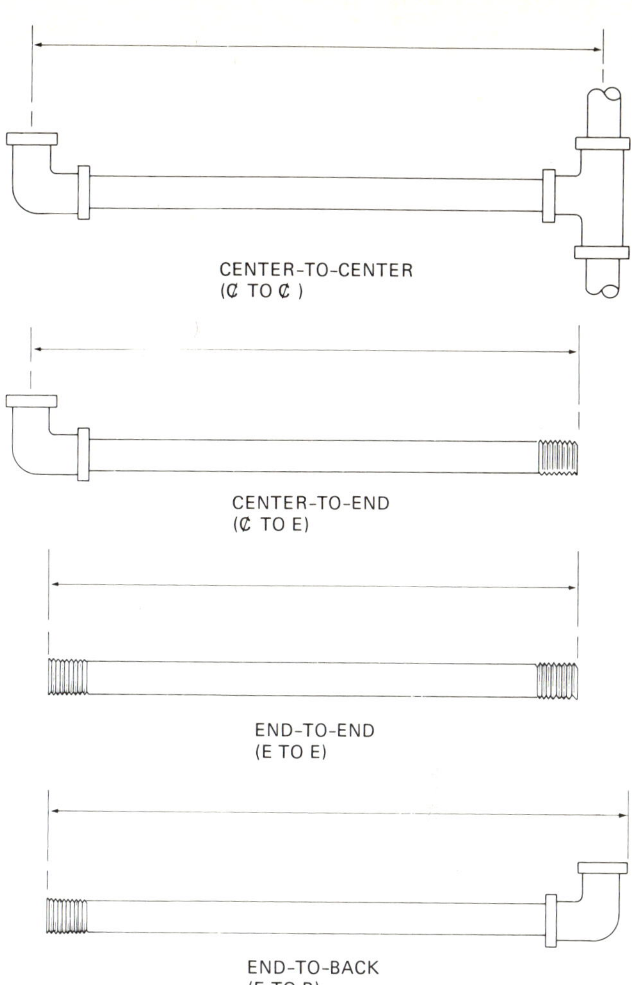

Figure 6-10. Pipe measurement

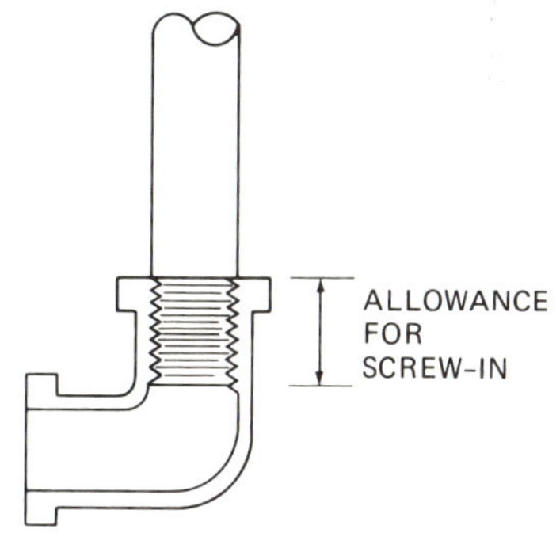

Figure 6-11. Thread allowance

TABLE 6-3 NORMAL ALLOWANCES FOR DIFFERENT PIPE SIZES

Pipe Size	Allowance
½″	½″
¾″	¾″
1″	⅝″
1¼″	⅝″
1½″	⅝″
2″	¾″

• QUESTIONS •

1. What is the measurement from the center of the ell to the end of the pipe?
2. What is the measurement from the center of the tee to the back of the ell?
3. What is the end-to-end measurement of the pipe?
4. What is the measurement from the back of the tee to the back of the ell?
5. If the center-to-center measurement of a section of pipe must be 22 inches, how long should the pipe be end-to-end?

• SPECIAL FITTINGS •

Occasionally fittings used in a water-supply system are there for special purposes. Some of these fittings allow a different type of pipe (copper, PVC, lead) to be connected to a steel-pipe system. Others allow a different size of steel pipe to be used. Figure 6-12 shows the more common special fittings.

Figure 6-12. Common special fittings

• THE PLUMBER'S TOOLS •

There are a variety of tools which are peculiar to steel-pipe installation. Figure 6-13 pictures one of each of the more common tools necessary to assemble and install steel pipe properly.

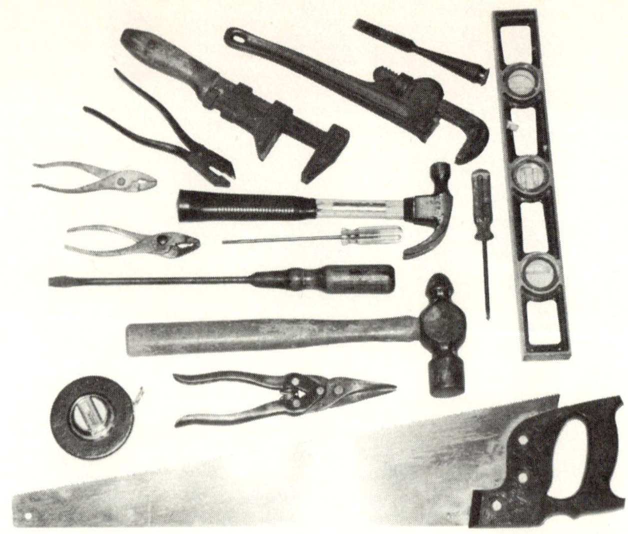

Figure 6-13. Typical plumber's tools

Pipe Wrenches

In discussing these tools and their proper use, we will begin with the pipe wrenches. Follow Figure 6-14 as we discuss the peculiar points of the wrench.

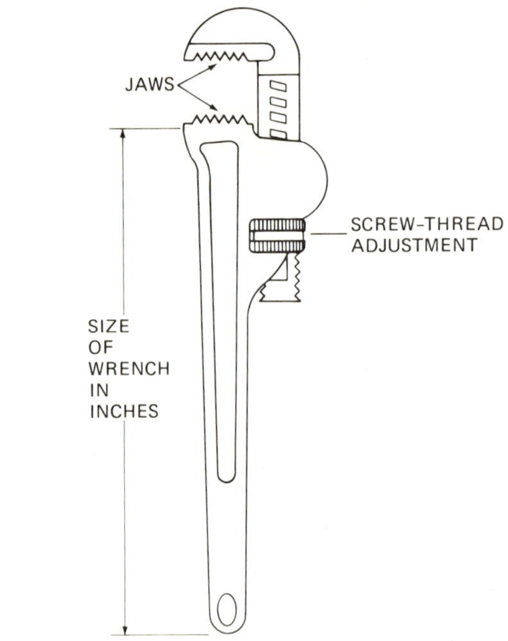

Figure 6-14. Pipe wrench

The most important difference between a pipe wrench and the other wrenches are the jaws, both the upper and lower. Pipe-wrench jaws are equipped with teeth to grip and hold the pipe as pressure is applied to the handle. A second important point is the ability of the jaws to adjust to the different sizes of pipe through the use of a screw thread and spring lock mechanism. The third, and most important to the proper use of the wrench, is its length from the lower jaw to the end of the handle. Each size of wrench is designed to be used with certain sizes of pipe (see Figure 6-15).

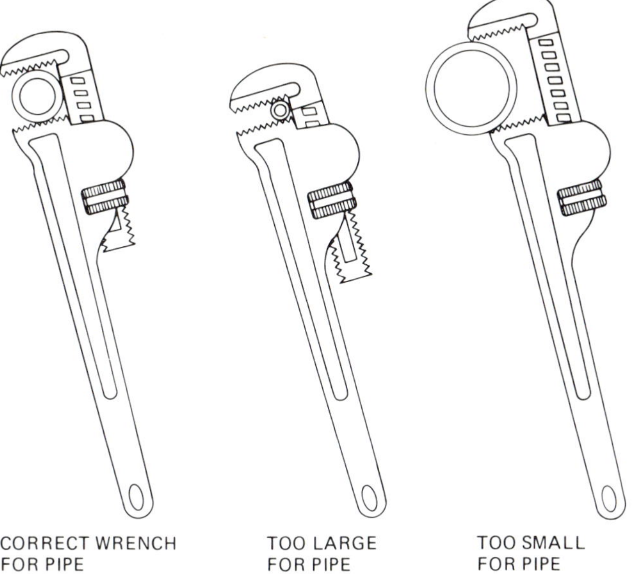

Figure 6-15. Pipe wrench selection

• CUTTING AND THREADING STEEL PIPE •

Earlier in this chapter, a brief statement introduced the process of threading steel pipe. Two other operations, cutting and reaming the pipe, are necessary before the threads can be cut. Figure 6-16 illustrates the two separate tools employed in these operations.

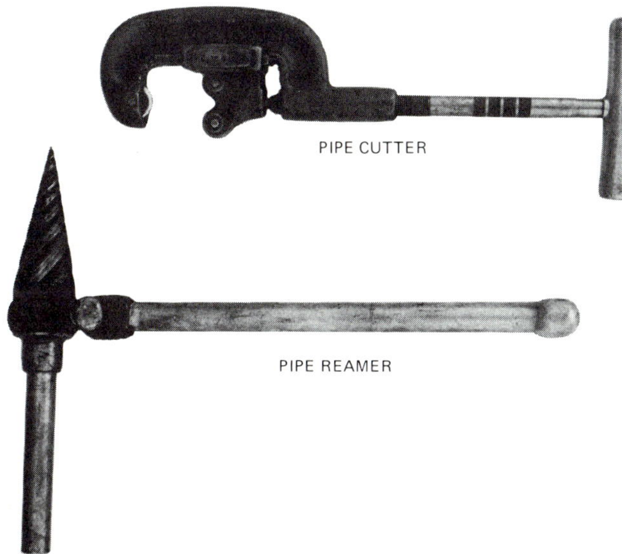

Figure 6-16. Steel-pipe cutting and reaming tools

The pipe cutter is composed of a rigid base with a round cutting blade at the top, a sliding lower jaw which contains two rollers, and a threaded shaft and handle. Operating the pipe cutter is simply a matter of opening the cutter so that the pipe will slide between the cutter wheel and the rollers, tightening the handle until the cutter blade and the rollers are in contact with the pipe, and then rotating the cutter around the pipe while applying pressure to the cutter blade by twisting the handle. This results in a smooth, even cut ready for the reaming and threading operations.

After the pipe is cut, a close look at the new edge will show a small, sharp lip around the inside diameter (ID) of the pipe. This sharp lip may trap rust and other very small particles and cause restriction of the water flow. Removal of this lip is the object of the reaming operation. The part of the reamer which cuts away this interior sharp lip is shown in Figure 6-17.

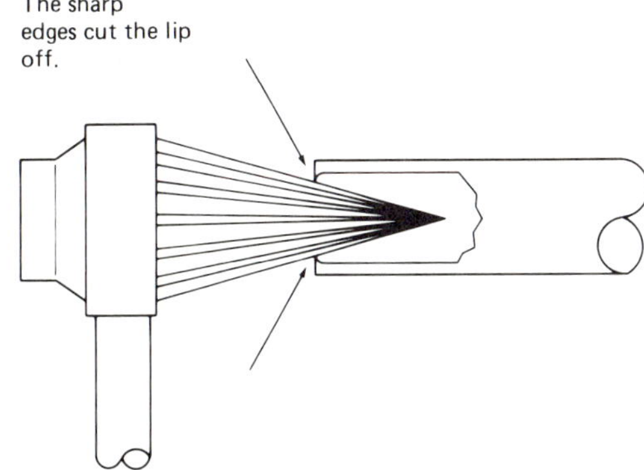

Figure 6-17. Reaming the pipe

The reamer is operated by placing the pointed portion into the inside diameter of the pipe and revolving it two or three turns. This will remove the sharp edge and eliminate the possibility of particle accumulation and restriction of water flow.

Threading the end of the pipe is accomplished through the use of a stock equipped with a set of dies. The primary function of the stock is to hold the dies and to transfer the force from the handle to the cutting edge of the dies. Figure 6-18 shows a cutaway picture of a typical pipe-thread die.

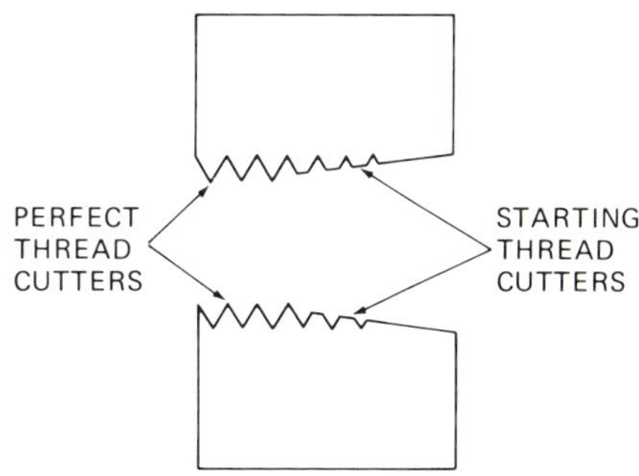

Figure 6-18. Pipe-thread die

Notice the three areas on the die: the starting thread cutters (flat top and bottom), the imperfect thread cutters (flat on the bottom), and the final perfect thread cutters (V shaped on both top and bottom). Cutting threads on a length of pipe requires two distinct applications of pressure. The first, when starting the thread, requires both forward and rotational pressure until the dies are started or caught. The second, after the dies are started, requires rotational pressure (Figure 6-19).

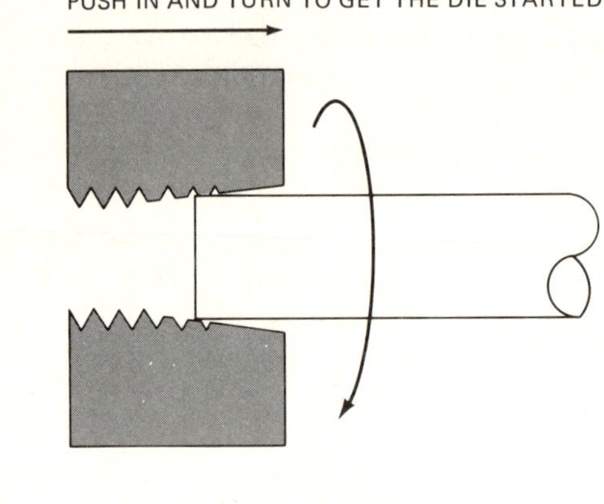

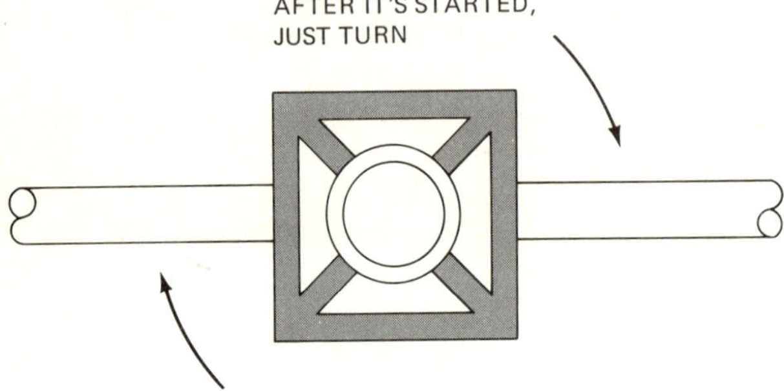

Figure 6-19. Thread cutting with a die

It is important to note the sequence of the reaming operation. Reaming should occur between the cutting of the pipe and the cutting of the threads. If the reaming is done after the threads are cut, the pressure applied by the reamer will expand the threaded end of the pipe, and the joint will probably leak (see Figure 6-20).

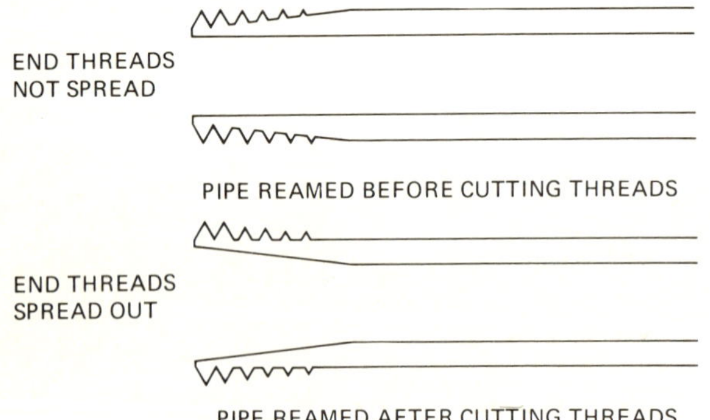

Figure 6-20. Threaded pipe spread by reamer

• WELDING STEEL PIPE •

A secondary method of joining steel pipe is by welding. This method is used when the system will be subjected to high pressure, as would a steam or gas line. The welded pipe joint is extremely strong and as durable as the pipe itself. Remember, high-pressure applications normally use the thicker-wall pipe.

There are two common methods of welding pipe, the oxyacetylene method and the electric-arc method. The oxyacetylene method burns two compressed gases, oxygen and acetylene, at a high temperature, while the electric-arc method uses electricity to melt the steel. The techniques of gas (oxyacetylene) and arc welding will be introduced in Chapter 8.

• QUESTIONS •

1. Pipe wrenches have three special features. What are they?
2. Five different sizes of pipe are listed below. Determine the proper size wrench to be used with each. There may be more than one wrench for each pipe listed.
   a. ¼-inch
   b. ½-inch
   c. ¾-inch
   d. 1¼-inch
   e. 2½-inch
3. Monkey and strap wrenches are used for chrome-plated fixtures or pipes. (T or F)
4. What are the three parts of a pipe cutter?
5. A pipe reamer is usually used after the threads are cut. (T or F)
6. Male threads cut on a length of pipe should have nine perfect threads. (T of F)
7. The two common methods of welding steel pipe are brazing and soldering. (T or F)

• COPPER PIPE •

In many new buildings copper pipe is used in the water-supply system instead of steel pipe. Copper pipe is manufactured by processing hot copper through a forming machine, resulting in a seamless pipe. Some of the advantages of copper pipe over steel are its lightness, its resistance to corrosion, its ability to allow liquid to flow more freely, and the use of swage or solder joints rather than threaded joints.

Of the advantages listed above, the last two are probably the most important. Comparison of the in-

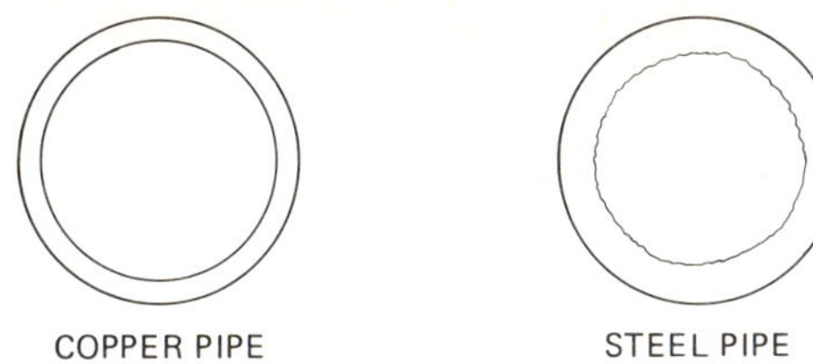

Figure 6-21. End view of copper and steel pipe

side surface of steel pipe with that of copper pipe (Figure 6-21) shows that the copper is much smoother. The small, rough bumps and pits in the steel pipe slow down the flow of liquid, while the smooth wall of the copper pipe does not. This fact generally allows the plumber to use a copper pipe one size smaller to replace steel pipe.

Types of Copper Pipe

There are two types of copper pipe in general use throughout the plumbing trade, a soft-wall copper tubing and a rigid-wall copper pipe. The difference between the two should be evident from their names. One can readily be bent; one cannot. A less obvious difference is the type of joint used with each. The soft-wall tubing normally uses a swage joint, while the rigid-wall tubing uses a soldered joint (Figure 6-22).

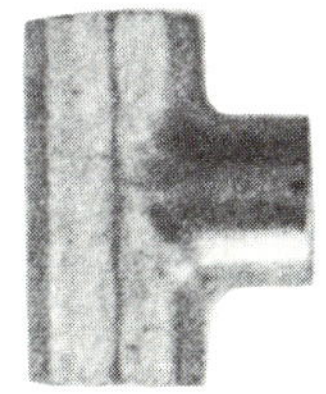

Figure 6-22. Copper-joint fittings

These joints do not require the additional tooling for thread cutting, nor do they require prethreaded nipples. This is another advantage of copper over steel pipe. The measurements used with rigid-wall

TABLE 6-4 NORMAL ALLOWANCES FOR DIFFERENT PIPE SIZES

Pipe Size	Allowance
½″	½″
¾″	¾″
1″	⅝″
1¼″	⅝″
1½″	⅝″
2″	¾″

copper pipe are the same as those used with steel pipe. Measurements with soft-wall copper tubing are similar but are less exacting because of the flexibility of the pipe. Table 6-4 shows the measurement allowances for rigid-wall copper pipe.

Cutting Copper Pipe

Copper soft-wall tubing or rigid-wall pipe is easily cut with either a hand hacksaw or a pipe cutter that is similar to but smaller than the steel-pipe cutter. The inside of the cut copper pipe must be reamed out in the same manner as the steel pipe. Copper-pipe reamers are usually attached to the cutter base. At other times, when a reamer is not available, a file or knife blade may be used.

Joining Copper Pipe

Since rigid-wall copper pipe does not readily bend, fittings are used to make changes of direction in the supply system. Though the fittings pictured in Figure 6-23 look like the steel fittings discussed earlier in this chapter, there are some major differences. They are lighter, appear smaller, and have no female threads.

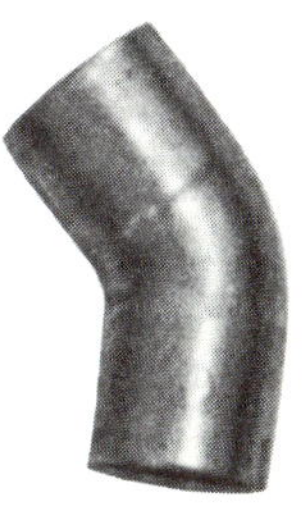

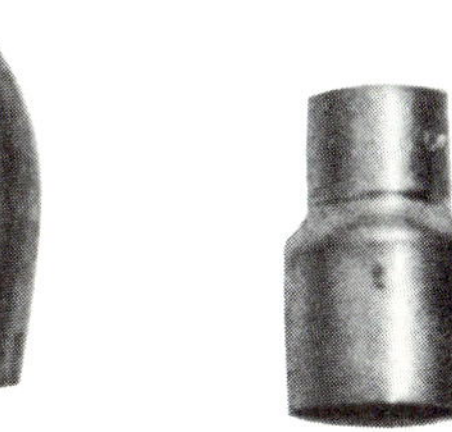

Figure 6-23. Rigid-wall copper-pipe fittings

Figure 6-24. Special rigid-wall copper-pipe fittings

Special fittings are manufactured to permit the joining of both types of copper pipes to a steel-pipe plumbing system. Figure 6-24 has pictures of many of these combination fittings.

Rigid-wall copper pipe is joined to the fitting by soldering. Solder is a soft metal composed of tin and lead. If solder is given the designation 50/50, it means that it is composed of 50 percent tin and 50 percent lead. If the designation is 60/40, it means that the solder contains 60 percent tin and 40 percent lead. This is important to understand because, as Table 6-5 shows, the more tin the solder contains, the lower the melting point will be.

TABLE 6-5 COMPOSITION OF SOLDER

% Tin	% Lead	Approximate Melting Temp (°F)
20	80	530
30	70	495
40	60	460
50	50	420
60	40	370

Preparing a soldered joint for copper pipe requires some type of abrasive material (sandpaper, emery cloth, or steel wool) and soldering flux (paste or liquid). Refer to Figure 6-25 as we discuss the steps in preparing and soldering the joint.

1. Clean about ½ inch of the end of the pipe with the abrasive material until it has a bright, shiny appearance.
2. Coat the shiny area with flux.
3. Repeat the cleaning and fluxing process on the inside of the fitting to be joined with the pipe.
4. Insert the fluxed end of the pipe into the prepared fitting opening, and using a twisting motion, seat the pipe to the back of the fitting.

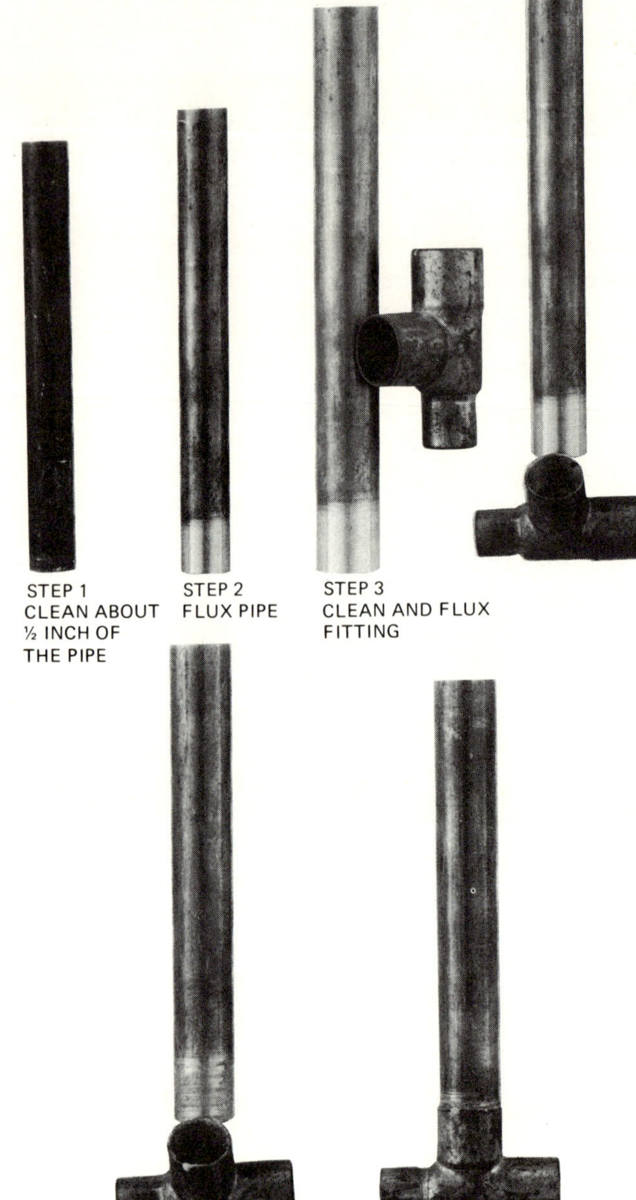

Figure 6-25. Soldering the copper-pipe joint

Completing the joint, or soldering it, is easy if you remember a few basic things:

1. Use an appropriate heat source. Most plumbers use a gasoline blowtorch, a liquefied-petroleum (LP) gas torch, or an oxyacetylene torch (Figure 6-26).

2. After the solder melts, it will only flow to the heat source. Therefore, just heating the pipe and not the fitting will not draw solder through the entire joint. Make sure to heat all sides of the pipe and fitting so that the solder can be drawn through the entire joint (Figure 6-27).

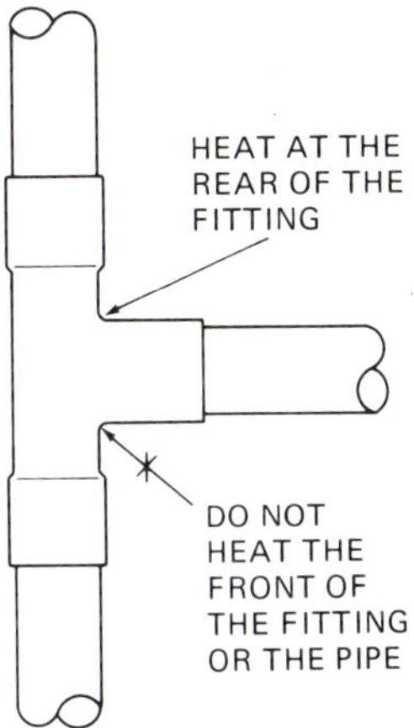

Figure 6-27. Heating the solder joint

3. Solders with different melting points should be used. Figure 6-28 shows a practical application of the use of two such solders. (See pg. 66)

Soft-wall copper tubing, as was mentioned earlier, can readily be bent to change the direction of the pipe and has many applications in the trade. It is generally used for short runs or when flexibility of

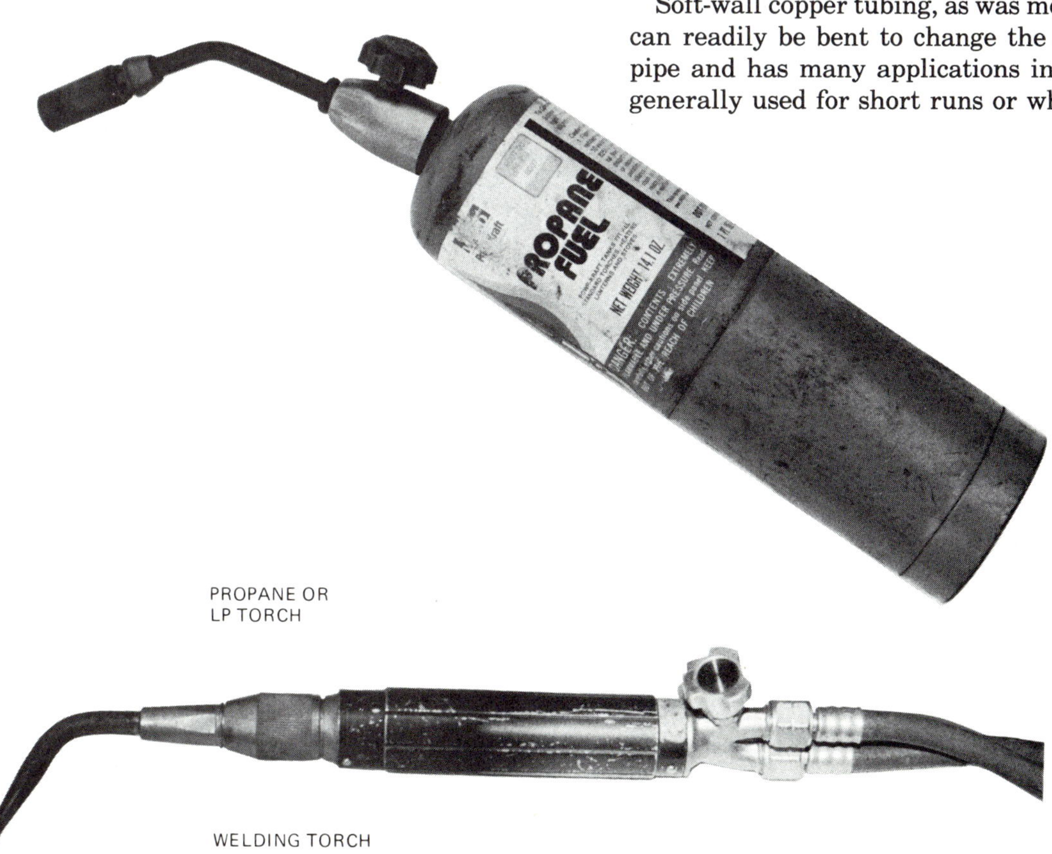

Figure 6-26. Heat sources for soldering

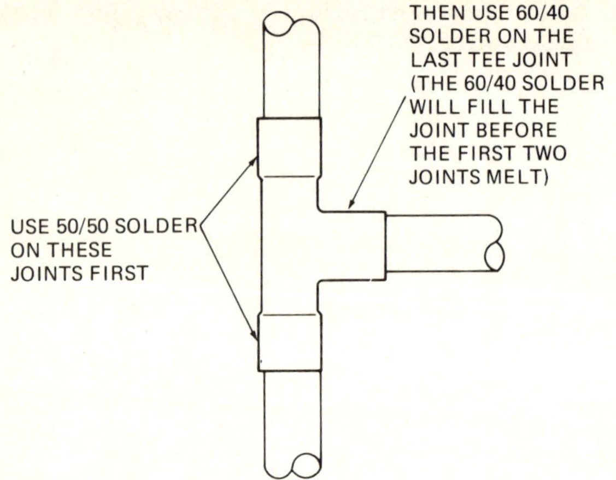

Figure 6-28. Using solders with two different melting points

a supply line is necessary. Although a change of direction can be made by curving the tubing, connections to existing steel or copper lines must be made using a swage or compression fitting. Figure 6-29 shows a cutaway view of each of these fittings.

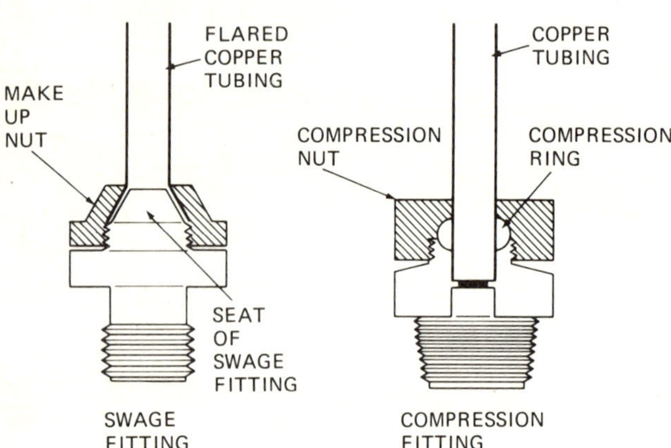

Figure 6-29. Soft-wall copper-pipe joints

Joining copper tubing with a swage fitting requires a special tool, called either a swaging tool or a flaring tool. As you noticed in Figure 6-29, the end of the copper tubing is flared out to fit the seat of the swage fitting. This flaring operation is done by the tool shown in Figure 6-30. The steps in the flaring operation are as follows:

1. Square and straighten the end of the tubing.
2. Slide the makeup nut over the end of the tubing.
3. Insert the tubing into the correct size hole in the flaring tool and tighten down the holding block.
4. Position the flaring block over the tubing and tighten down until the copper tubing is flared out on the seat of the holding block. Remove the tubing from the flaring tool.
5. Seat the flared tubing on the swage fitting and tighten down the makeup nut for a watertight joint.

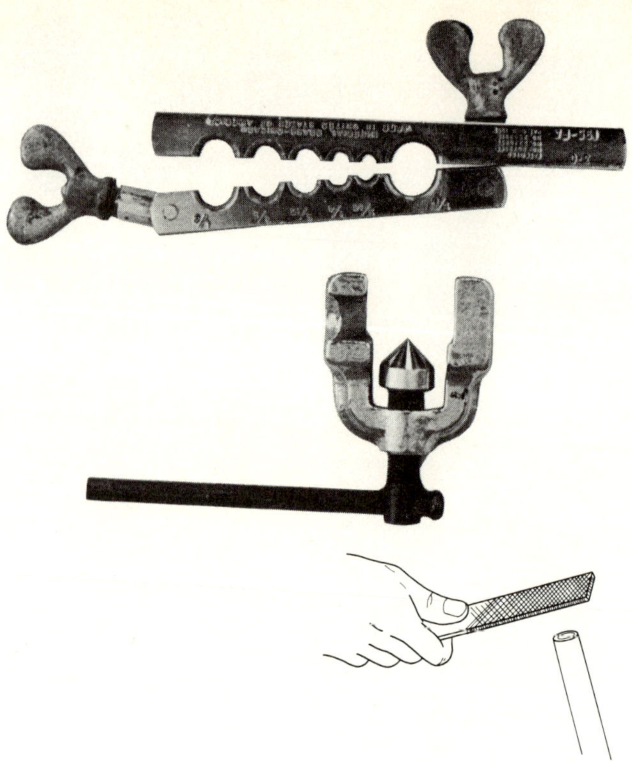

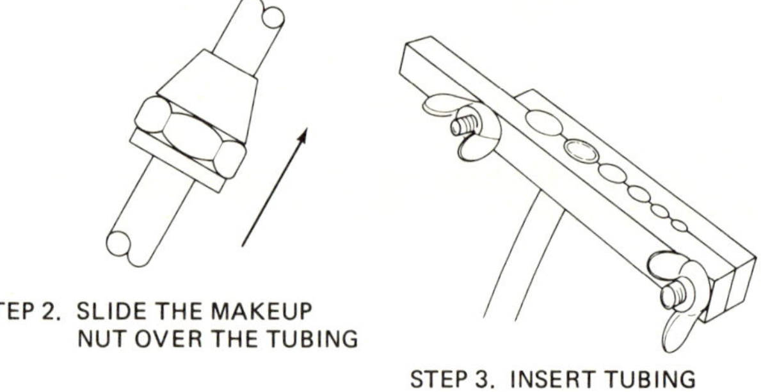

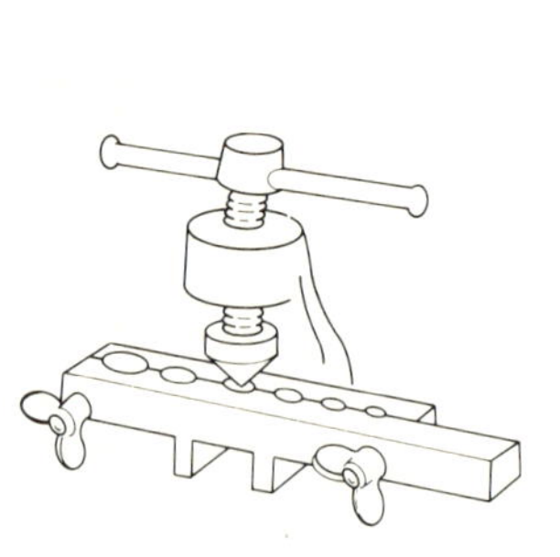

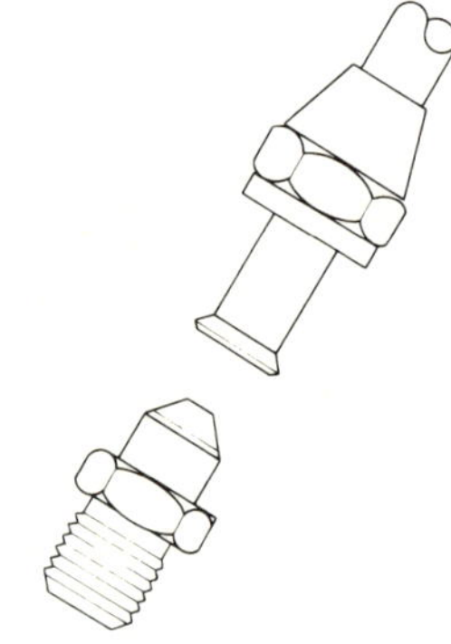

Figure 6-30. Making a swage joint

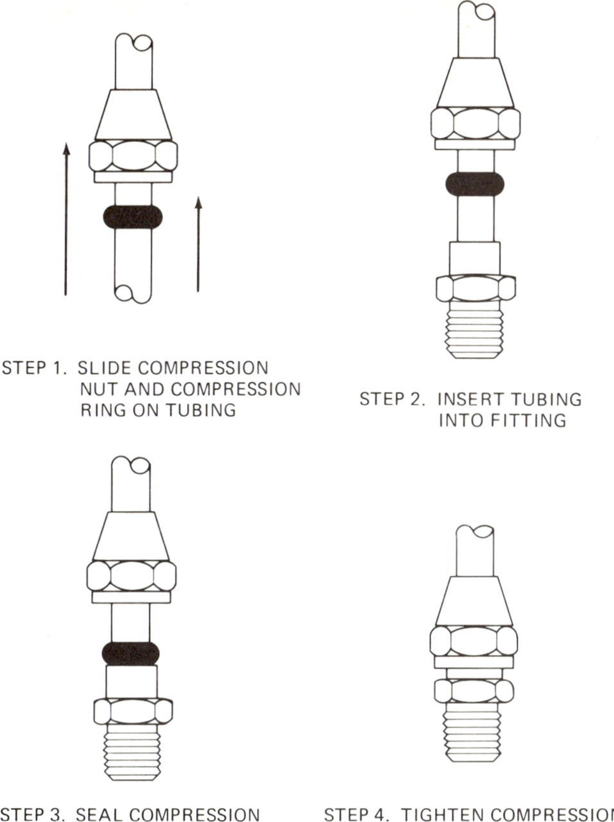

Figure 6-31. Making a compression joint

The compression fitting uses no special tools. Follow the series of steps in Figure 6-31.

1. After the tubing is cut and reamed, the compression nut and then the compression ring are slid onto the tubing.
2. The tubing is then inserted into the fitting.
3. The compression ring is brought down and seated.
4. The compression nut is then brought down and tightened.

• QUESTIONS •

1. What are the two main advantages of copper pipe?
2. A ⅜-inch-ID copper pipe could be used to replace a section of ½-inch-ID steel pipe. (T or F)
3. How much tin does 60/40 solder contain?
4. 40/60 solder will melt before 60/40 solder. (T or F)
5. What are the three steps in preparing a proper solder joint?
6. Heating the fitting is sufficient to ensure a well-soldered joint. (T or F)

• PLASTIC PIPE •

Many varieties of plastic pipe are on the market today, but the type used most frequently in the plumbing trades is made of polyvinyl chloride (PVC). PVC pipe is manufactured in much the same manner as seamless steel and copper pipe, by means of a forming machine. It is a hard-wall (nonflexible) pipe and is used in many different applications. In addition to being used in some water systems, it is used for lawn-sprinkler and swimming-pool systems because of its resistance to corrosion or chemical erosion and its ability to flex with ground settling.

Cutting Plastic Pipe

Any of the plastic pipes are readily cut with a handsaw, and the hard-wall PVC pipe can also be cut with a pipe cutter. The edge of the pipe must be squared, either by using a miter box during the cutting operation or by using a rasp file afterward. As with steel pipe, the ID of the plastic pipe must be reamed to smooth out the interior edge of the pipe. The measurements used with hard-wall PVC are basically the same as those used with steel and hard-wall copper pipes. Other plastic pipes are flexible, and the measurements are not as exacting.

Plastic Fittings and Nipples

Fittings and nipples for plastic pipes are available in the same range and almost the same sizes as steel pipe. Combination fittings (like those used in steel-to-copper conversions) are all but eliminated by the use of a plastic screw-thread nipple or fitting. Copper-to-plastic conversions are made using the copper-to-screw-thread fittings.

Joining Plastic Pipe

Hard-wall and some flexible plastic pipes are joined to the fittings by a gluing or cementing process. Though this is relatively simple, certain steps must be followed to ensure a watertight joint. The first two of these steps were covered earlier, squaring the end of the pipe and reaming the inside diameter to remove the burrs.

The most important point in joining any plastic pipe is use of the correct adhesive or cement. Each particular plastic pipe may require a special cement, so be sure that the correct cement is used. The different types of plastic pipe and the type of cement used with each are listed in Table 6-6.

A watertight plastic pipe joint can be made following the steps outlined:

1. Accurately measure, cut, and ream out the pipe.

TABLE 6-6 PLASTIC PIPE JOINTS

Type of Pipe	Type of Joint
PVC, CPVC, ABS Styrene, CAB	Solvent cemented
FRP	Adhesive bonding (bell and spigot or butt strap)
Polyethylene Polypropylene Polybutylene	Heat fusion (butt, electrical, or socket)

2. The end of the pipe and the inside of the fitting should be cleaned using an abrasive material in the same way the copper pipe was cleaned.
3. The proper cement must then be applied to the pipe and the fitting. The cement must be smoothly and quickly applied to each part; short, choppy strokes take too long, and the cement will harden before the two parts can be joined.
4. After the cement is applied, the two parts are slipped together and given a slight turn to spread the adhesive. Many of the cementing agents used harden very quickly, so care must be taken to put them together in the correct position. These steps are covered in Figure 6-32 at right.

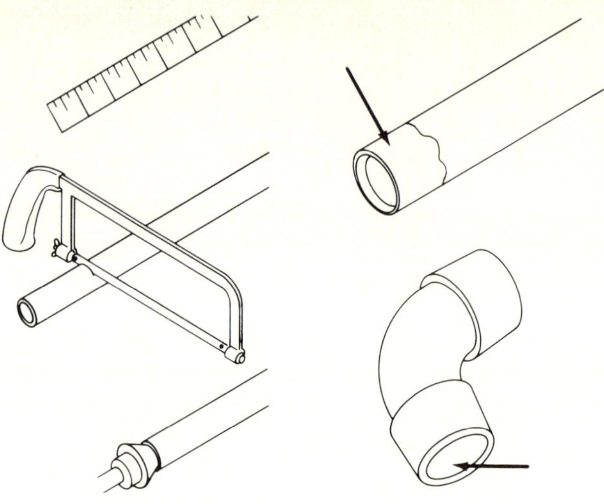

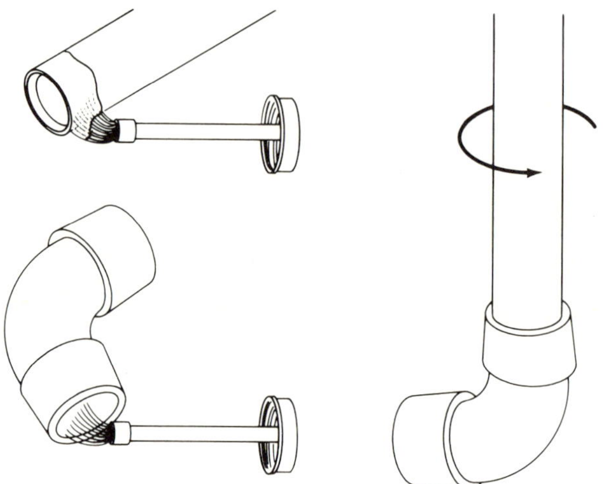

Figure 6-32. Cementing the plastic-pipe joints

• CONTROL VALVES •

We have discussed and worked with the three types of pipes commonly used in water systems. There are many other components of a supply system, the most important of which are the valves. Many different types of valves are used throughout the supply system to control the flow of water. Just imagine how often the water would have to be turned off if each building did not have a separate control valve. Every time a faucet or pipe needed repair, the whole city system would have to be turned off.

In this section, we will discuss the various supply-system control valves. This will not include faucets, which are only terminal or outlet points.

In Chapter 3, we examined the community water-distribution system and discovered that this underground system was composed of water mains. We further learned that these water mains had outlets to each building site or lot served by the community supply and that an outlet ended with a street cock. A street cock is a ground key valve. This is the first type of valve we will discuss. A drawing of a ground key valve and its different parts is shown in Figure 6-33 at right.

As you can see from the picture, this valve works on a very simple principle. When the top bar is turned across the flow, the holes in the stem are

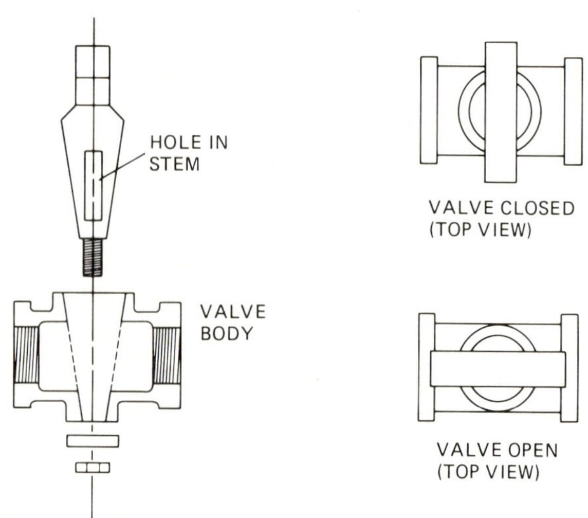

Figure 6-33. Ground key valve

not lined up, and the flow is stopped. When it is turned so that the bar is lined up with the pipe, the hole in the stem is lined up with the openings, and the flow of water is not blocked. This valve is normally made of brass, is resistant to corrosion and chemical action from the soil, and has a long underground life. These valves do not get much use and very seldom develop a leak, but when they do, it requires a ditch large enough so that a person can repair or replace the valve.

The street cock is the primary control point for the water system of any building. It is always ahead of the water meter which counts the number of gallons of water used. In most buildings there is another control valve either just before or just after the meter. Figure 6-34 is a diagram of a closed and open building control valve.

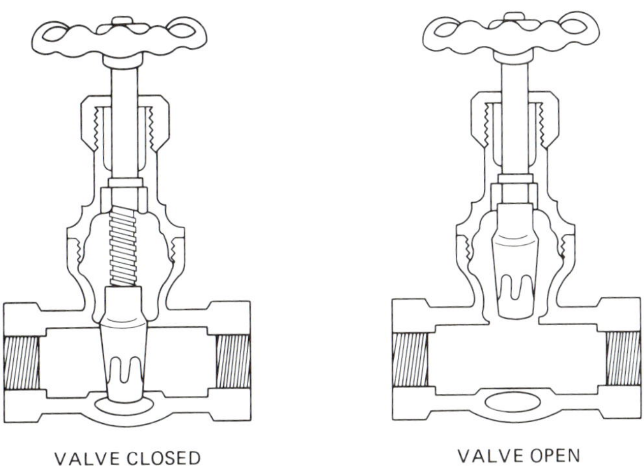

Figure 6-34. Gate-valve operation

This second control point normally makes use of a gate valve. The gate valve is easily recognized by the raised part directly under the handle. It operates much like an overhead garage door. Turning the handle lowers a brass gate across an opening in the body of the valve, closing it off and stopping the water flow. A gate valve allows full flow of water and should be in either the fully open or the fully closed position. The position of this valve, right next to the water meter, is important because it is used to close off the water supply to the building when repairs to the system must be made. Figure 6-35 pictures a type of gate valve.

The most common valve in the water-supply system is the in-line compression valve. This in-line valve is often referred to as a globe valve because of its rounded body. The globe valve can be used to regulate the flow of water because it uses a rubber or neoprene washer which is compressed onto the valve seat to close off the supply. Pictures of a compression valve and its major parts are shown in Figure 6-36 A&B.

Figure 6-35. Gate valve

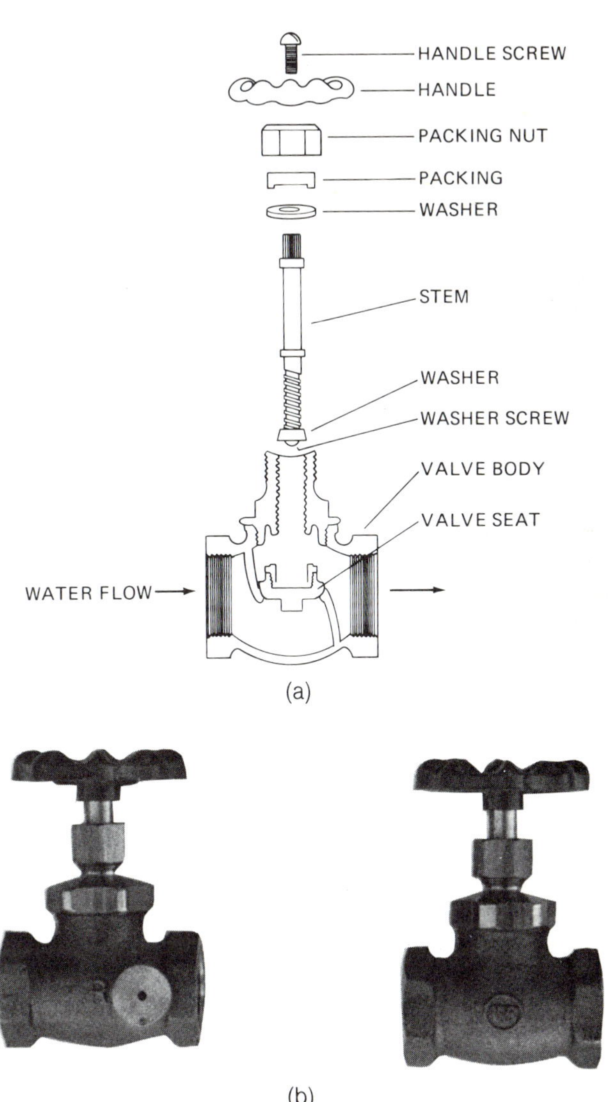

Figure 6-36. Compression valve

The last type of water-system valve we will discuss is the check valve. This valve allows water or other liquid to flow in one direction only. Check valves are used in supply systems for large buildings and in hot-water heating systems, and they have many applications in sanitary systems (Chapter 7). Figure 6-37 shows a cutaway view of a typical check valve.

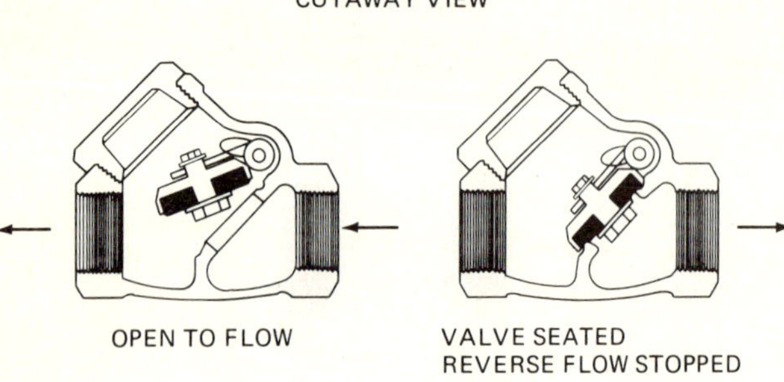

Figure 6-37. Check valve

• SECONDARY WATER SYSTEMS •

In Chapter 3, we looked at the various ways in which water could be supplied: (1) the municipal system, with its central purification and pumping system; (2) the industrial system, a smaller model of the municipal system; and (3) the individual system, which supplies water from a well or other source to a single building or farm. This section will outline the two basic systems of a building and discuss some of the differences between the municipal supply source and the individual system source. Although all these systems use basically the same pipes and fittings, there will be some differences in the way the pipes or lines are routed in the building.

• THE HOT-WATER SYSTEM •

So far we have discussed the pipes, fittings, and valves used in the water-supply system and have made no distinction between the different types of water systems in a building. In this chapter we will only discuss the human use or consumption systems, not the recirculating systems used in heating or air conditioning, which will be discussed in later chapters of this book. Most buildings have two human use or consumption systems, one for cold water and one for hot water. These two systems are supplied by a common source which is divided through the use of fittings and is controlled by valves.

The cold-water lines supply cold water to all plumbing facilities, the kitchen, bath, the laundry, and hose connections, while the hot-water lines supply hot water only to the kitchen, bath, and laundry areas. As you will see in the next section, while there may be more than one subsystem of the cold-water line, there is rarely more than one hot-water line. Figure 6-38 shows the two supply lines in a typical one-story dwelling.

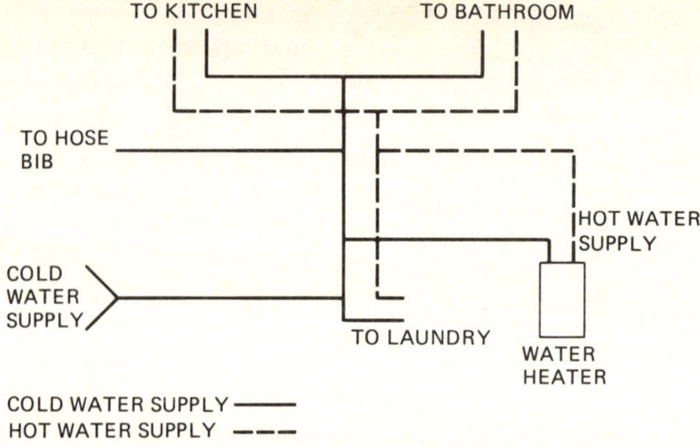

Figure 6-38. Hot-water and cold-water supply lines

• DUAL COLD-WATER SYSTEMS •

At one time in your life, you may have been visting or out camping somewhere and found that it was difficult to rinse all the soap off your face or hands. If you had asked about it, you would probably have been told that it was due to the soft water. This does not mean that the water would be good for a water bed. It just means that many of the dissolved minerals had been removed from the water. Most wells or other individual water supplies (and some municipal systems, as well) furnish water that contains many dissolved minerals. The mineral might be iron, alkali, sulfur, or some other mineral. This type of water is referred to as hard water, and though it can be used in a water-supply system, it is likely to cause some problems. Iron-rich water will stain clothes and plumbing fixtures, sulfur can cause the water to have an unpleasant odor or taste, and alkali- or phosphate-rich water can cause faucets and showers to clog up.

Many people, when facing a problem caused by hard water, purchase a water softener and have it installed in their system. The purpose of a water softener is, just as you suspected, to remove the dissolved minerals from the water and thereby eliminate the problems they cause. The water softener uses a salt (or saline) solution and a filter process to remove impurities from the water supply. A drawing of a typical water softener is provided in Figure 6-39.

The problem that a water softener poses for the plumber is not a matter of connecting it. That is relatively easy. The problem is that not all the water used in the building has to be softened. Water for sprinkling the lawn or garden would not have to be softened, and believe it or not, many people do

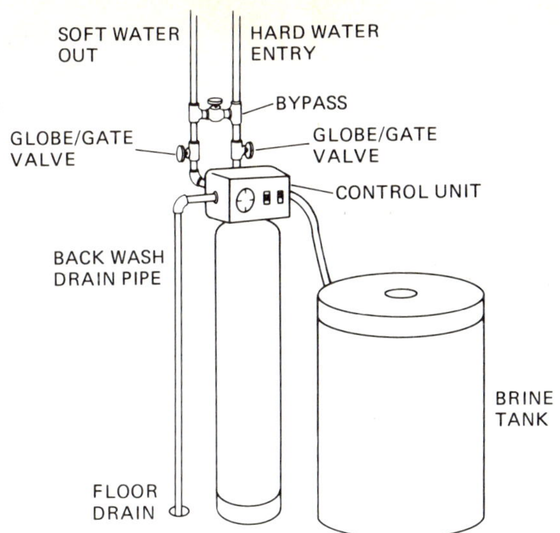

Figure 6-39. Water softener

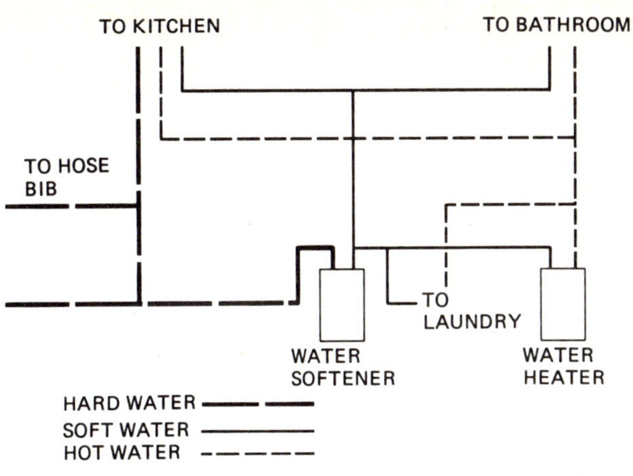

Figure 6-40. Water supply lines using a water softener

not care for the taste of softened water. To eliminate problems like this, it is sometimes necessary to include two separate supply lines for cold water, one for soft water and one for hard water. Figure 6-40 shows a cold-water system with two separate lines.

Another possible problem with an individual supply is purification. The municipal and industrial systems of supply include the purification step, but the water from a well or other source is not always pure enough for human use. There are many commercial water purifiers on the market for use in an individual supply system. As with the water softener, the connection is no problem, but it may be necessary to install separate supply lines to the hose connections and the water closets (commodes).

• SUMMARY •

In this chapter, we have studied the water-supply system of a typical dwelling. We have learned about pipe and the material used in its manufacture, the different methods of connecting pipe, the special tools used primarily for installing pipes, and the different kinds of water and water supplies.

• WORDS PLUMBERS USE •

polyvinyl chloride (PVC)
malleable steel
rupture
glavanized pipe
black pipe
inside diameter (ID)
outside diameter (OD)

thread joint
external
internal
pipe vise
pipe fitting
nipples
pipe wrench

pipe cutter
reamer
threading die
solder
swage joint
combination fitting
flaring tool

miter box
valve
street cock
ground key valve
gate valve
in-line compression valve
globe valve
check valve

7
THE SANITARY SYSTEM

As we noted in Chapter 1, the present-day system of plumbing came about because of the large numbers of people living in cities and towns. You can imagine the problems which could occur if provisions for waste removal were not made. Rodents and insects would run wild, disease would be widespread, and the water supply could become contaminated.

The sanitary or sewer system of a building accumulates and transports the human waste products to a treatment or processing area. It is important to remember that this treatment or processing area can be a simple septic system or a complex municipal sewage plant. The end result is the same, a sanitary method of disposing of human waste.

The purpose of this chapter is to introduce the student to the many different materials, tools, and processes associated with the installation of a building's sanitary system.

After completion of this chapter, you will be able to:

- Identify the tools, materials, and procedures used in constructing the sanitary system of a typical dwelling.
- Construct a segment of a typical dwelling's sanitary system using the appropriate tools, materials, and procedures.

• SANITARY-SYSTEM MATERIALS •

There are many different materials used in the manufacture of pipes for a sanitary system. Table 7-1 is a list of the different types.

TABLE 7-1 MATERIALS USED IN A SANITARY SYSTEM

Component	Material
Soil pipe	Cast iron
Steel pipe	Malleable steel
Plastic pipe	PVC
Cement pipe	Cement reinforced with steel
Vitrified clay tile	Baked and glazed clay
Orangeburg pipe	
Clay drainage tile	Baked clay
Steel-pipe fittings	Malleable steel or cast iron

Notice the wide range of pipe sizes, all the way up to 36 inches in diamter [92.5 mm]. The larger pipes are normally used in the municipal sanitary systems and carry the waste from private, commercial, and industrial buildings. The sanitary system for a single dwelling normally uses 4-inch pipe for the main sewer (horizontal) and stacks (vertical) with a 2-inch pipe for supplemental sewage (sinks, tubs, washbasins, etc.).

• CAST-IRON PIPE •

One of the most common sewage pipes used in the interior of the building is made of cast iron and is called soil pipe. It is manufactured by the centrifugal casting process and is normally made in lengths of 5 feet. Soil pipe is usually made with a large bell or hub at one end, called the hub end. The other end has a small bead and is called the spigot end, although some lengths have a bell or a hub on each end. The difference between these two types is shown in Figure 7-1. Soil pipe also comes in two wall thicknesses, standard and extra heavy.

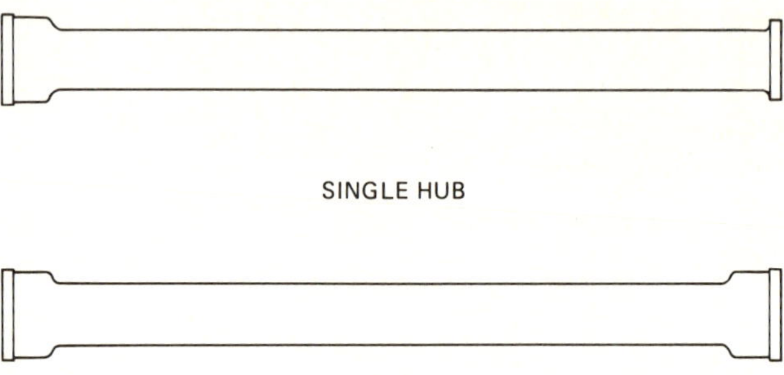

Figure 7-1. Cast-iron soil pipes

Measuring and Cutting Cast-Iron Pipe

As shown in Figure 7-1, the 5-foot length of pipe is measured from the base of the hub to the end of the spigot. The hub is not included in the measurement because it is used to join the pipes together. Figure 7-2 presents a cutaway view of two joined sections of soil pipe.

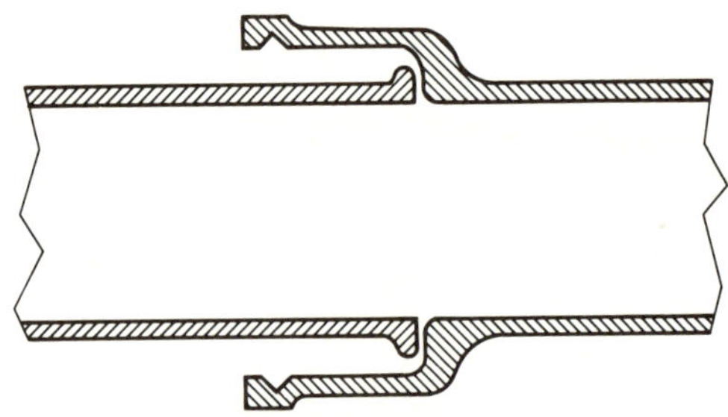

Figure 7-2. Cast-iron soil pipe joint

End-to-end measurement of soil pipe must be made from the base of the hub to the end of the pipe. Other ways of measuring are outlined in Figure 7-3 on the next page.

After you have determined the length of soil pipe that you need, you must cut it squarely to seat it in the hub of the adjoining pipe. There are many

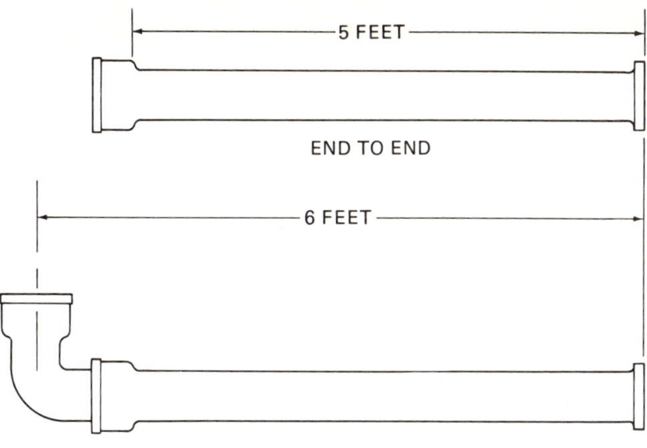

Figure 7-3. Measuring soil pipe

methods of cutting soil pipe, but the method most commonly used by plumbers is the hammer-and-chisel method. This method consists of the following steps:

1. Measure the length of soil pipe needed, starting at the base of the hub, and, using a chalk or soapstone marker, draw a line around the entire pipe where it must be cut.
2. Support the pipe at the marked line by placing a wood block (2 × 4) beneath it. This wood block is the secret of success in cutting soil pipe by the hammer-and-chisel method.
3. Now, using a sharp chisel, align the chisel blade with the chalk mark and, using moderate strokes, strike the head of the chisel with the hammer.
4. Repeat this procedure around the entire pipe, making sure that the pipe does not roll off the support block during the process.
5. Once the pipe has been weakened enough by the chisel blows, it will crack and break along the marked line. Do not expect it to break the first time around. It sometimes takes two or three times around the pipe to weaken it sufficiently.

Figure 7-4 shows the proper sequence of cutting soil pipe using a hammer and chisel.

There are many other methods used to cut soil pipe. Some take longer but assure a better and more even cut.

Another method involves the use of a hacksaw. This procedure is as follows:

1. Using the hacksaw, cut a small groove (about ⅟₁₆ inch deep) all the way around the pipe to weaken it.
2. Place a support block under the section of pipe to be used.

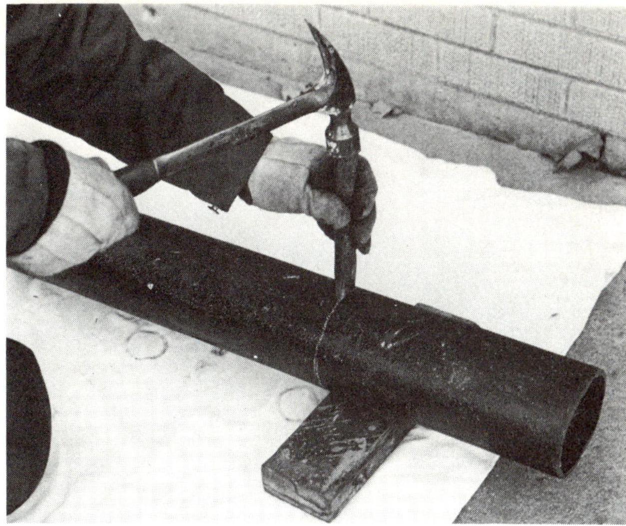

Figure 7-4. Cutting soil pipe, hammer and chisel method

3. Using a hammer, strike lightly on the sawed line until the pipe cracks and breaks.

Figure 7-5 illustrates this method. (See below and page 76.)

Figure 7-5. Cutting soil pipe, hacksaw method

A third method of cutting soil pipe requires a special tool called a soil-pipe cutter. This tool is also known as a chain cutter or pipe cracker. The use of this tool is shown in Figure 7-6. The chain portion of the cutter is wrapped around the soil pipe at the measurement mark and tightened slightly, then the levers are brought together, cutting the pipe evenly and instantly.

Figure 7-6. Cutting soil pipe, chain cutter method

Though a square or even cut is preferred, small irregularities are common. It is suggested that you attempt to square the large irregularities of the cut of soil pipe using a wrench, as demonstrated in Figure 7-7.

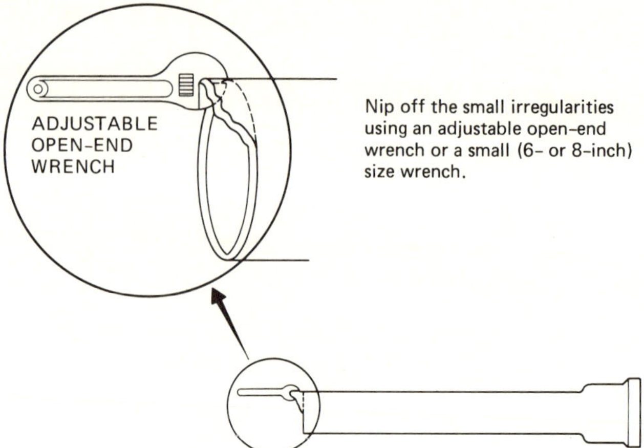

Figure 7-7. Breaking off irregularities, wrench method

Joining Soil Pipe

Before we discuss the caulked lead joint used with soil pipe, it is important that we identify the special tools used in the process. There are two distinct types of tools used in making this joint, and though they look very much alike, there are some very special differences. The two types are yarning irons and caulking irons shown in figure 7-8.

The differences between the two are obvious. The yarning iron has a longer blade to enable the plumber to reach the bottom of the hub joint. The tip of this iron is flat rather than beveled like the tip of a caulking iron. Notice also that caulking irons are beveled two different ways. The inside irons are sharpest at the inside curve, and the outside irons are sharpest at the outside curve. The reason for this will become apparent when we discuss finishing the caulked joint.

The caulked lead joint uses two sealing agents: oakum, a loosely twisted treated hemp rope, and molten lead. After the pipe is cut to size, it is placed in position with the cut end resting on the hub of another pipe. One or two layers of oakum are then tamped down into the hub and, using the yarning irons, tightly packed between the hub and the cut end of the inside pipe (Figure 7-9*a*). The two pipes are then checked for straightness or alignment (Figure 7-9*b*), and the oakum is repacked if an adjustment is made. On vertical points, molten lead is then ladled into the hub directly from the lead pot, filling the remaining space to the top (Figure 7-9*c*).

Horizontal pipes require another step, the use of

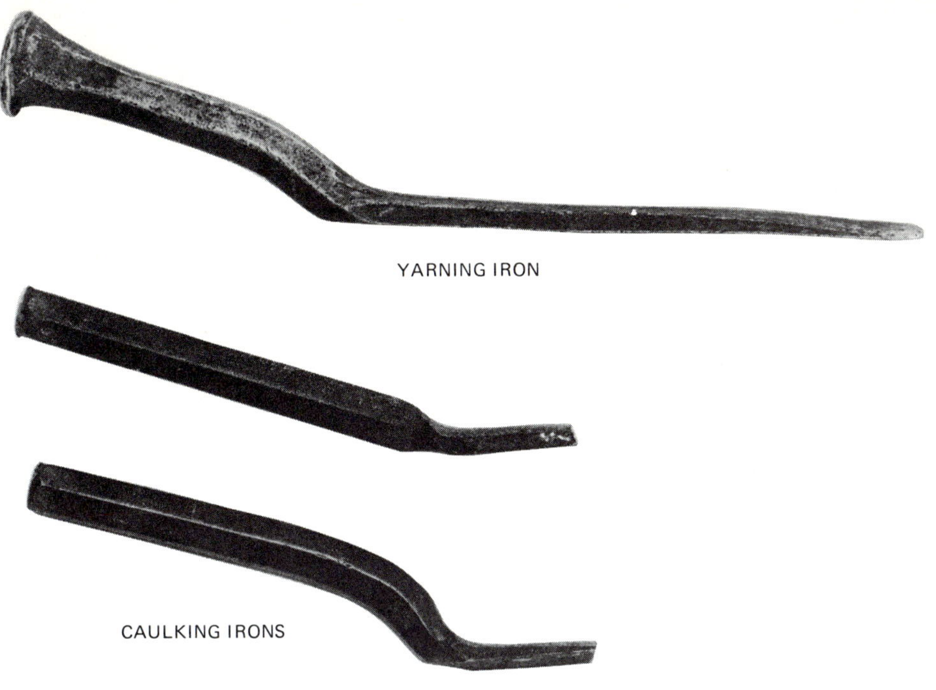

Figure 7-8. Soil-pipe joint tools

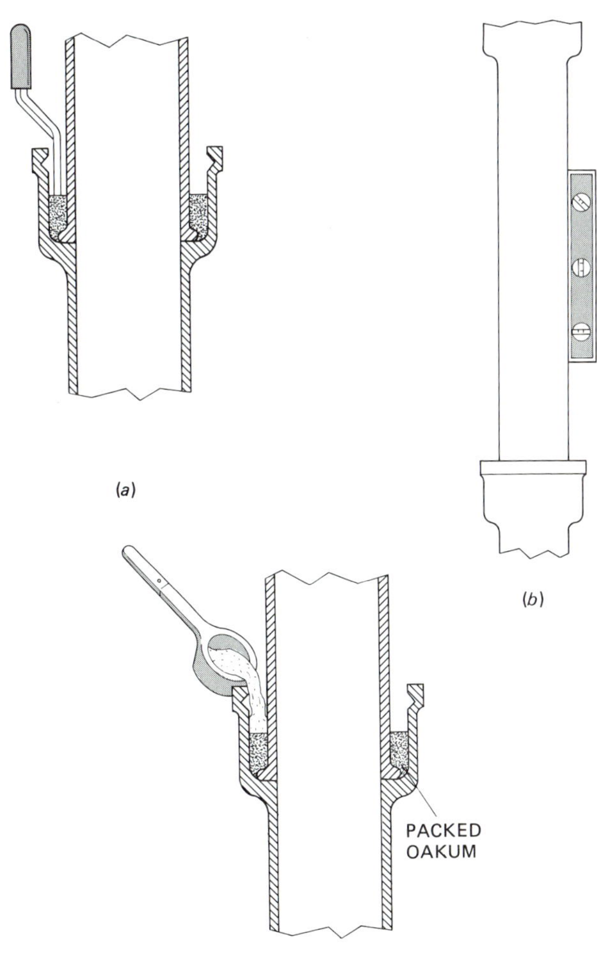

Figure 7-9. *(a)* Packing oakum in the soil-point joint; *(b)* Using a level to check vertical alignment; *(c)* Pouring the molten lead

a joint runner rope, a 1-inch-diameter rope made of asbestos about 2 feet in length. The joint runner rope is wrapped around the pipe, clamped at the top, and then tamped around the edge of the hub to seal off the bottom and sides (Figure 7-10). Molten lead is then poured into the opening at the top and fills the space left in the hub.

One final operation is necessary to ensure the watertightness of this joint. This last step is the caulking of the poured lead seal. Lead is a very soft metal, able to be poured into any shape when melted, but like any metal, as it hardens, it shrinks. As the lead in the hub cools and hardens, it shrinks and leaves very slight cracks between the pipe and the lead seal. Since these cracks could allow liquid or gas to escape, the caulking process is used to pack down the soft lead and close off those small cracks.

Figure 7-10. Preparing a horizontal soil-pipe joint

Figure 7-11 shows a cross section of a properly caulked joint, and identifies the following sequence of steps:

1. Using the outside caulking iron as shown, pack down the lead seal around the entire outside edge of the joint.
2. Using the inside caulking iron as shown, pack down the lead seal around the entire inside edge of the joint. Remember the difference between the inside and outside caulking irons and you will be able to finish a caulked lead joint that looks just like the example.

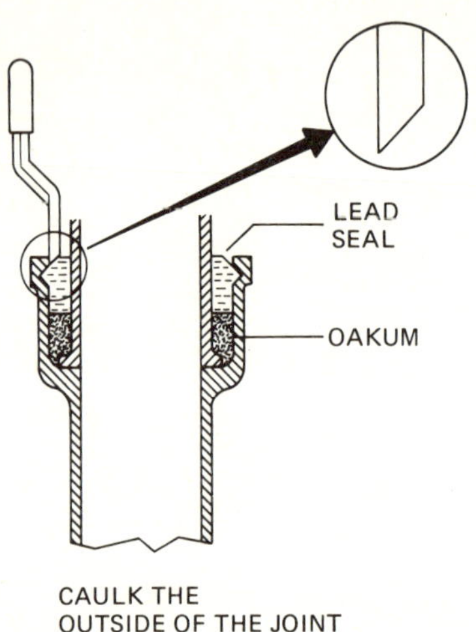

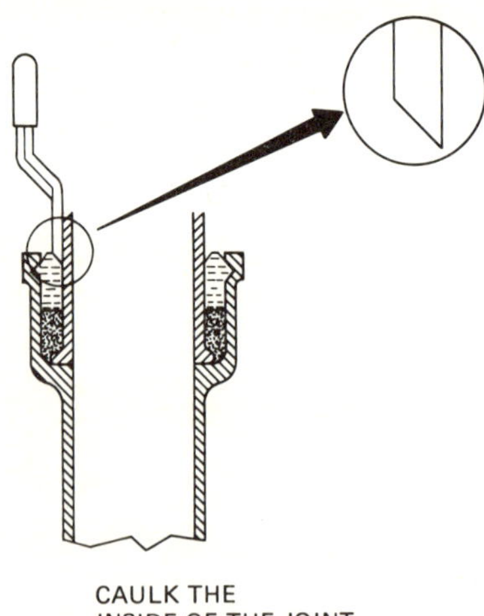

Figure 7-11. Caulking the poured lead joint

Safety with Molten Metals

Molten metal, even the lead used for a caulked joint, can be very dangerous. Any moisture which comes in contact with it will immediately turn to steam. The resulting pressure will act as an explosive and will *blow* the molten metal out of the container. Even a damp ladle can cause this explosion, and the hot metal can seriously injure anyone in the area if it gets on their skin or into their eyes.

A good way to make sure the ladle is dry is to place it over the lead pot as in Figure 7-12. This will serve a twofold purpose: the ladle will dry out and it will get hot. The ladle must be dry so that there will be no moisture present when it is dipped into the lead pot. The ladle should also be hot so that as the lead is removed from the pot, it will stay hot longer and not cool as rapidly.

Figure 7-12. Melting the lead and drying the ladle

Gasket Joints

Some local codes permit the use of a neoprene gasket joint in place of the caulked lead joint. This gasket is sized to fit tightly into the hub of a length of soil pipe or fitting after being lubricated. Then when the spigot end of the connecting pipe is forced in, it forms a watertight seal.

No-Hub Joints

Another method of joining soil pipe is the relatively new no-hub joint. This joint is made up of a neoprene sleeve and gasket, a stainless-steel sleeve, and two stainless-steel clamps. Figure 7-13 is a cross-sectional drawing of a no-hub joint and its parts.

Joining pipe using the no-hub joint process is quick and easy, but some local codes do not permit its use.

Soil-Pipe Fittings

So far, we have only discussed soil pipe in a straight-line application, from one point to another. Curved sections of soil pipe are made so that turns

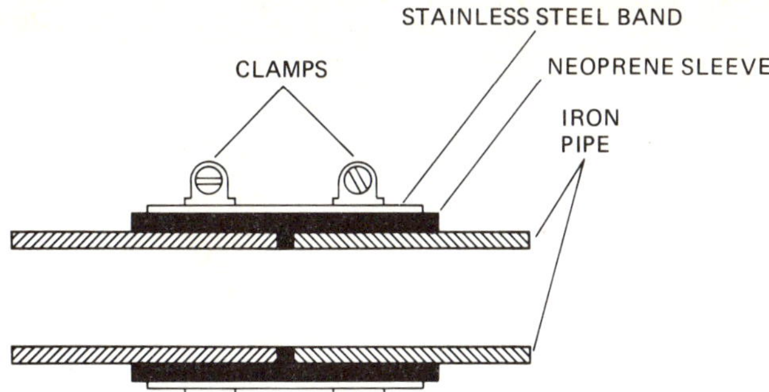

Figure 7-13. No-hub soil-pipe joint

can be made, and other pipes are cast so that two or more pipes can be run together. These special pipes are called fittings and come in a variety of shapes and in all the sizes of pipe. Figure 7-14 contains drawings of most of the common soil-pipe fittings.

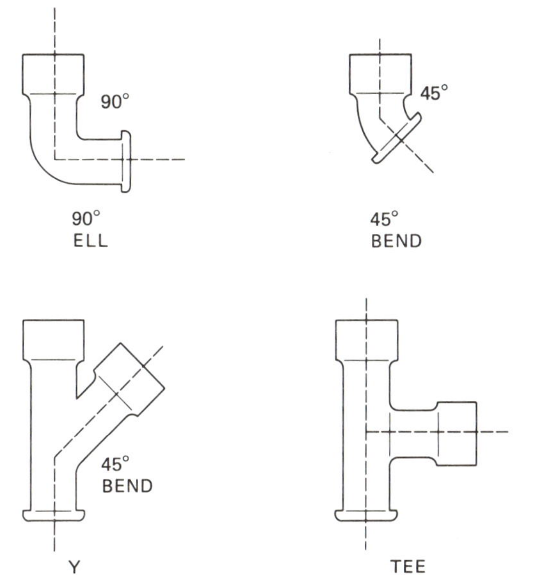

Figure 7-14. Common soil-pipe fittings

As you can see from Figure 7-14, all the fittings have a hub and spigot so that straight pipe and these fittings can be joined using the caulked lead joint.

• MALLEABLE-STEEL PIPE •

Soil pipe is seldom the only type of pipe used in a sanitary system. Earlier in this chapter it was stated that some portions of the sanitary system used 2-inch pipe rather than 4-inch pipe. These smaller pipes serve fixtures such as a kitchen sink, a washbasin, a tub or shower, or some other fixture primarily designed to dispose of liquid waste. Two-inch soil pipe is available, and in some cases it is advantageous to use. In other cases, however, it is more economical to use malleable-steel pipe. The steel pipe used in the sanitary system is never subjected to high pressure; therefore, standard lap-weld pipe is used in such applications. Since it is being subjected to water flow, galvanized pipe rather than black pipe is normally used.

Joining Steel Pipe

Most of the steel-pipe joints used in the sanitary system are the threaded joints discussed in Chapter 6. Cutting 2-inch pipe and threading it requires the same steps and the same types of tools as the water-system pipes. One step does become more important, the reaming of the pipe. It is imperative that the ID of the sanitary line have no sharp edges to accumulate hair, grease, or other materials and cause a stoppage. Stoppages must be cleaned out, and if correct provisions were not made initially, cleaning out a sewer line can be time-consuming and costly.

A second method of joining steel pipe in a sanitary system is the caulked lead joint method. Since we have just discussed this method on 4-inch pipe, there should be no reason to go over it all again.

Drainage Fittings

Threaded steel fittings used in the sanitary system are essentially of two different types, standard fittings and drainage fittings. There is a basic difference between the two, as shown in Figure 7-15.

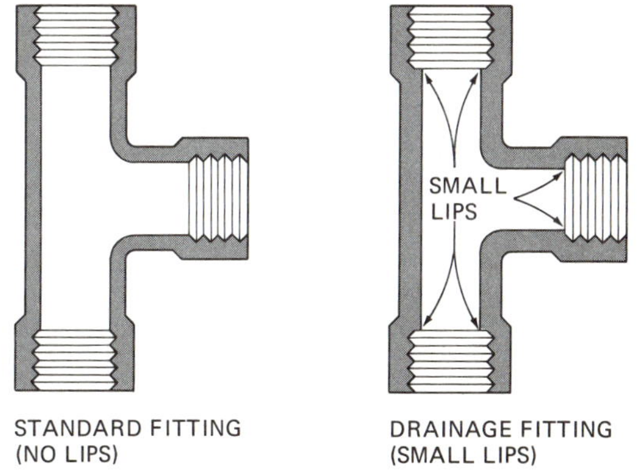

Figure 7-15. Standard fitting and drainage fitting differences

The cutaway views of the standard fitting and the drainage fitting show the important difference, the small lip at the base of the thread in the drainage fitting. This small base tends to give the draining water a smooth flow from the pipe into the fitting and out again. This eliminates the little ledge which is evident in the standard fitting and does not allow the waste material to build a stoppage point. Any time there is liquid flow in a steel-pipe portion of a sanitary system, drainage fittings must be used.

• VENT-SYSTEM MATERIALS •

There are certain areas of a sewer system in which there is no liquid flow. The vent system is an example. A subsystem of the sanitary system, the vent system is necessary for two reasons. The first is that the waste materials transported by the sanitary system can generate a volatile gas (methane gas). If there were no opening or vent to the outside to let this gas escape, it could build up a great amount of pressure and cause a fire hazard or damage the sewer system.

The second reason for a vent system was demonstrated in Chapter 5 when water seals (traps) for the fixtures were introduced. Remember, in that discussion traps were used to seal off the odors in the sewer system from the interior of the building, and the demonstration showed that if the traps were not vented, the water seal would siphon off. Figure 7-16 covers this.

Since only air or sewer gas passes through the vent system, there is no need to use the specially designed drainage fittings. Therefore, ordinary galvanized fittings are normally used.

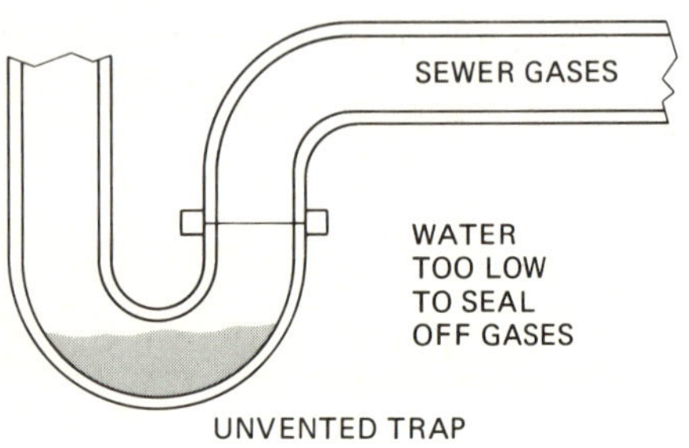

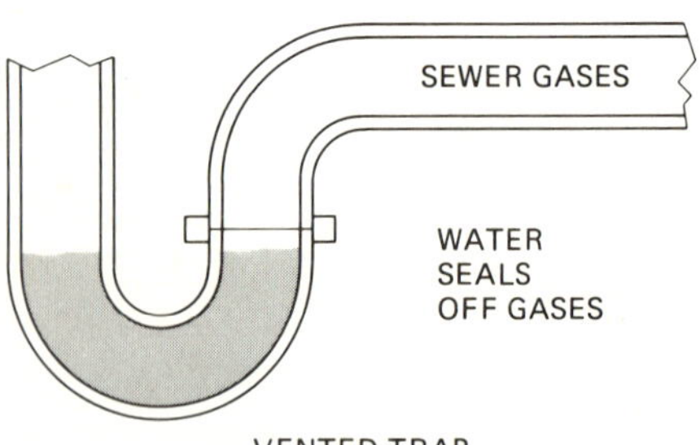

Figure 7-16. Vented and unvented trap seals

• STEEL-PIPE TOOLS •

Many of the tools described in Chapter 5 are also used in working with the sewer system. The larger pipe wrenches, strap wrenches, hammers, chisels, caulking and yarning irons are the mainstays of the plumber's toolbox. In some sewer systems, the large-diameter pipes used require a different type of pipe wrench. In these cases, a chain wrench is used. This chain wrench works on the same principle as the strap wrench. A picture of one is shown in Figure 7-17.

Figure 7-17. Chain wrench

• QUESTIONS •

1. What is soil pipe made of?
2. Soil pipe comes in 6-foot lengths and has a bell or hub on one end. (T or F)
3. Oakum is a loosely woven tarred hemp rope. (T or F)
4. A common soil-pipe joint is made using oakum and cement. (T or F)
5. The hub of the soil pipe is included in the measurement of the piece of pipe. (T or F)
6. Soil pipe must be reamed after it is cut by the hammer and chisel method. (T or F)
7. Why are drainage fittings manufactured with a small lip at the base of the threads?

• VITRIFIED-CLAY PIPE •

Vitrified-clay pipe, often referred to as glazed-tile pipe, is made by molding moistened clay into the shape and size of the pipe desired, removing it from the mold, applying the glazing material, and then firing (baking) the entire section of pipe in a kiln (oven) to make it hard and moisture-resistant. One end of the glazed-tile pipe is made in the same shape as the bell or hub end of the cast-iron pipe. Though there is no raised bead at the spigot end of the tile pipe, there is a small groove about an inch from the edge. A picture of a section of glazed-tile pipe is shown in Figure 7-18. Table 7-2 shows the various sizes and lengths of tile available.

Figure 7-18. Glazed tile pipe

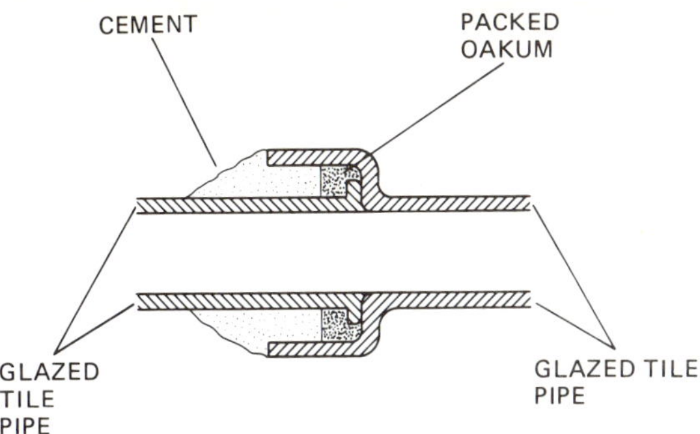

Figure 7-19. Glazed-tile joint

TABLE 7-2 VITRIFIED CLAY TILE SIZES

I.D. (Inches)	Length (Feet) (less hub)
4	2 to 5
6	2 to 5
8	2, 2½, 3
10	2, 2½, 3
12	2, 2½, 3
To 42″ in 3″ increments	

Cutting Glazed-Tile Pipe

Glazed-tile pipe is cut using the hammer-and-chisel method used in cutting soil pipe. The measurement is taken from the base of the hub or bell, a chalk mark is made around the pipe at the point of measurement, a wood block is placed under the point of cut, and then the chisel and hammer are used in the same manner as in cutting the soil pipe.

Joining Glazed-Tile Pipe

The procedure for joining sections of glazed-tile pipe is the same as that used in joining soil pipe. The spigot end is placed into and seated in the bell or hub end of another pipe. Oakum is then packed in to help seal and hold the joint. Now, instead of using lead to make the permanent seal, wet cement is packed around the joint and fills the remaining hub space. Once the cement has dried and cured, the joint is strong and waterproof. Figure 7-19 shows a cutaway view of a glazed-tile pipe joint.

There are commercial substances available as substitutes for cement, and their use is strictly a matter of personal choice or economics.

Many manufacturers of glazed tile have marketed their pipe or fittings with a polyvinyl chloride gasket around the spigot end of the pipe. This greatly reduces the amount of work and time required to lay glazed-tile pipe. The gasket is lubricated with a solution of liquid soap, then the gasketed spigot end of one pipe is forced into the hub of another. The PVC gasket forms a watertight and permanent joint and allows the tile to be slightly flexible in case of earth settling. This innovation is shown in Figure 7-20.

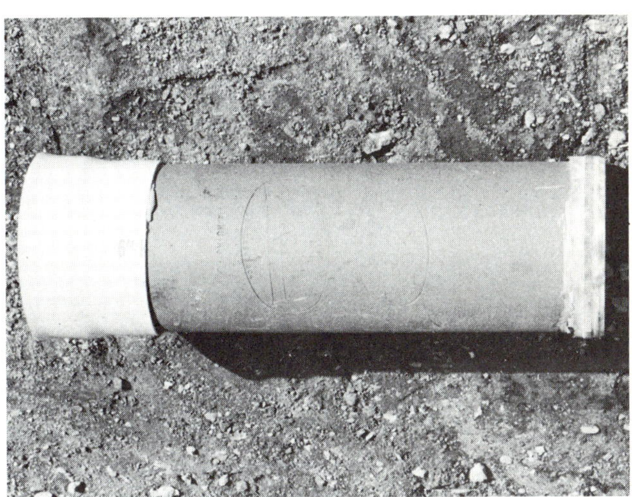

Figure 7-20. Glazed tile with PVC gasket

Glazed-Tile Pipe Fittings

Fittings for use with glazed-tile pipe are available in basically the same bends as soil pipe. All joints are made in the manner described above.

Joining Glazed-Tile Pipe to Soil Pipe

The main sewer line must extend a given distance beyond the foundation of the building before it can be connected to a glazed-tile sewer line. The joint made at this connection is basically the same as one which joins two sections of tile, using oakum and cement. A minor difference in the appearance of the two joints is that the cast iron-to-tile joint should have a larger cemented area, as shown in Figure 7-21.

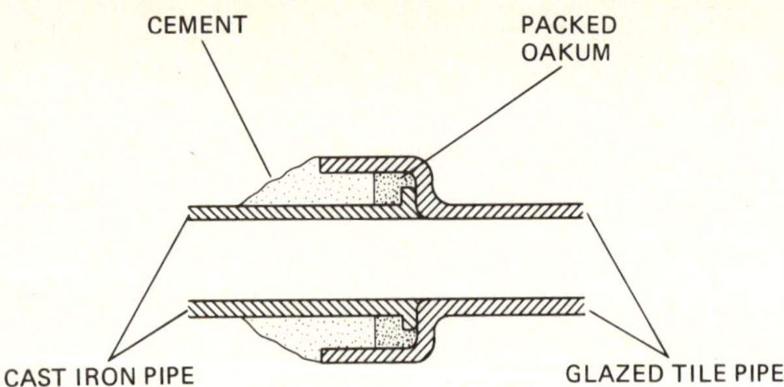

Figure 7-21. Cast iron-to-glazed tile joint

TABLE 7.3 CONCRETE PIPE SIZES

I.D. (Inches)	Length (Feet)
6	3 to 5
8	3 to 5
10	3 to 5
12	3 to 5
To 36″ in 3″ increments	

• QUESTIONS •

1. What is glazed-tile pipe also called?
2. Oakum and lead are used in making glazed-tile-pipe joints. (T or F)
3. Some companies manufacture glazed-tile pipe with a neoprene seal which will make a watertight and slightly flexible joint. (T or F)
4. What tools are used to cut glazed-tile pipe?

• CEMENT AND CONCRETE PIPES •

Cement and concrete pipes are manufactured in much the same way as the vitrified-clay tile. That is, they are formed of cement or concrete in a molding process, and then, instead of being glazed and fired, they are dried and allowed to cure (harden). These pipes are used for many of the same purposes as glazed-tile pipe outside the building and for the underground movement of sewage material.

Sizes of Cement and Concrete Pipe

Cement and concrete pipes are manufactured in lengths from 2 to 4 feet, and the ID of the pipe ranges from 4 to 24 inches, as shown in Table 7-3.

Cutting and Joining Cement and Concrete Pipe

Cement or concrete pipe may also be cut by the hammer-and-chisel method. It is usually not necessary to cut this pipe, however, since a little proper planning in selecting the pipe lengths will result in the correct amount of pipe for the sewer run.

Cement or concrete pipe is also joined in the same manner as vitrified-clay pipe, using oakum and cement to provide a watertight and solid joint.

One disadvantage of cement or concrete pipe is that it may deteriorate when used in a sewer line. The acids formed by the waste material attack the pipe and decompose it. Vitrified-clay pipe is better suited for most sewer applications.

• COPPER PIPE •

Large-diameter copper pipe is also used in sanitary systems. Its noncorrosive nature suits this application well, but the cost of copper may put it out of reach for a sanitary system. Copper pipe is often used in sewer lines in remodeling a building because of its lighter weight and the ease with which it can be assembled.

Large-diameter copper pipe is cut and joined in the same manner as the hard-wall copper pipe studied in Chapter 6. Sweat solder joints of large-diameter copper pipe may present a problem with heat control for the beginner.

• PLASTIC PIPES •

In Chapter 6, we discussed the use of plastic pipe in the water-supply system. The sanitary systems of many new buildings are entirely of plastic pipe. These plastic systems are resistant to deterioration, are easily installed, have sufficient strength to withstand any of the pressures encountered, and are more economical.

Cutting Plastic Pipe

Large-diameter plastic pipes are cut by any one of the methods discussed in Chapter 6. A saw or pipe cutter will ensure a square edge, and since this is a sanitary-system pipe, the reaming operation is absolutely necessary.

Plastic Fittings

Plastic fittings for sanitary systems are available in the same sizes and bends as soil-pipe fittings. Most of them are shown in Figure 7-22.

Plastic fittings are made so that the ID of the fitting is the same as that of the pipe. This means that there is no space between the pipe and the fitting in which waste material can accumulate. If such an accumulation were to occur it could cause a stoppage in the flow of the pipe. Figure 7-23 is a cutaway view of a completed plastic joint.

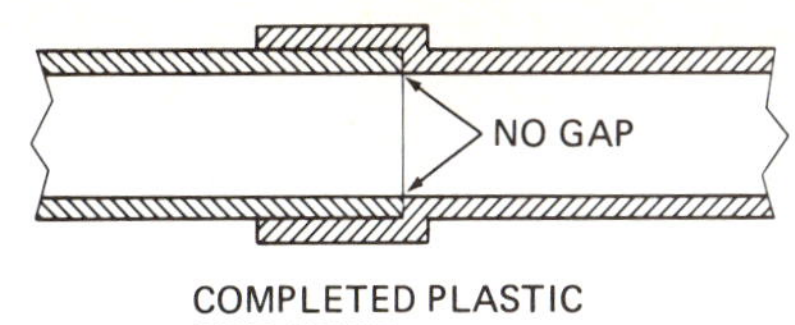

COMPLETED PLASTIC
PIPE JOINT

Figure 7-22. PVC fittings for sanity systems

Figure 7-23. Plastic drainage fittings

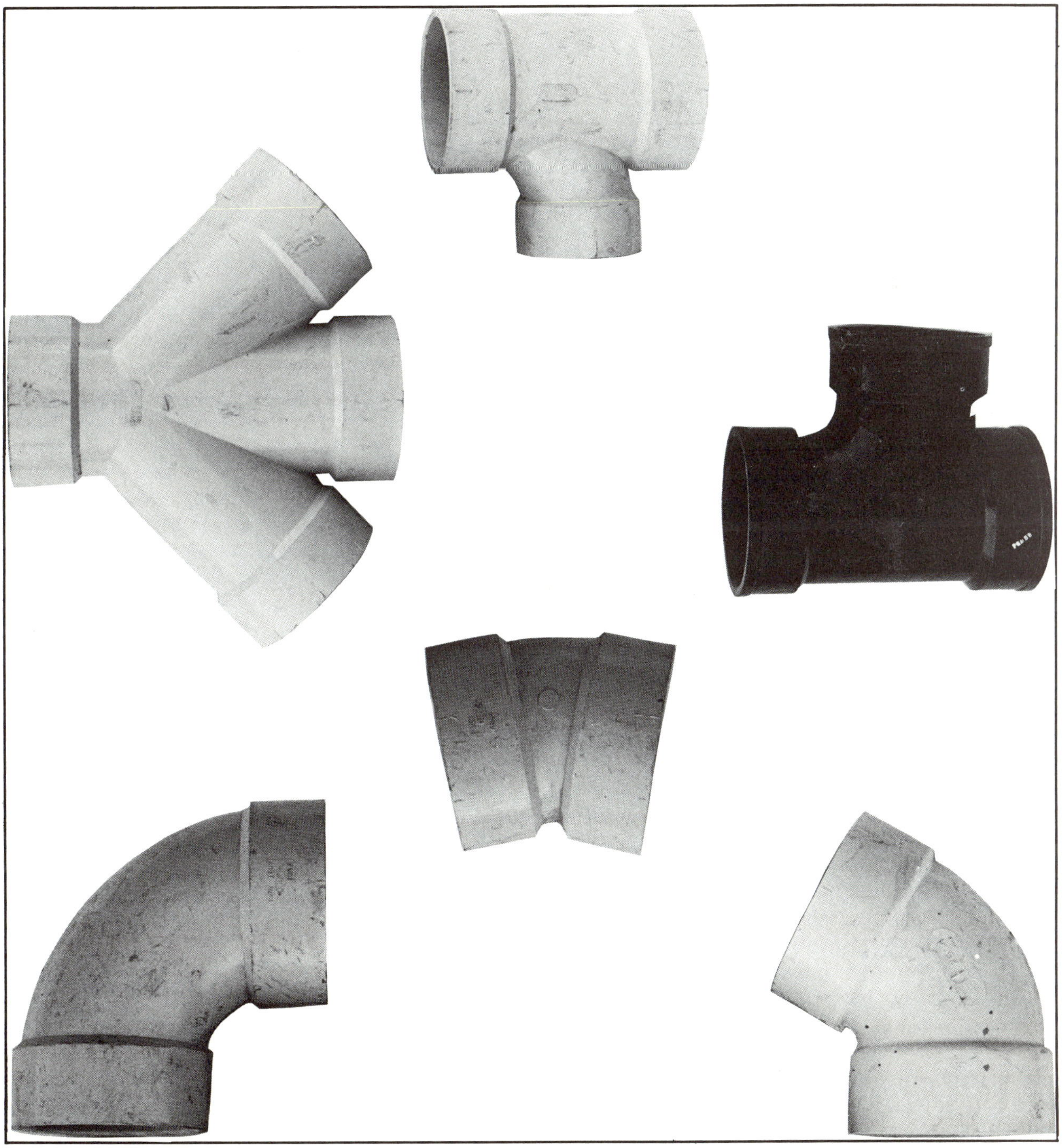

Joining Plastic Pipe

Again, the basic procedures for joining plastic pipe were covered in Chapter 6. However, here is a brief review of the cement or adhesive used with each type of plastic. Look over Table 7-4 and refresh your memory.

TABLE 7-4 PLASTIC PIPE JOINTS

Type of Pipe	Type of Joint
PVC, CPVC, ABS Styrene, CAB	Solvent cemented
FRP	Adhesive bonding (bell and spigot or butt strap)
Polyethylene Polypropylene Polybutylene	Heat fusion (butt, electrical, or socket)

Always work swiftly when applying the cement to the pipe and the fitting, because some of the adhesives dry very quickly.

Plastic pipe can be joined to soil pipe, malleable-steel pipe, vitrified-clay pipe, or cement or concrete pipe. The joints between plastic pipe and soil pipe, vitrified-clay pipe, cement pipe, or concrete pipe are made using the oakum and cement joint we previously discussed. Special threaded fittings or nipples are required to join plastic to malleable-steel pipe or fittings.

Plastic pipe has other applications within the sanitary system besides those in the sewer lines. Some plastic pipe comes with small, evenly spaced holes for use in a septic system's drainage field or for use as drain tile around the footing or under the basement of a building. These applications will be covered in a later section of this chapter.

• FIBER PIPE •

Fiber pipe, also called bituminous pipe or Orangeburg pipe, is also used in some sanitary systems. Fiber pipe is cut with a handsaw or a hacksaw, and a special tool is used to taper the male end of the pipe. This cut end must be tapered so that it can be joined using the coupling process outlined in Figure 7-24.

Fiber-Pipe Fittings

Fiber-pipe fittings are made in the same range and sizes as other sewer-system materials. The fitting joints are tapered so that the end of the pipe can be seated to form a watertight joint as in Figure 7-24.

Like plastic pipe, fiber pipe is also made to be

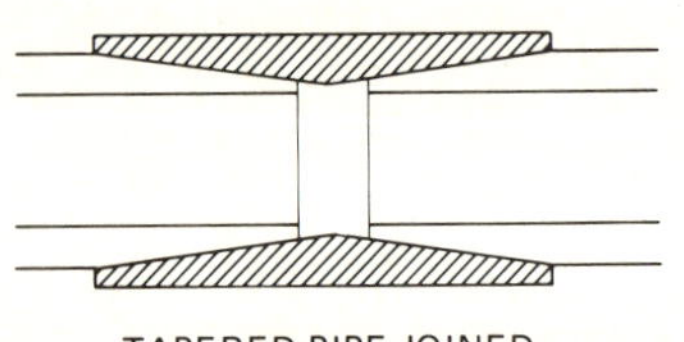

Figure 7-24. Fiber (bituminous) pipe and fitting

used in the drainage field of a septic system or in the drainage system around or under a building. It also has small holes evenly spaced along its length to allow liquid to drain away.

• SEPTIC SYSTEM •

A septic system is an on-site sewage-treatment system which combines a large-capacity tank (septic tank) with a network of perforated pipes (drainage field). Septic tanks may be made of various materials, such as concrete, plastic, Fiberglas, and metal. The septic tank is designed to hold solid waste and allow the liquid waste to move into the drainage field. Natural or added bacteria decompose much of the solid waste into gas, which escapes through the vent system or the soil, and sludge, which settles to the bottom of the tank. Figure 7-25 shows the layout of a typical septic system.

Sizes of Septic Tanks

The number of bedrooms or persons living in the building dictates the size or capacity of the septic tank (in gallons) necessary to treat all the sewage anticipated. The more bedrooms or people, the larger the septic tank needed. Table 7-5 lists the recommended sizes of septic tanks.

TABLE 7-5 RECOMMENDED SIZES OF SEPTIC TANKS

Number of Bedrooms	Number of Persons	Size of Septic Tank in Gallons
2	4	600–750
3	5	750–900
4	8	900–1000
5	10	1000–1300
6–8	12	1300–1500

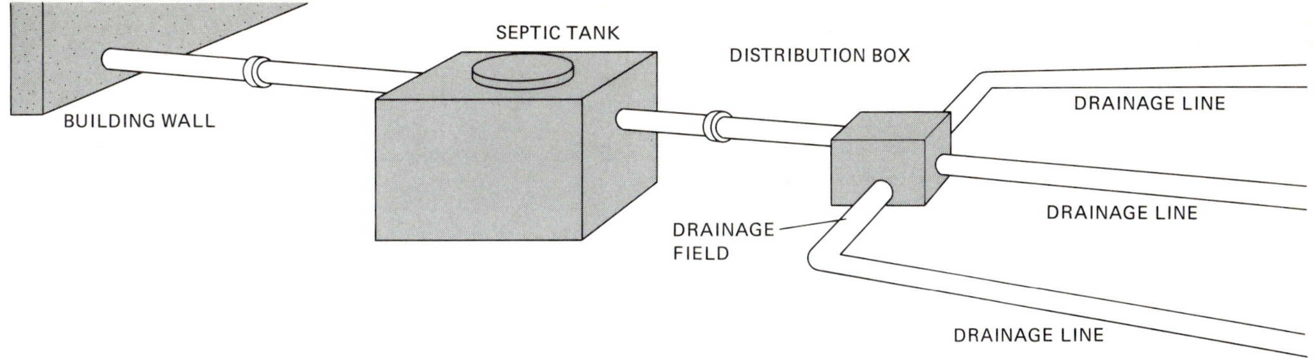

Figure 7-25. Septic-system layout

The Drainage Field

As shown in Figure 7-25, the drainage field is laid out away from the house and has three or more branches. The number of branches and their respective lengths are determined by the local code and the type of soil present. Generally, very sandy soil will absorb liquid waste faster than clay soil and shorter drainage fields are used.

The types of pipe used in the drainage field are governed by local codes. This pipe may be made of clay tile (not vitrified), plastic, or other material. Some of the pipes used (called perforated pipes), which have regularly spaced holes to permit easy drainage from the pipe to the soil, were discussed earlier in this chapter. Figure 7-26 shows some of these perforated pipes.

• QUESTIONS •

1. Are cement and concrete pipe very resistant to chemical deterioration?
2. Are plastic pipes and fittings well suited to sanitary system applications?
3. Plastic fittings, like drainage fittings, do not have a gap between the interior surface of the fitting and the pipe. (T or F)
4. Fiber pipe is also known as bituminous pipe and Orangeburg pipe. (T or F)
5. What types of pipe may have small equally spaced holes along their length?
6. Perforated pipe is used in the drainage field of a septic tank. (T or F)

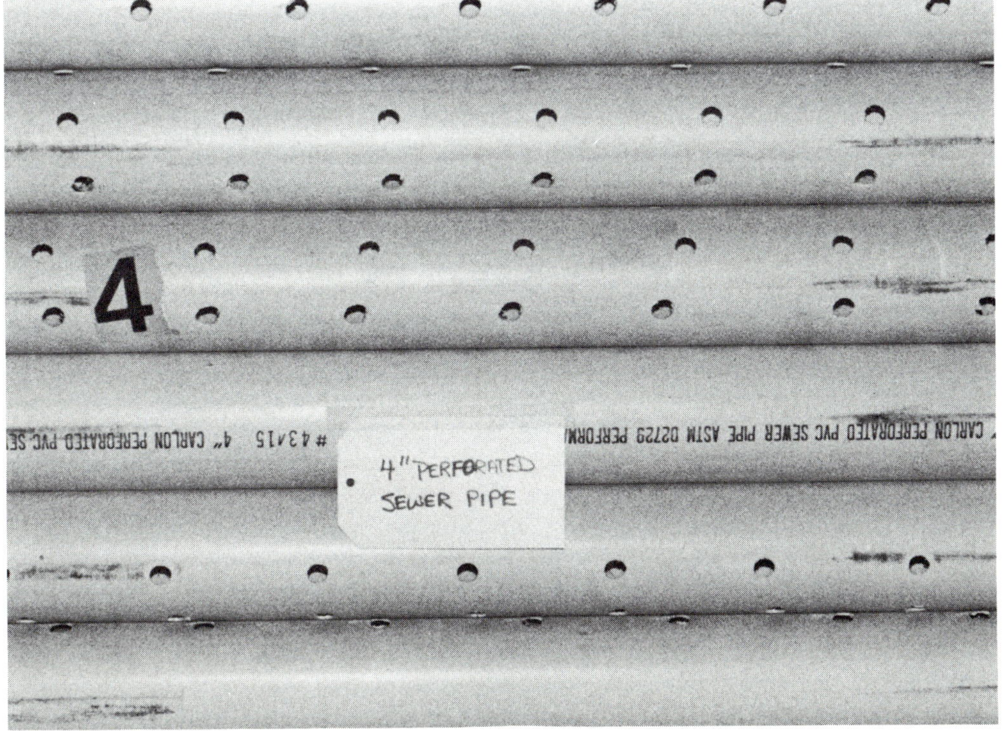

Figure 7-26. Perforated drainage pipe

7. A 500-gallon septic tank would serve a building with four bedrooms or six persons. (T or F)
8. The septic tank allows bacteria to decompose the solid waste matter. (T or F)
9. What two products are formed by the decomposition process in the septic tank?
10. Sandy soil requires a larger drainage field than clay soil. (T or F)

• THE SEWER AND VENT SYSTEM •

Up to this point, the materials used in the sanitary system have been discussed and the procedures have been discussed and practiced. This section of Chapter 7 will discuss and practice some simple applications of the materials and procedures studied. These exercises will be developed using simple floor plans (blueprints) and side views (elevations) of the areas which contain sanitary system pipes.

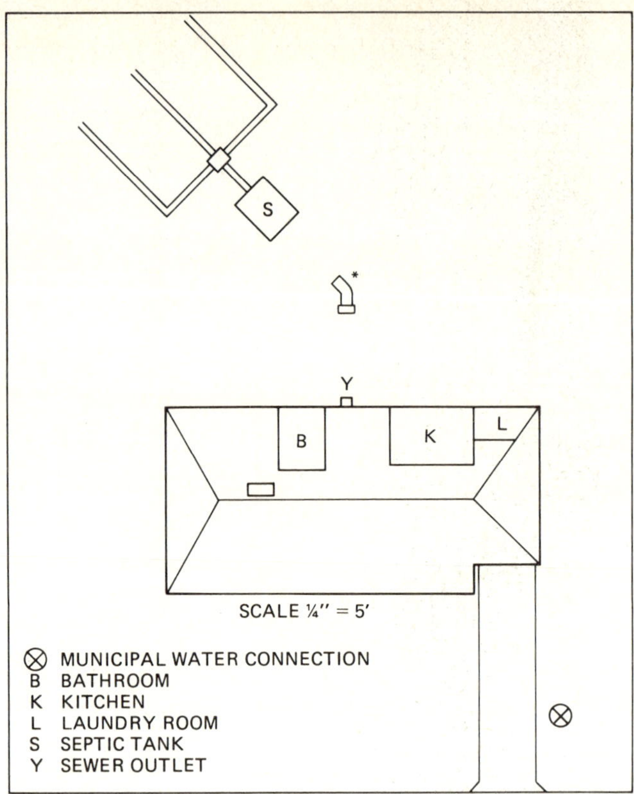

List of Materials		
Item	No.	Type
*45° Bend	1	Vitreous clay
5′ Section		
2′ Section		
		Soil pipe
Cement		
Oakum		
Lead	5#	

Figure 7-27. Site plan and list of materials

The Building Sewer

Earlier in this chapter we made reference to the building sewer. This is the portion of the sanitary system which extends from the building out to the sewer main or the on-site treatment point. Local plumbing codes establish the different types of pipes which can be used in these building sewers and how close these sewers can come to the foundation of the building. The depth at which the sewer is placed depends on the depth of the sewer main and the depth to which the ground freezes in the winter (frost line).

Problem 7-1

Figure 7-27 is a drawing of where a building will be constructed on the lot. This is called a site plan or a plot plan. The site plan furnished here shows the location of the sewer main and the location of the bathroom, the kitchen, and the laundry room. The local code permits vitrified-clay pipe in this case, and the building sewer cannot be closer than 10 feet from the foundation of the building. You have all the information necessary to determine the location of the proposed building sewer and to order the materials necessary to construct it. The list of materials has been partially filled out. Determine the location of the building sewer and complete the list of materials needed to build the line.

The Main Sewer Line

The main sewer line of the building is that section of the sewer which gathers the waste material from the vertical stacks and transports it beyond the foundation of the building to the building sewer. As you can tell from the description, there is a difference between the sewer main (outside) and the main sewer line (inside). There are basically two types of main sewer lines: the underground line and the hanging line. The difference between these two is shown in Figure 7-28.

Most single-story buildings without a basement use the underground line, as do many buildings with a basement and connections to a municipal sewer

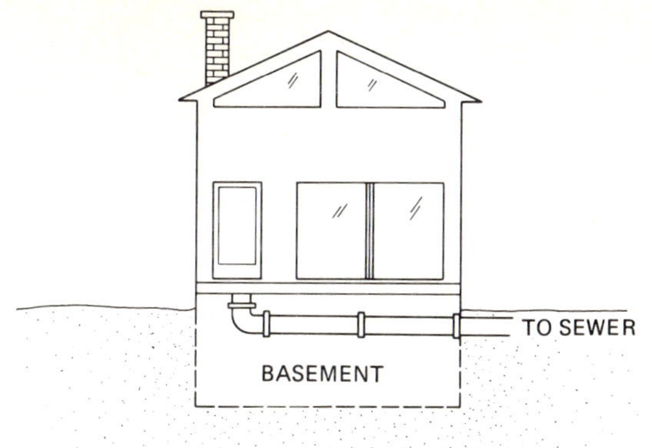

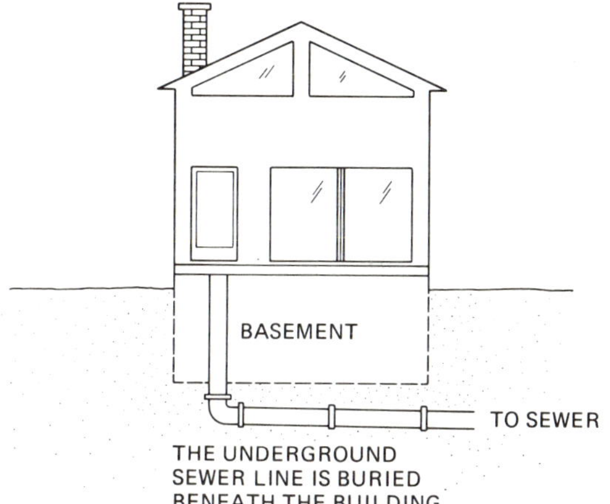

Figure 7-28. Building sewer placement

main. Buildings with a basement and a septic system may have to use a hanging sewer because of the depth of the basement and the depth of the septic system. Figure 7-29 illustrates this.

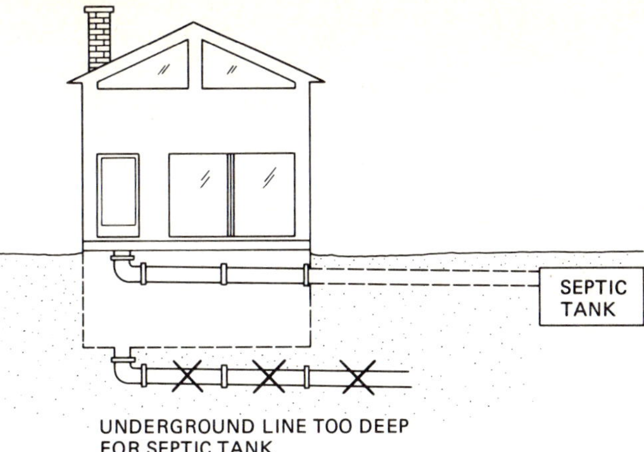

Figure 7-29. Hanging sewer line

Problem 7-2

Figure 7-30 is a simplified floor plan (blueprint) of the building in Problem 7-1. The local code calls for 4-inch cast-iron pipe to be used in an underground main sewer line. The building sewer is located at point *A*, the bathroom stack at point *B*, and the kitchen and utility stack at point *C*. Determine the location of the main sewer line, connecting point *A* with points *B* and *C*, and complete a list of materials needed to run this line.

Cleanouts

In planning main sewer lines for a building, attention must be given to the possibility of a stoppage of the line. Planning for this problem includes the positioning of capped or plugged openings to the sys-

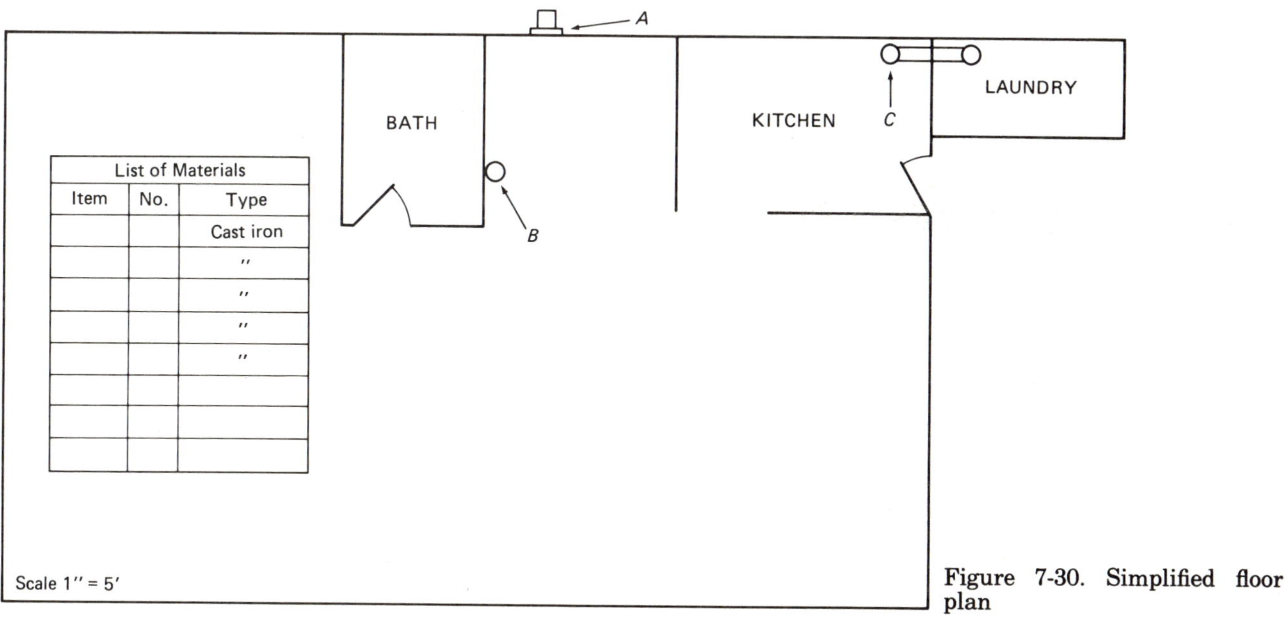

List of Materials		
Item	No.	Type
		Cast iron
		″
		″
		″
		″

Figure 7-30. Simplified floor plan

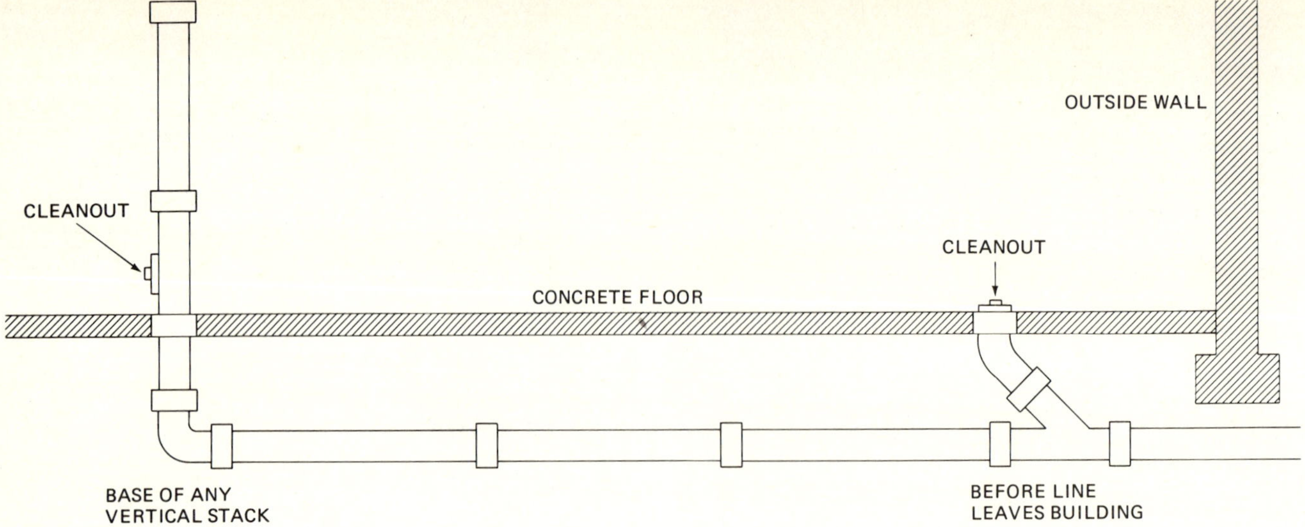

Figure 7-31. Cleanout positions

tem. Figure 7-31 is a drawing of the different areas where stoppages may occur and points where cleanouts should be placed.

Note the placement of the cleanouts for the underground main sewer line. They are located at the base of the last stack in the line. This placement affords a simple access for cleaning the entire length of the main sewer from one point. Each branch line, like the kitchen branch in Problem 7-2, should also have a cleanout at its base.

Stacks and Vents

The stack and vent of a sanitary system serves two purposes. First, it transports the waste material from the upper floors to the main sewer, and second, it provides air circulation throughout the entire system. One of the blockage points in Figure 7-31 was at the base of a stack which used a short quarter bend (90-degree fitting). Figure 7-32 points out some other fitting combinations which could be used to eliminate those stoppages. In general, the more gentle the curve at the base of a stack, the less the potential for stoppage.

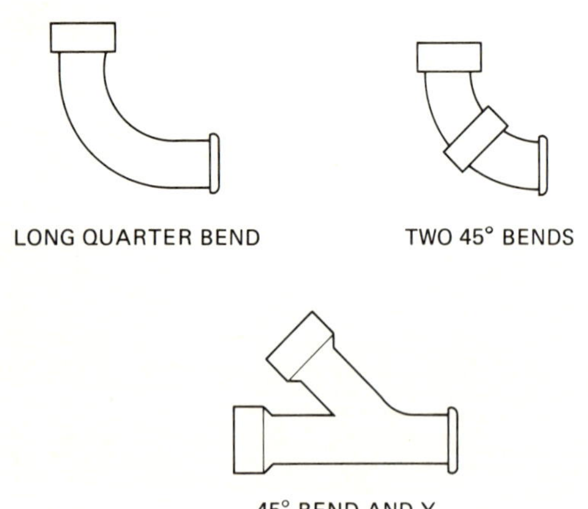

Figure 7-32. Blockage prevention

Problem 7-3

Figure 7-33 diagrams a stack and vent combination which could be used to serve the bathroom of the house in Problem 7-2. Look this diagram over carefully, noting any apparent problems and correct them by determining the location of new sewer stack or vent lines. Remember, each fixture should be vented to overcome trap siphoning, and the vent must extend through the roof of the building.

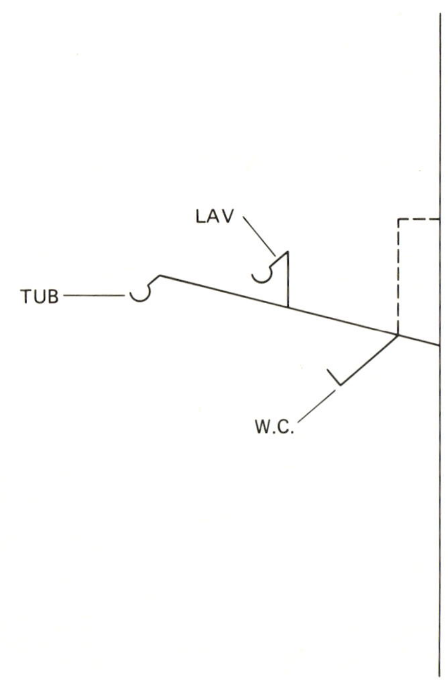

Figure 7-33. Multiple-fixture stack (bathroom)

Single-Fixture Stack

A single-fixture stack would serve one fixture in an area away from the main or multiple-fixture stack. If the local code permits, this stack may consist of 2-inch malleable-steel pipe which extends through the roof for venting purposes. Figure 7-34 shows a single-fixture configuration serving a kitchen fixture.

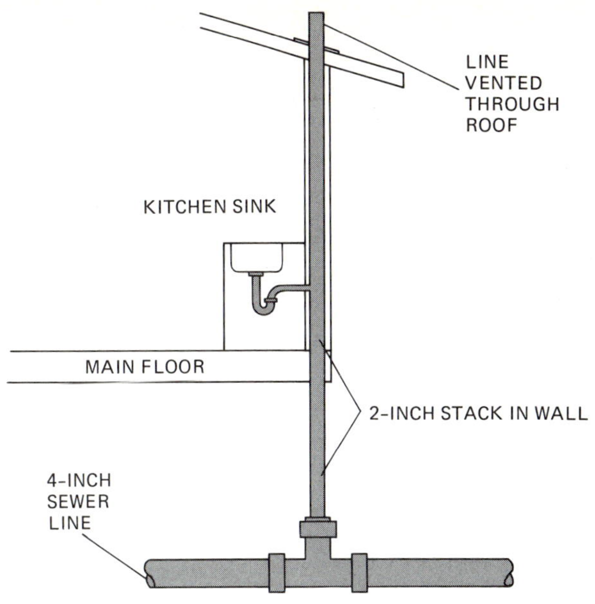

Figure 7-34. Single-fixture stack (kitchen)

• SUMMARY •

This chapter has provided a broad overview of the materials, procedures, and layouts of a sanitary system of a building. There are many additional points that are not covered in this book. Knowledge of drainage slopes, grades, relief vents, grease traps, sump pumps, etc., in addition to the basic concepts provided here is necessary to good plumbing practice.

• WORDS PLUMBERS USE •

stack
soil pipe
hub end
spigot end
soil-pipe cutter
yarning irons
caulking irons
caulked lead joint
oakum
joint runner rope
neoprene
gasket joint
no-hub joint
drainage fitting
vitrified
glazed-tile pipe
fiber pipe
septic system
perforated pipe
drainage field
underground line
hanging line
cleanout

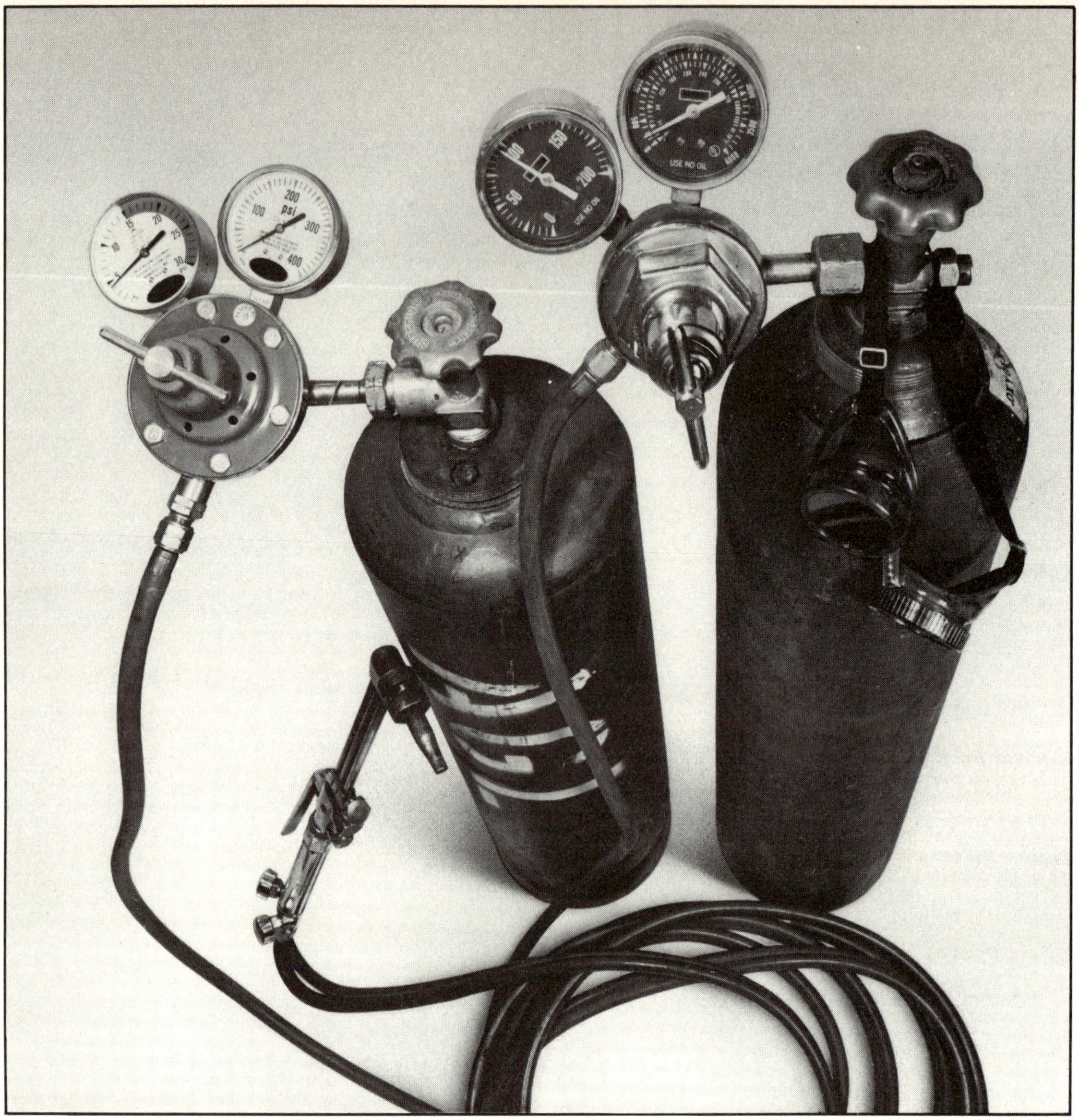

8

WELDING PROCESSES

The purpose of this chapter is to introduce you to the different equipment, materials, and procedures used by plumbers and other persons involved in joining steel products. Many plumbing jobs involve providing pipe pathways for very high-pressure fluid flow. The ordinary screw-thread joint was not designed to withstand these very high pressures, and a stronger method of joining these pipes must be used. Most of these very high-pressure joints are made using one of the welding processes which we will discuss.

At the completion of this chapter, you will be able to:

- Identify the equipment and materials used in the following processes:
  - Oxyacetylene welding
  - Arc welding
  - Oxyacetylene cutting
- Select the appropriate welding rod, prepare the material for welding, and, using the oxyacetylene welding process, successfully weld steel pipe together.
- Select the appropriate electrode, prepare the material for welding, and using the ac arc-welding process, successfully weld steel pipe together.
- Prepare the oxyacetylene welding equipment for cutting steel and successfully make straight, curved, and leveled cuts.

• OXYACETYLENE WELDING •

The term oxyacetylene is a short form of the names of the two gases used in this process. Oxy stands for oxygen, a gas which we all breathe and a gas which a fire needs to burn. Oxygen is about one-fifth of the air around us, but the oxygen which we will be using is pure. It is stored in a green or yellow tank under high pressure (about 2400 pounds per square inch) and should be handled with great care. A picture of an oxygen tank is shown in Figure 8-1.

Figure 8-1. Oxygen tank

Acetylene, the second gas used in this welding process, is a very combustible, almost explosive gas. It is normally *not* present in the air around us and must be manufactured. Like oxygen, acetylene is stored in a tank, but the tank is painted red or black. Acetylene is also under pressure in the tank, but only up to 210 pounds per square inch. Acetylene gas in a tank with a higher pressure would explode the tank like a bomb. Figure 8-2 shows a typical acetylene tank.

Figure 8-2. Acetylene tank

Both the oxygen and the acetylene tanks are used, together with regulators, hoses, and torches, to create a flame. This flame burns at about 6000°F, and most steel will melt at about 2800°F. A picture of the tanks, regulators, hoses, and a torch which make up a typical oxyacetylene welding set is shown in Figure 8-3 on the next page.

Figure 8-3. Oxyacetylene welding set

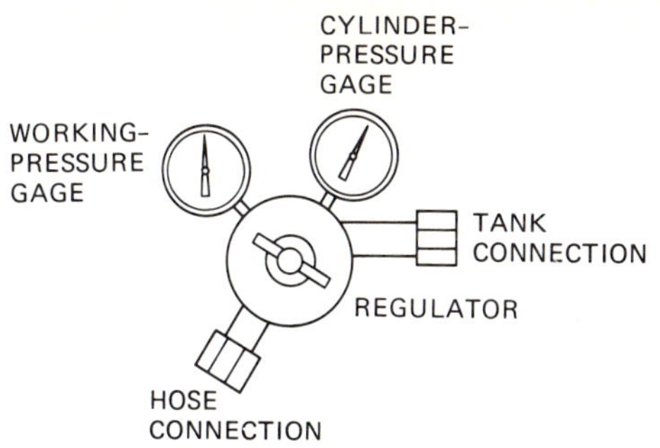

Figure 8-4. Typical gas regulator

• PRESSURE REGULATORS •

Since both the oxygen and the acetylene gases are under pressure in tanks, a system of dropping that high pressure to a usable low pressure is needed. Both the oxygen regulator and the acetylene regulator have the same duty, to lower the high pressure at the tank valve so that it can be used in a welding or cutting procedure.

There are many different types of regulators on the market today, but the most common is the two-gage single-stage regulator. This means that the first gage will measure the gas pressure in the cylinder and the second gage will measure the gas pressure at the outlet of the regulator. You will see the importance of this second pressure as we discuss lighting and adjusting the torch later in this chapter. Figure 8-4 is a drawing of a typical regulator.

The regulator shown in Figure 8-4 had no figures on the gages. They were purposely left blank because the oxygen regulator has one set of figures and the acetylene regulator has another.

The Oxygen Regulator

The main difference between the oxygen and the acetylene regulators is the pressure. Remember, the oxygen in the tank is normally at 2400 pounds per square inch. This means that the first gage, the one nearer the tank, must go up to at least 2400.

The second gage, which measures the pressure at the outlet of the regulator, will show the working pressure of the oxygen in the hose and at the torch. Some oxyacetylene processes (cutting, for one) require a higher oxygen working pressure. For that reason, the second gage will normally be able to indicate up to 100 pounds per square inch. Figure 8-5 shows a typical oxygen regulator with gage pressures.

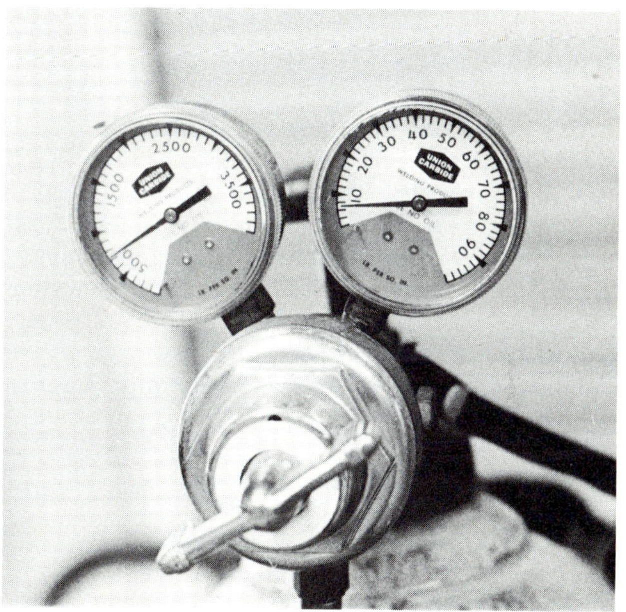

Figure 8-5. Oxygen regulator

The Acetylene Regulator

The pressure in a full acetylene tank is normally about 210 pounds per square inch. Therefore, the first acetylene pressure gage must be able to read at least that high. The second gage is the most important one, though. Take a close look at the picture in Figure 8-6 and note that a red line begins at 15 pounds per square inch. This is a danger signal. Acetylene gas at a pressure of more than 15 pounds per square inch, in any place other than the *special tank* designed for it, may blow up. Be *very careful* when using acetylene.

Figure 8-6. Acetylene regulator

• HOSES •

As you noticed in Figure 8-3, there was a hose which led from each regulator to the torch. Each hose is color-coded. The oxygen is green and the acetylene is red. The threads on these hoses as well as the threads on the tank and regulators are different, so that it is impossible to make a mistake when connecting a new tank, regulator, or torch. The oxygen equipment has right-handed threads (just like a nipple or pipe), while the acetylene equipment has left-handed threads.

• TORCHES •

The oxyacetylene torch has two threaded connections, two valves, a mixing chamber, and a tip with a small hole, or orifice. Each threaded connection provides a supply of gas (either oxygen or acetylene) to a valve. When the valves are opened, the gas flows to the mixing chamber, and the acetylene and oxygen are mixed so they will burn properly at the tip. The tip of the torch allows the mixed gases to flow out through the orifice, where they burn. Figure 8-7 shows a diagram of an oxyacetylene torch.

There are many different sizes of oxyacetylene torch tips. The sizes are numbered from 00 to 15, and the larger the number, the larger the hole or orifice. Of course, the larger the orifice, the more gas pressure required, and the larger the flame. With a larger flame, you can heat a bigger area. Table 8-1 shows the different tip sizes and the pressures used with each tip.

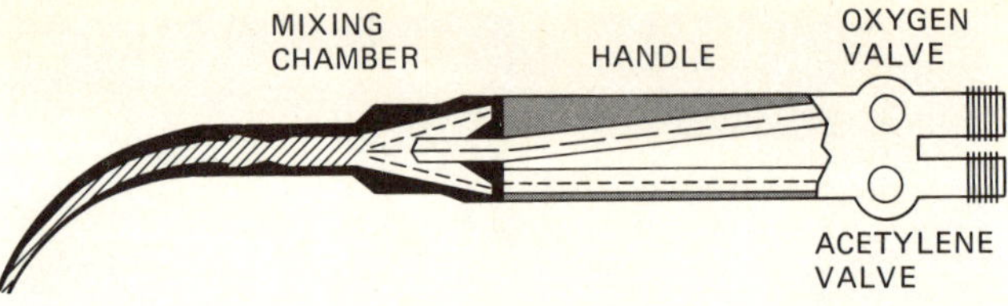

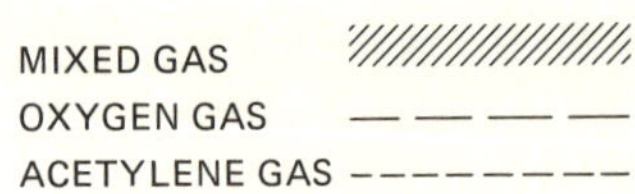

Figure 8-7. Oxyacetylene torch

TABLE 8-1 TORCH TIP SIZES AND PRESSURES

Tip Size	Pressure	
	Oxygen	Acetylene
1	1	1
2	2	2
3	3	3
4	4	4
5	5	5
6	6	6
7	7	7
8	8	8
9	9	9
10	10	10
11	10	10
12	10	10

Lighting the Torch

There are three basic items which you must know before you attempt to light the torch. They are:

1. The pressures of each gas required at the torch valve
2. The sequence of lighting the torch
3. The three different types of flames which can be adjusted from any pressure setting

• ADJUSTING THE PRESSURE •

The oxyacetylene welding set in your shop or laboratory will be ready to turn on. This section will explain how to turn on the tanks and set the required pressure for both the oxygen and acetylene gases. Figure 8-8 shows the oxygen pressure when opening and adjusting the torch. Use the following steps to set the pressure when you begin to weld.

1. Read the number size of the tip on the torch you will be using.
2. Check the valve below the second gage (pressure regulator valve) to be sure it is backed out or loose.

3. Open the tank valve *all the way* (this makes sure that there is no gas leaking around the valve stem).
4. Slowly turn the pressure regulator valve (below the second gage) until the needle points to the same number as the tip you have on (for a No. 3 tip, the pressure needle should point to 3). If you have a 00 or a 0 tip, use 1 for the pressure required.
5. Open the oxygen valve on the torch (connected to the green hose) and reset the needle to the proper number. (This is working pressure.)
6. Close the oxygen valve on the torch.

Figure 8-8. Properly adjusted oxygen regulator

There is one important change in adjusting the acetylene pressure, and that is in step 3. Since acetylene gas is so flammable, the tank valve should be opened only one-fourth of a turn. This allows enough gas to flow through the regulator and makes it easy to shut off the gas if an emergency occurs. Step 2 also becomes more important. You must be sure to have the pressure regulator valve backed off or loose. Remember, acetylene gas is very dangerous at 15 pounds of pressure or more.

If you follow the same steps you used in adjusting the oxygen pressure, making the third step one-fourth of a turn instead of full open, you will have no difficulty in adjusting the acetylene pressure.

• LIGHTING THE TORCH •

Lighting the oxyacetylene torch becomes a safe and easy operation if you remember the following steps and safety rules. Follow through these steps:

1. Put on the dark safety glasses.
2. Open the acetylene valve (the valve with the red hose) on the torch very slightly, about one-fourth of a turn.
3. Place the striker at the tip of the torch and squeeze the handles together. This will cause sparks to ignite the acetylene gas. (Safety note: Never use a match to light the gas.)
4. Adjust the acetylene gas until you have a bright flame with no black smoke.
5. Slowly open the oxygen valve on the torch (the valve on the green hose) until you can see three different-colored flames, a small bright blue one at the tip, a lighter blue flame about 1 inch long, and a very faint blue flame at the end.
6. After you have identified the three different flames, continue to open the oxygen valve slowly. The light blue (middle) flame will move toward the tip of the torch and get smaller until it is exactly the same size as the bright blue one. This is the correct flame for welding and brazing. Figure 8-9 shows a properly adjusted acetylene regulator.

Figure 8-9. Properly adjusted acetylene regulator

To turn off the torch, turn off the acetylene gas valve on the torch first. This may cause a small popping noise, which is normal. It is always important to turn off the acetylene first to prevent an afterfire in the tip of the torch.

There are three types of flames which can be used in welding. You have seen two of them now. The one with the three blue flames is called a carburizing flame, one which has more acetylene gas than oxygen. The one which has one bright blue flame and a very faint flame is called a neutral flame, one with equal amounts of oxygen and acetylene. The third type of flame is called an oxidizing flame, one which has more oxygen than acetylene. This third flame, the oxidizing one, is made by opening the oxygen valve past the neutral flame position. Try it with your torch and you will see that the bright blue cone of flame at the top of the torch will get lighter and more pointed. See Figure 8-10 for drawings of these three flame types.

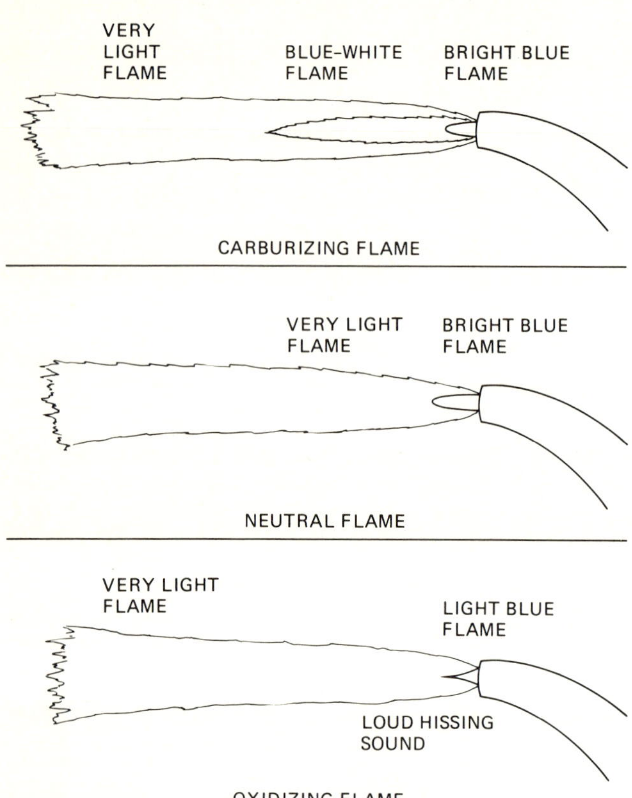

Figure 8-10. Three oxyacetylene flame patterns

• QUESTIONS •

1. What are the two colors that an oxygen tank is likely to be painted?
2. What are the two colors that an acetylene tank is normally painted?
3. The oxygen tank is shorter than the acetylene tank. (T or F)
4. Acetylene gas can be explosive at more than 15 pounds of pressure. (T or F)
5. Always turn on the oxygen gas first when lighting the torch. (T or F)
6. The acetylene gas is always turned off first when shutting off the torch. (T or F)

• OXYACETYLENE WELDING PROCESS •

This welding process uses heat supplied by the burning gases to melt the edges of two pieces of metal together and make a single piece of metal. This is called a cohesive combination. Supplies needed are an oxyacetylene welding setup, two pieces of steel (either steel pipe or flat steel), and some mild-steel welding rods.

Before you start welding two pieces of metal together, it is very important that you know what to look for in this welding process. After you have the torch lit and the metal ready, move the torch flame down to the area you want to weld. The tip of the bright blue neutral flame should be brought down until it almost touches the metal (Figure 8-11).

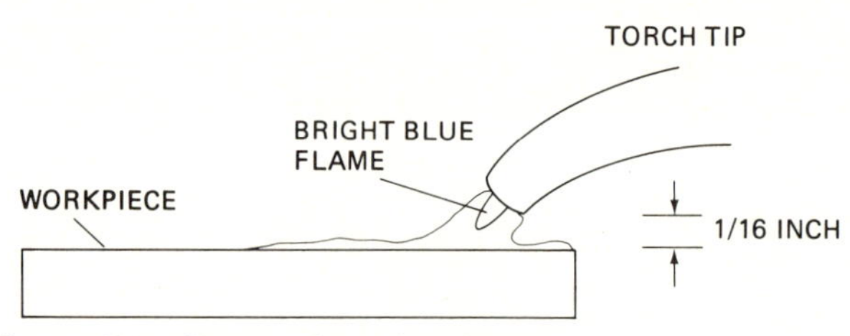

Figure 8-11. Proper flame height

Soon after the flame is close to the metal, a small, bright, shiny spot will appear right under the flame. When this happens, begin to move the tip of the torch in a small circular motion (about a ¼-inch circle). This motion will help to spread the torch heat over a wider area. Shortly, the steel will begin to melt and form a small pool or puddle. The size of the puddle of molten steel must be controlled by moving the torch, and since we want to make a weld along a line, the torch must be moved in a certain direction. Different ways to move the torch are shown in Figure 8-12. A picture of the melt strip that results is shown in Figure 8-13.

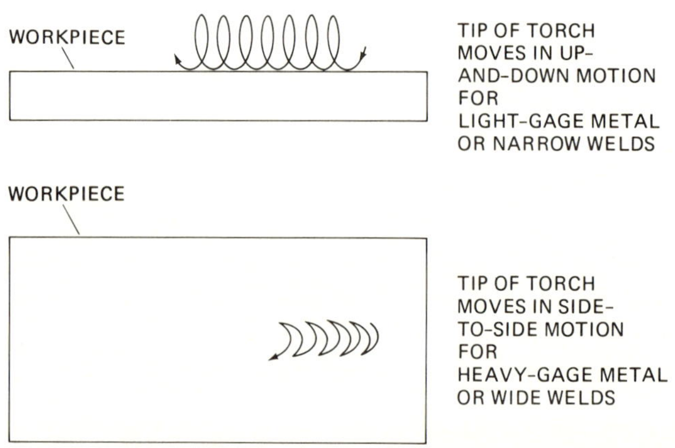

Figure 8-12. Torch movement

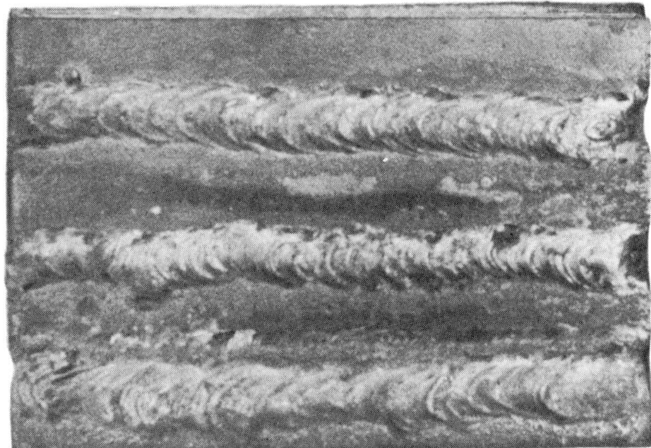

Figure 8-13. Oxyacetylene melt strip

• SELECTING TORCH TIP AND WELDING ROD •

A welding rod must be selected so that it will supply the proper amount of steel to the weld. Welding rods normally come in various diameters ranging from $\frac{1}{16}$ to $\frac{5}{32}$ inch, although larger rods are available for special welding jobs.

When selecting the correct-sized rod, you must consider the thickness of the metal to be welded. This will determine the size of torch tip which you will be using and the correct welding rod to use. Table 8-2 will help you select the correct tip and welding rod for each job.

TABLE 8-2 WELDING TIP AND ROD SIZE

Tip Size	Rod Size	Plate Thickness
1	$\frac{1}{16}$	22–16 gage
2	$\frac{1}{16}$–$\frac{1}{8}$	$\frac{1}{16}$–$\frac{1}{8}$ inch
3	$\frac{1}{8}$	$\frac{1}{8}$–$\frac{3}{16}$ inch
4	$\frac{3}{16}$	$\frac{3}{16}$–$\frac{1}{4}$ inch
5	$\frac{3}{16}$	$\frac{1}{4}$–$\frac{3}{8}$ inch
6	$\frac{1}{4}$	$\frac{3}{8}$–$\frac{5}{8}$ inch
7	$\frac{1}{4}$	$\frac{1}{2}$–$\frac{3}{4}$ inch
8	$\frac{1}{4}$	$\frac{5}{8}$–1 inch
9	$\frac{1}{4}$	1 inch +
10	$\frac{1}{4}$	1 inch +
11	$\frac{1}{4}$	1 inch +
12	$\frac{1}{4}$	1 inch +

• USE OF WELDING ROD •

The steel welding rod, also called filler rod, is used to add extra metal to the weld. As you may have noticed when you practiced welding, the metal which was melted was lower than the metal on each side. The welding rod is used to add metal to the puddle so the welded area will be slightly higher than the original metal. This adds strength to the weld.

• WELDING PIPE •

The only difference between welding flat metal and welding pipe is that you are always welding on the top of the flat metal. You will have to weld all the way around on pipe. This takes much practice and some special techniques which your instructor will demonstrate.

Preparing the Pipe

Pipe must be prepared for welding. The edges of the pipes to be joined should be ground down so that there is a small V when they are placed together. The drawing in Figure 8-14 shows this.

Welding these two pipes together will require that enough welding rod be used to fill that small V. The drawing in Figure 8-15 shows what the completed weld should look like.

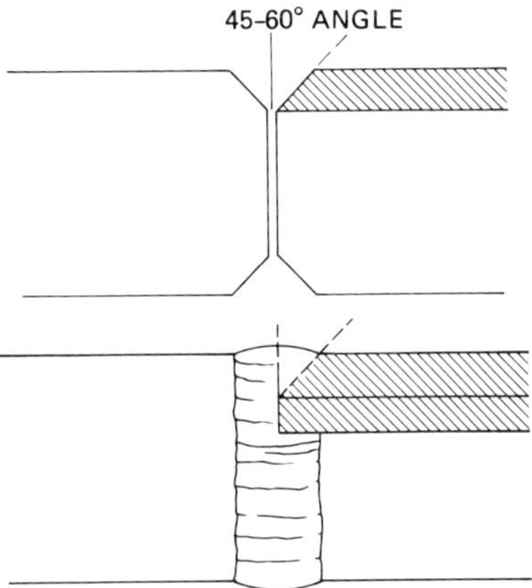

Figure 8-14. Preparing the welded joint

Figure 8-15. The finished pipe weld

Out-of-Position Welding

Earlier, the comment was made that you would have to weld all around the pipe. This means that you will have to be able to weld down each side and underneath the pipe when you are out on a job. This is called out-of-position welding. There are many different ways of welding out of position. Figure 8-16 shows just a few.

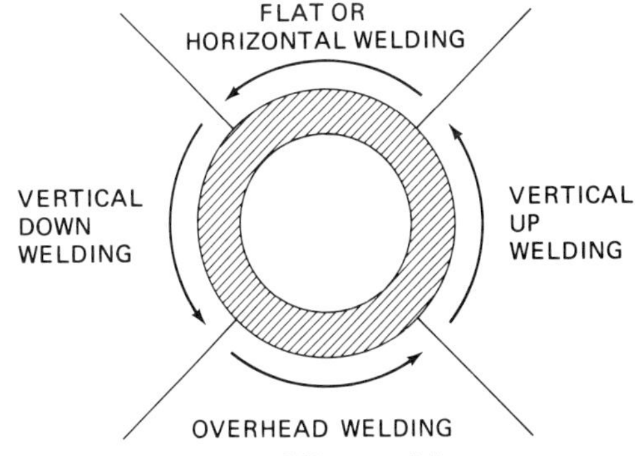

Figure 8-16. Different welding positions

Your instructor will demonstrate these different ways and instruct you in the techniques which are best.

• ARC WELDING •

Oxyacetylene welding uses the heat from two burning gases to melt the steel, while arc welding uses the heat from electricity jumping an air gap. Just as electricity jumping across the gap of the spark plug in your car causes enough heat to ignite the gas in the cylinder, the heat generated by the electricity from the arc welder jumping the gap between the electrode and the metal causes the metal to melt.

Figure 8-17 is a picture of a typical alternating-current (ac) arc-welding set. Become familiar with all the parts and their names.

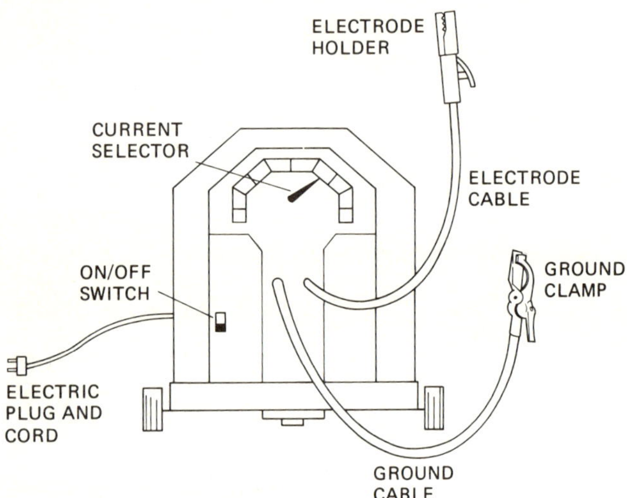

Figure 8-17. Arc-welding machine

• THE AC ARC-WELDING MACHINE •

The ac arc-welding machine uses alternating current, the same type of electricity used in most ordinary houses. While the normal current in a house is 110/120 volts, the ac welding machine normally uses the same current as an electric stove or dryer, 220/240 volts.

This welding machine is a transformer type, which changes the voltage (from 220/240 to 30) and amperage (from 20 up to 200) of the electricity so that it can be used for welding purposes. Figure 8-18 is a picture of an ac arc-welding machine.

The figures on the front of the machine indicate the different amperages which can be selected. A good rule of thumb is "the thicker the metal being welded, the higher the amperage needed." There are other things which indicate the amount of amperage needed, such as the electrode type, the electrode size, and the kind of metal being welded. Since we are only concerned with welding steel, we shall identify and discuss only those items which pertain to welding steel and steel pipe.

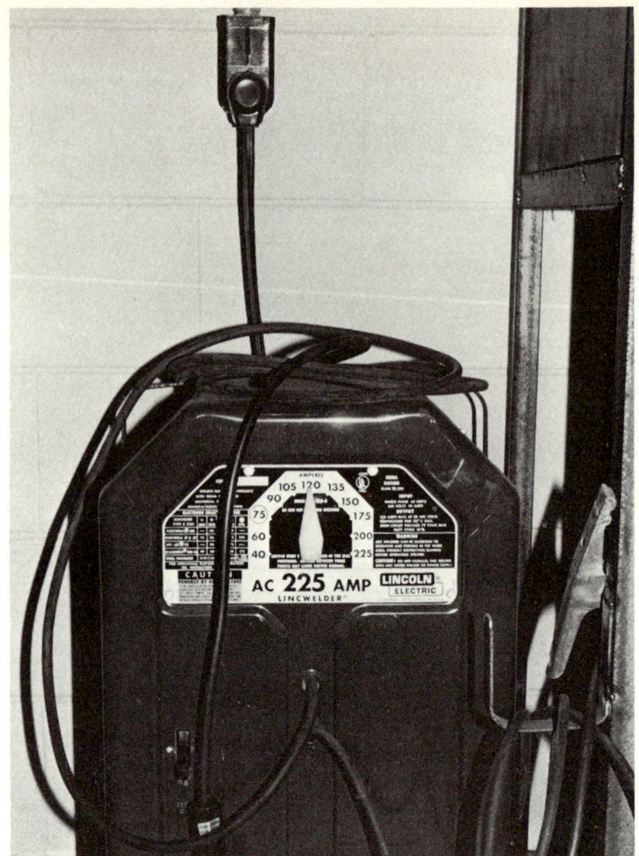

Figure 8-18. AC arc-welding machine

The other parts of the welding machine are the electrical cables which conduct the current to the work area. They are called the ground cable and the electrode cable. Figure 8-19 shows a close-up of the two cables.

Figure 8-19. Welding cables

Notice that the ground cable ends in a spring clamp. This clamp is called a ground clamp and is clamped to the metal which is to be welded. You should also notice that the electrode cable ends in an electrode holder. This is also called a "stinger" and holds the electrode which is used in arc welding. The ground cable and the electrode cable must both be used to complete the electric-arc-welding circuit.

• ELECTRODES •

The electrode is a long (about 13 inches) round piece of steel coated with a material called flux. The electrode has three purposes in arc welding. They are:

1. To conduct electricity to the point to be welded.
2. To supply molten steel to the weld to increase the thickness of the weld.
3. To supply flux to the weld to clean the molten metal and form a protective cover after the weld is made.

Figure 8-20 is a drawing of a typical electrode. Electrodes are made from many different types of steel. We will only discuss the three which would normally be used in the plumbing trade. Each electrode is named by a series of numbers. The three electrodes which we will discuss are the 6011, 6013, and 7014. These numbers are stamped on one end of the electrode for easy identification. While each of the numbers means something to the journeyman welder, it is only necessary for us to know what these three electrodes can do. Table 8-3 shows what each rod is designed to do. Read it carefully.

You will have to remember this information about the different electrodes. If you select the wrong type, you may wind up with a leaky joint. You should also pay close attention to the amperage used with each rod, since this is important in setting the welding machine before you begin to weld.

• SAFETY EQUIPMENT •

Safety equipment is very important when you are arc welding. The light from the arc can blind you or cause a bad burn on exposed skin. The arc is almost as bright as the sun, and it is much closer to your eyes and skin, so use the personal safety equipment furnished. Check the equipment shown in Figure 8-21 and notice what the well-protected and well-dressed welder wears.

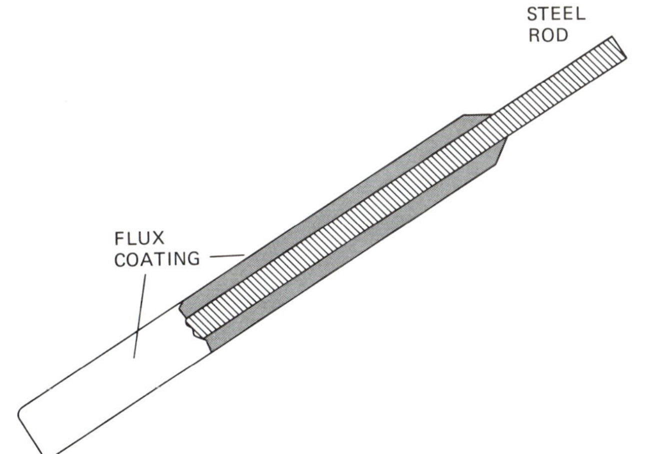

Figure 8-20. Arc-welding electrode

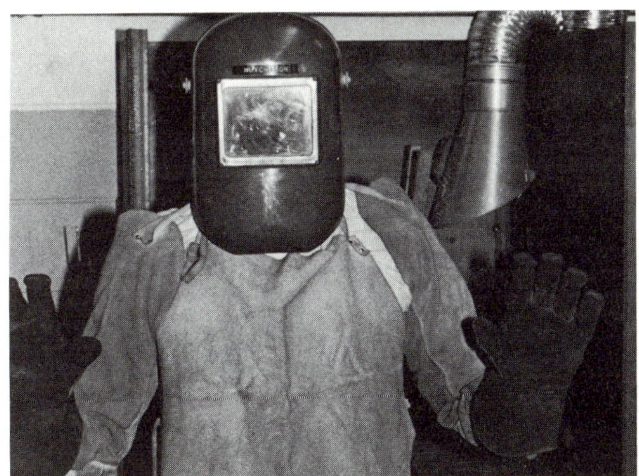
Figure 8-21. Arc-welding safety equipment

TABLE 8-3 COMMON WELDING ELECTRODES

Electrode	Welding Position	Current	Penetration	Tensile Strength, Pounds per Square Inch
6011	All	dcRP ac	Deep	62,000
6013	All	dcRP dcSP ac	Mile	67,000
7014	All	dcRP dcSP ac	Medium	70,000

• PREPARING TO WELD •

The following steps will help you set up the machine for practice welding. Figure 8-22 shows a set-up arc welding machine.

1. Be sure the ac welder is off before you plug it in.
2. Select the correct electrode for the job you are going to do.
3. Set the correct amperage on the welding machine.
4. Connect the ground clamp to the workpiece.
5. Put the electrode in the electrode holder.
6. Put on all safety equipment. Raise face shield.

If you followed through the steps above, you are now ready to strike the arc.

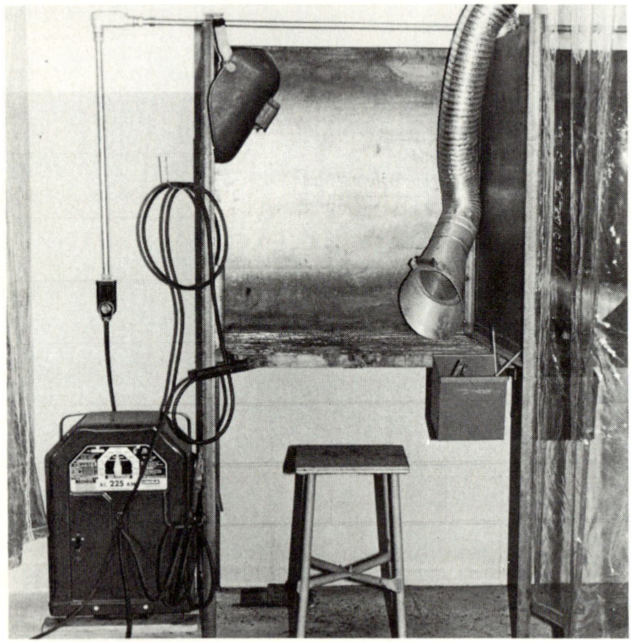

Figure 8-22. Arc-welding machine setup

• STRIKING THE ARC •

Striking the arc can be accomplished in different ways. Two of the most popular ways are by scratching or tapping the tips of the electrode across the workpiece. Follow the steps shown in Figure 8-23 to practice striking the arc.

1. Turn the welder power switch *off.*
2. With the electrode in the holder (step 5 above), move the tip of the electrode to about ½ to ⅜ inch above the workpiece.
3. Then scratch the top of the electrode across the workpiece as you would light a match, or tap the tip of the electrode against the workpiece as if it were a light hammer.
4. At the end of the scratch or tap, the tip of the electrode should be about ⅛ to ¼ inch above the workpiece.

• MAKING THE WELD •

Making the weld, or running the bead, as it is also called, is a matter of keeping the arc going, watching the puddle of molten metal, and moving that puddle down a straight line just as you did with the oxyacetylene puddle. The most important thing to remember is that the electrode melts and gets shorter as you are welding. This means you have to lower the electrode continually or the arc will break and stop. The tip of the electrode must always be ⅛ to ¼ inch above the workpiece.

• QUESTIONS •

1. How many volts of electricity does the ac arc welder use?
2. What does the ground cable attach to?
3. What are the three most common electrodes?

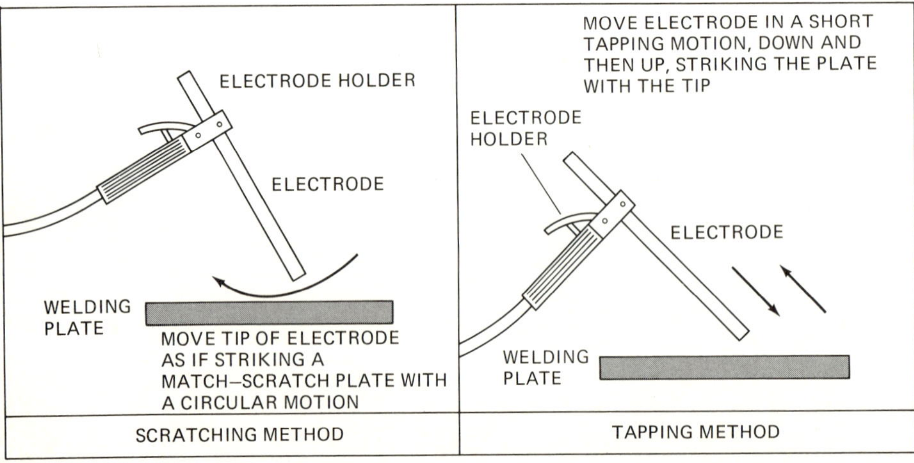

Figure 8-23. Striking the arc

4. When arc welding, the light from the arc can cause blindness or sunburn. (T or F)
5. The ground clamp is also called a "stinger." (T or F)
6. What are the three purposes of an electrode?
7. What amperage should be used for the rods listed below?
   7014
   6011
   6013

• WELDING PIPE •

In using the arc-welding process for pipe, you will find that there is the same requirement for "out-of-position" welding that we had in oxyacetylene welding. You will weld up, down, and underneath the pipe. Remember, electrodes that are made to do out-of-position welding should be used. Your instructor will demonstrate the different techniques for out-of-position arc welding. Learn them and practice until you can make good out-of-position welds.

As with oxyacetylene welding, the edges of the pipe to be welded must be ground. Figure 8-24 shows pipe which has been properly prepared for arc welding.

45-60° ANGLE

Figure 8-24. Weld preparation

• OXYACETYLENE CUTTING •

Cutting with the oxyacetylene flame uses all the same equipment as oxyacetylene welding with one exception, the torch. Figure 8-25 shows a drawing of an oxyacetylene cutting torch with the different parts identified.

Notice the differences between the cutting torch and the welding torch. The cutting torch is larger. It has three valves and an oxygen lever, three tubes carry the gases from the body to the tip, and the tip is entirely different. Let us start at the lower valves and discuss each of the components of the cutting torch. The two main valves, the ones near the hose connection, are just like the ones on the welding torch. One allows oxygen flow into the body of the torch, where the other two oxygen valves control it. The other main valve controls the acetylene. The *lever valve* in the middle of the torch body *controls the oxygen used in the cutting process,* while the *upper oxygen valve only controls the oxygen which is mixed with acetylene.* Each of the three tubes going from the body to the tip of the torch allows gas flow from one of the three gas control valves, the acetylene valve, the cutting oxygen lever valve, and the mixing oxygen valve. The cutaway drawing in Figure 8-26 shows these control valves and how the different gases flow to the tip.

A close look at the torch tip in Figure 8-27 will show you that there are a series of small holes around a larger center hole. These smaller holes are flame holes, and each one produces a small oxyacetylene flame when the torch is lit. The center hole allows the cutting oxygen (from the lever valve) to flow out between the small flames.

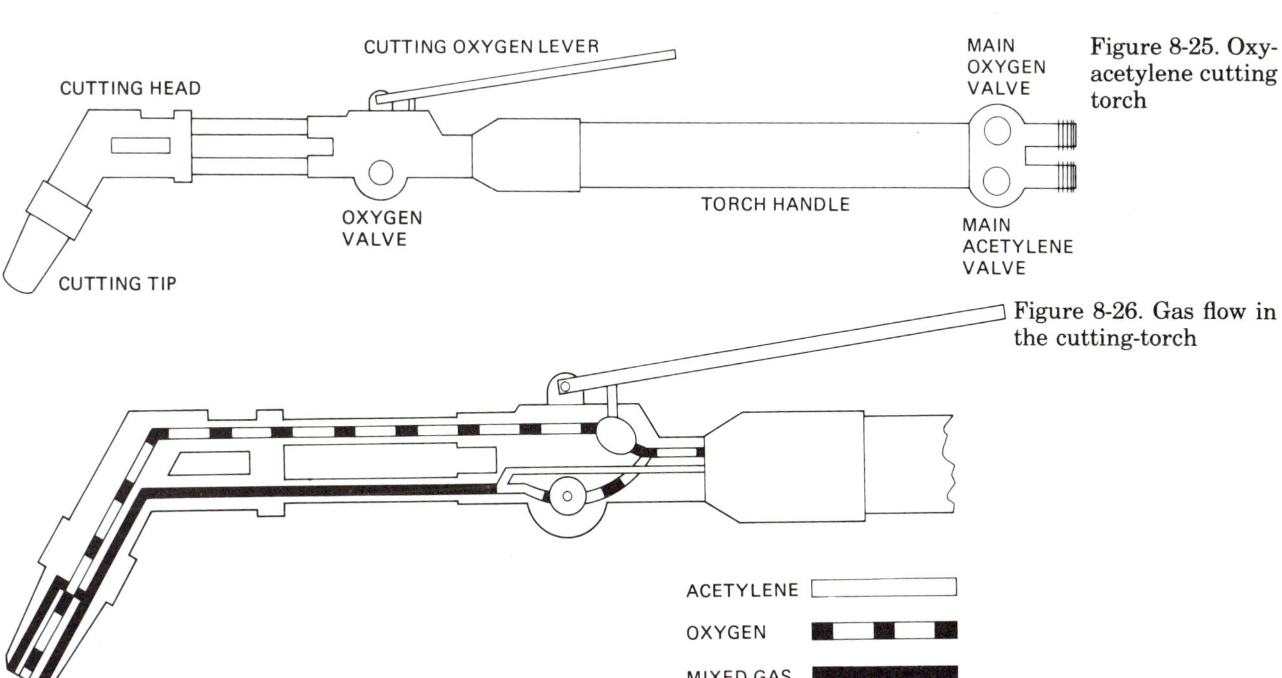

Figure 8-25. Oxyacetylene cutting torch

Figure 8-26. Gas flow in the cutting-torch

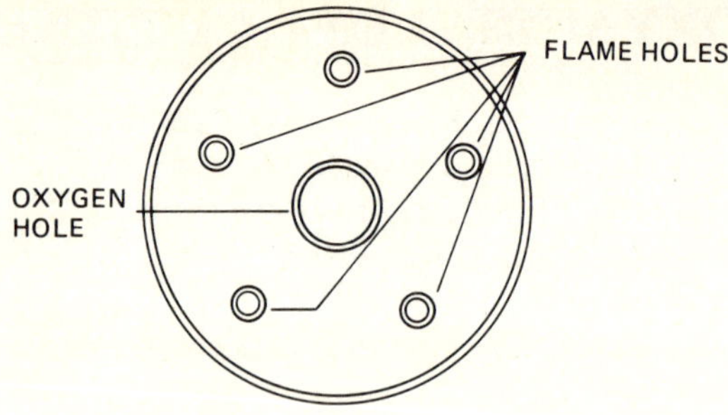

Figure 8-27. Cutting-torch tip

Oxyacetylene Cutting Principles

Remember, when we were discussing steel pipe, we referred to rusting as oxidization. That is, the steel and oxygen (in water or air) combine to form rust. This oxidizing process is the base for oxyacetylene cutting.

It works this way: The small flame holes around the oxygen hole are used to heat up the steel until it is red hot. Then the oxygen lever valve is opened and the red-hot metal is oxidized so rapidly that it burns. Once the cut has been started, the torch can be moved, and the oxygen will continue to burn through the metal.

Cutting Operations

Before we can begin using the oxyacetylene cutting torch, there are some equipment and pressure adjustments which must be made. The basic equipment change was stated earlier—the welding torch must be removed from the tank hoses. Be sure the main tank valves are off before attempting to disconnect the torch, and remember that the acetylene fittings always have left-hand threads.

Pressure Adjustment

After you have connected the cutting torch to the hoses, the size of the cutting tip must be determined. For general-purpose pipe cutting, a No. 1 tip should be used. The acetylene pressure should be adjusted the same way as the welding pressure, 1 pound per square inch for a No. 1 tip. Oxygen pressure should then be adjusted to a minimum of 20 pounds per square inch. Figure 8-28 shows the properly adjusted pressures for cutting. Follow through the steps below to adjust the oxyacetylene pressures for cutting.

1. Open the oxygen tank valve and set the pressure-regulator valve for 20 pounds per square inch.
2. Open the acetylene tank valve (one-fourth turn, remember) and set the pressure-regulator valve at 1 pound per square inch (for a No. 1 tip).
3. Turn the main oxygen valve on the torch to the full open position. Then squeeze the oxygen lever valve just long enough to check for oxygen flow at the tip.

Figure 8-28. Proper cutting pressures

4. Open the main acetylene valve slightly, and using a striker, light the gas at the tip of the torch.
5. Adjust the acetylene flame as with the welding torch. Then, using the upper oxygen valve, adjust for neutral oxyacetylene flame at the tip.
6. Squeeze the oxygen lever valve. You should see a clear path of gas through the flames at the tip (Figure 8-29). You are ready to use the cutting torch.

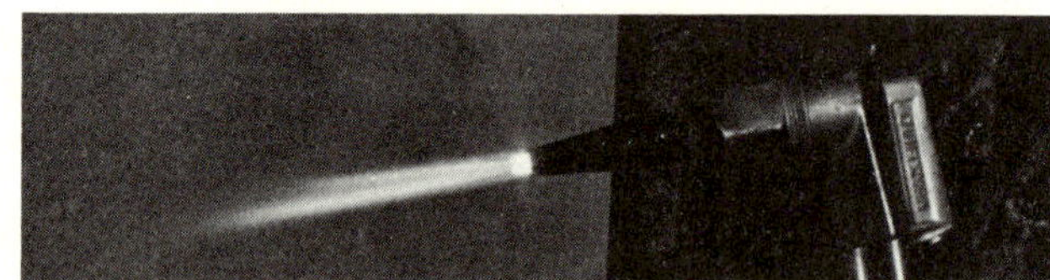
Figure 8-29. Adjusted cutting-torch flame

Cutting Procedures

There are two things which have a direct bearing on the cut which you will make: the distance from the tip of the neutral flame to the metal, and the speed at which the torch is moved. The diagram in Figure 8-30 will show you the approximate height of the flames.

The speed at which the cut can be made varies with the thickness of the metal being cut. You will soon determine this after making a few practice cuts.

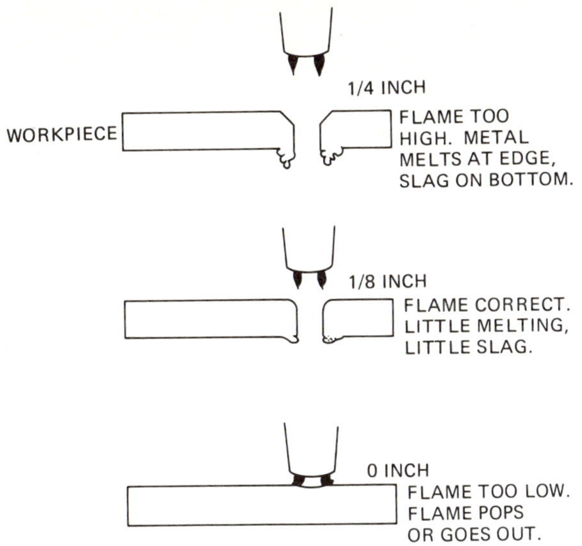

Figure 8-30. Cutting flame height

• SAFETY EQUIPMENT •

The safety equipment for oxyacetylene cutting is the same as that for oxyacetylene welding: welding goggles and gloves. In addition, since this cutting process produces many sparks, it is recommended that pants cuffs be rolled down to keep sparks from causing burn holes or a fire.

• QUESTIONS •

1. The oxyacetylene cutting torch has two oxygen valves. (T or F)
2. What gas flows through the large center hole in the tip of the torch?
3. The acetylene pressure is set at 10 pounds per square inch for a No. 3 tip. (T or F)
4. Oxygen pressure is set at a minimum of 20 pounds per square inch. (T or F)
5. The most important item of safety equipment is the welding goggles. (T or F)

Figure 8-31 shows common errors made when using the cutting torch.

• SUMMARY •

This chapter introduced oxyacetylene and arc-welding equipment, practices, and procedures. It was designed to provide the plumber trainee with basic knowledge and skill in both welding mediums as well as indoctrination into the theory and practice of oxyacetylene cutting.

SIDE	FACE	TOP	PROBLEM
			NONE
			PREHEAT FLAMES TOO HIGH/HOT (MELTING AT TOP)
			CUTTING SPEED TOO FAST
			CUTTING SPEED TOO SLOW

Figure 8-31. Common cutting errors

• WORDS PLUMBERS USE •

oxygen
acetylene
oxyacetylene welding
gas regulator
orifice
welding rod
arc welding
transformer
welding electrodes
oxyacetylene cutting torch

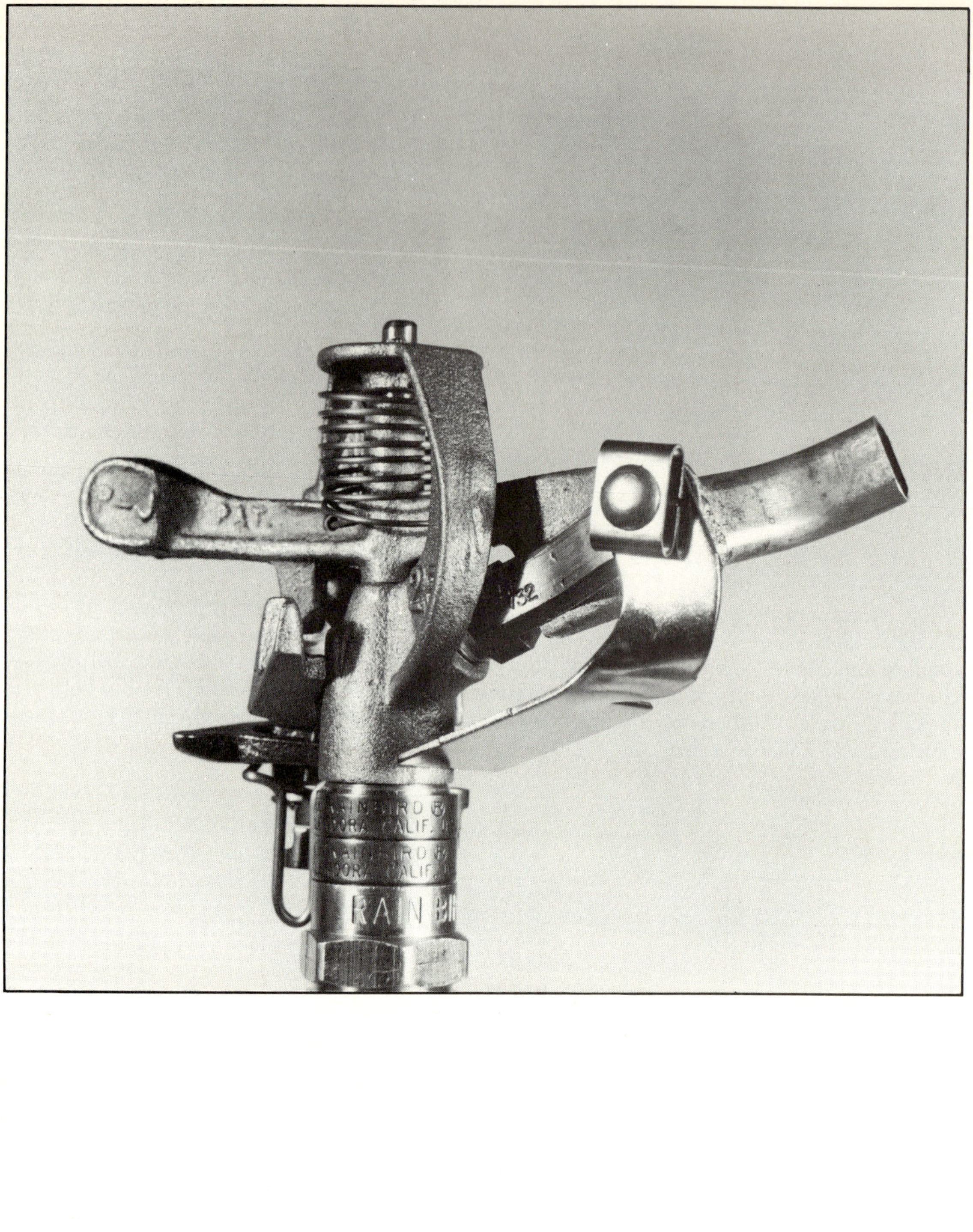

9 PLUMBING ACCESSORIES

Up to this point, we have covered the various fixtures and systems of a typical dwelling. This chapter will discuss some important accessories which may or may not be used in a building's plumbing system. Many of these accessories are for comfort or work-saving reasons. This chapter will acquaint you with some of the plumbing-system accessories you may encounter in the course of your trade.

At the completion of this chapter, you will be able to:

- Identify four different types of hot-water heating devices.
- Identify the different common accessories that require connection to the plumbing system.

• HOT-WATER SYSTEM •

The first water-system accessory which we will discuss is the hot-water heater. The hot-water heater is the most common accessory used today. Very few buildings are without some form of hot-water heater.

There are basically four types of hot-water heaters on the market. They are the electric, gas, solar, and heat-exchange types. The electric, gas, and solar heaters are independently operated. That is, they are not connected to the building's heating system. The heat-exchange type (an older type) uses the building heating system to heat the water. We will discuss each of these hot-water heaters individually.

Most hot-water heaters available today are automatic. That means that the water will be heated to a certain temperature and then the heater will turn off. As the heated water is used, cold water comes into the heater, lowering the temperature of the water in the tank. When the water in the tank cools off, the heater senses this, and the heat element automatically turns on to bring the water in the tank back up to the correct temperature.

• THERMOSTATS •

The device which senses the temperature of the water and turns the heat on and off is called a thermostat. A thermostat is used in all of the automatic heaters, electric, gas, and solar. There are even times when a thermostat might be used with the heat-exchange type of water heater.

When metal is heated, it expands or gets larger. The thermostat makes use of this principle of metal expansion by using a thin strip of metal which acts as a spring. As the water in the tank reaches the correct temperature, the metal in the thermostat expands and breaks a contact, turning the heating element off. As the water cools, the metal also cools and shrinks back to its original size. This closes the contact, and the heater element turns on. Figure 9-1 is a drawing of a simple thermostat.

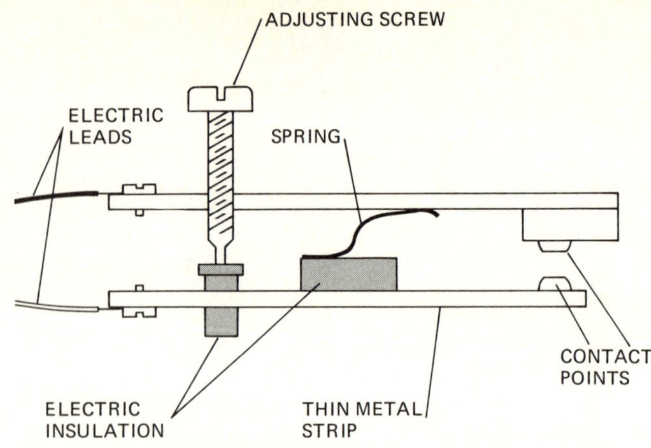

Figure 9-1. Simple thermostat

• ELECTRIC HOT-WATER HEATERS •

There are many different electric hot-water heaters available today. They are available in many different sizes or capacities (the amount of water they will hold), with the 30- to 40-gallon size most commonly used in a dwelling. A picture of a typical 40-gallon electric hot-water heater is shown in Figure 9-2.

Figure 9-2. Automatic electric water heater

Since the electric hot-water heater does not make any smoke or fumes, it is ideally suited for use in small or closed-in areas, such as under a sink counter or in a closet. The electrical elements which heat the water are enclosed in the tank. Figure 9-3 is a cutaway drawing of an electric water heater.

Notice the cold-water inlet pipe. It has a filler pipe that forces the cold water down to the bottom of the tank. This filler pipe is there for two reasons. First, it prevents the cold water from mixing with the hot water near the top of the tank. Second, it directs the cold water to the bottom, where the ther-

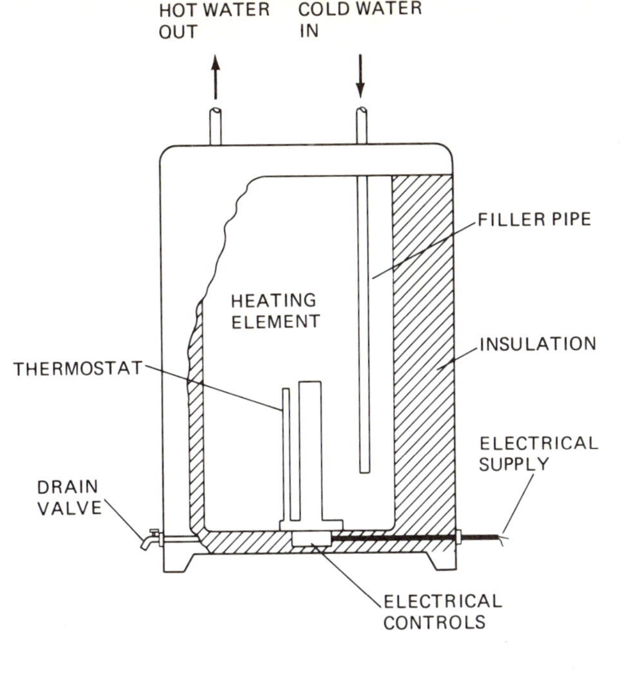

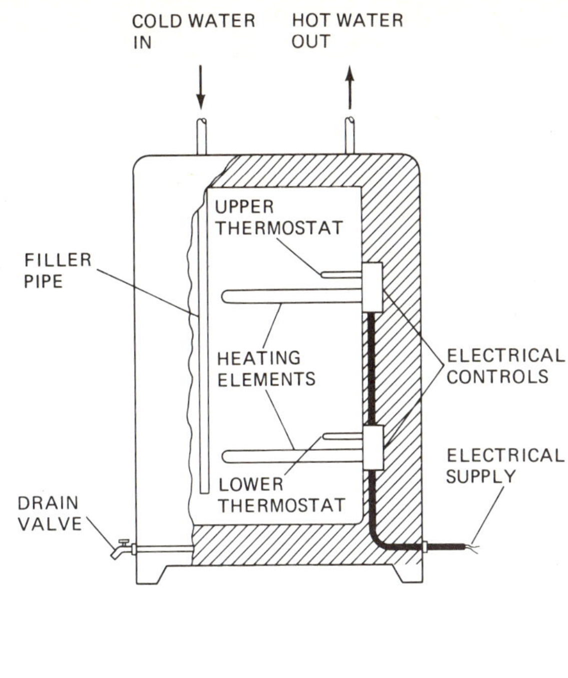

Figure 9-3. Automatic electric water heater components

mostat and heating elements are, so that it may be heated more rapidly.

The plumber normally connects the hot- and cold-water systems to the electric heater. The electrical components are usually connected by an electrician.

• GAS HOT-WATER HEATERS •

There are two types of gas hot-water heaters, the manual and the automatic. There are still manual hot-water heaters on the market today, but they are seldom used. Figure 9-4 is a drawing of a manual hot-water heater and the hot-water storage tank.

In a system with a manual gas hot-water heater, the gas is turned on and lit each time hot water is needed. It is left on until the water in the storage tank reaches the correct temperature, and then the heater is turned off manually.

The automatic gas water heater works on the same principle as the electric heater. The only difference is that gas instead of electricity is used as the fuel. It uses a thermostat to control the gas and has a pilot light to ignite the gas as the thermostat turns it on. A typical automatic hot-water heater is shown in Figure 9-5 on page 108.

These gas water heaters also come in a variety of sizes. Normally, the 30- or 40-gallon size is used for a typical dwelling. Figure 9-6 is a cutaway view of an automatic gas water heater. (See page 108.)

The cold-water inlet and the hot-water outlet are the same as in the electric heater. There are some

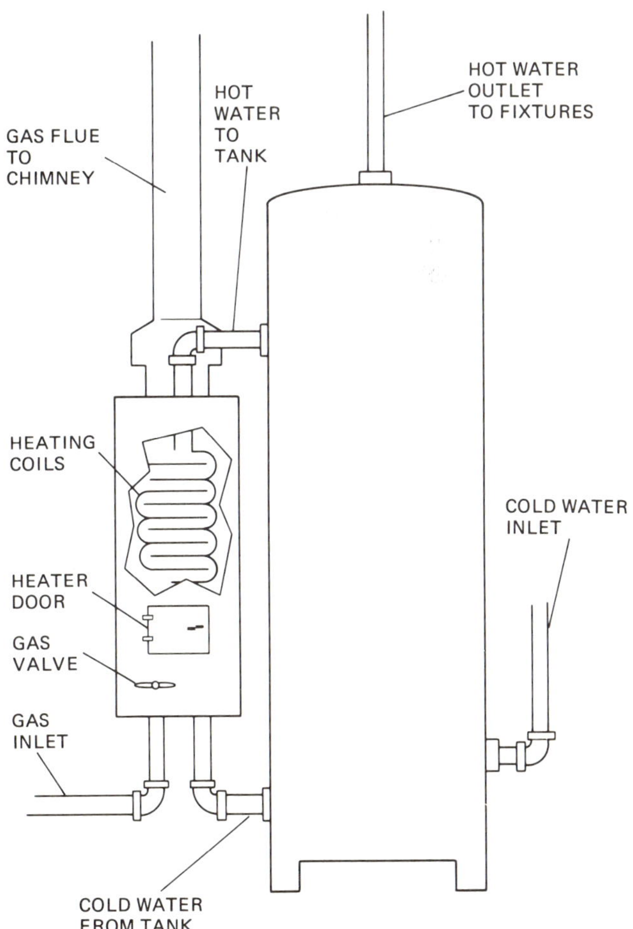

Figure 9-4. Manual water heater and tank

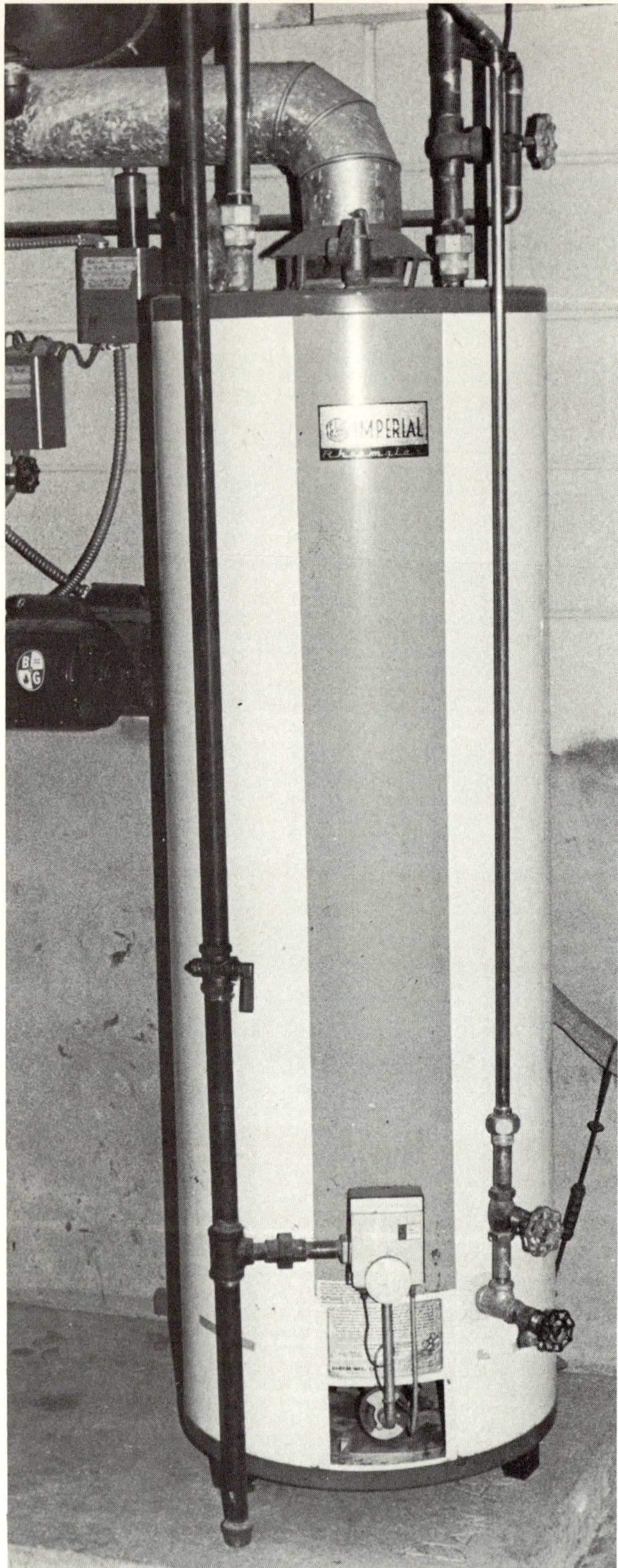

Figure 9-5. Automatic gas water heater

important differences in connecting the heater, since the plumber also must connect the gas line and must connect the flue to the chimney. This heater burns gas to heat the water, and the fumes and smoke from the burned gas must be vented to the chimney.

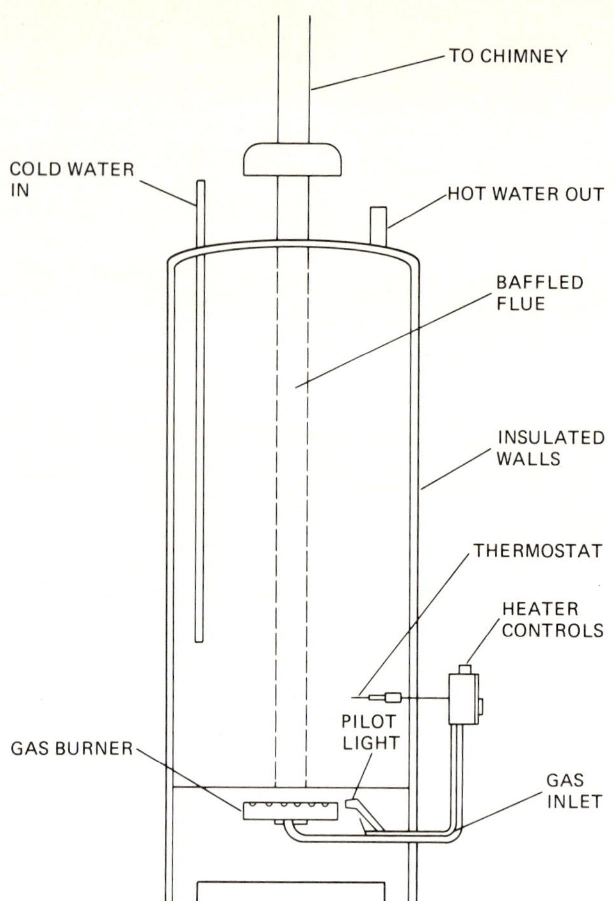

Figure 9-6. Automatic gas water heater components

• INSTALLING THE AUTOMATIC HOT-WATER HEATER •

Automatic heaters, either electric or gas, require the same connections to the water system. The cold-water inlet should have a control valve installed so that the heater can be turned off and drained periodically. It will be necessary to use a union between the cold- and hot-water pipes and the heater. Figure 9-7 shows a typical method of connecting the water lines to an automatic heater.

Notice the safety or relief valve on the hot-water outlet line. This valve protects the hot-water heater and the water system in case of a thermostat failure. If the thermostat failed to turn the heat off, the water in the heater would boil (212°F, 100°C) and cause enough pressure to blow up the tank. The safety or relief valve opens at a temperature below 180°F [80°C] and releases this extra pressure. The pipe connected to the safety valve simply directs the water to the floor.

After the water lines are connected to the electric heater, the plumber's job is done. The gas heater requires two more steps. The plumber must connect

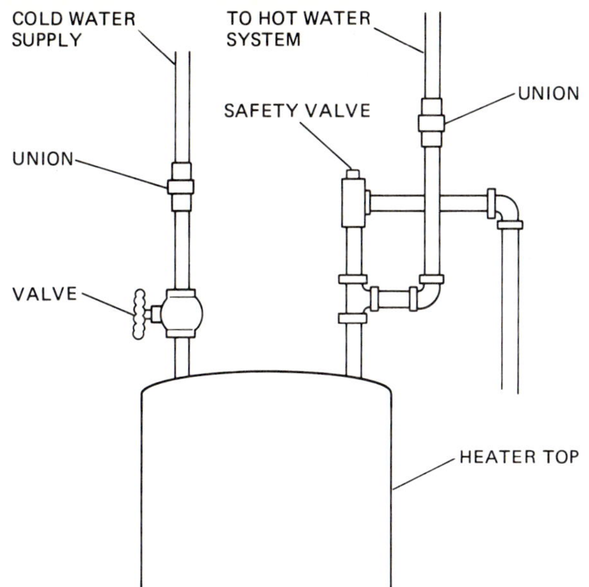

Figure 9-7. Connecting the automatic water heater

the gas line and run the flue into the chimney. Figure 9-8 shows a standard gas-line connection.

Notice the capped nipple labeled "rust trap" on the bottom of the vertical gas line. This trap catches small bits of dirt and rust which could clog the gas control unit.

The flue connection to the chimney is a matter of placing the heater hood (normally furnished with the heater) in position on the top of the heater. The

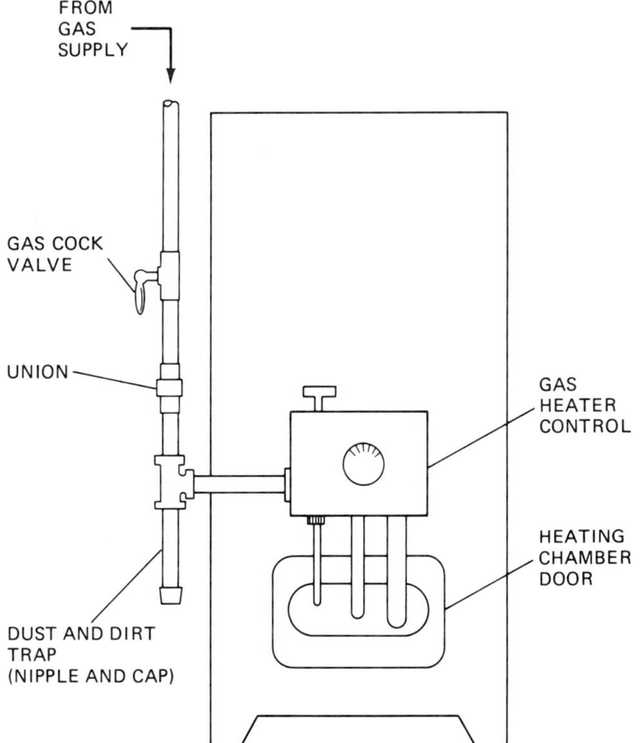

Figure 9-8. Gas-line connection

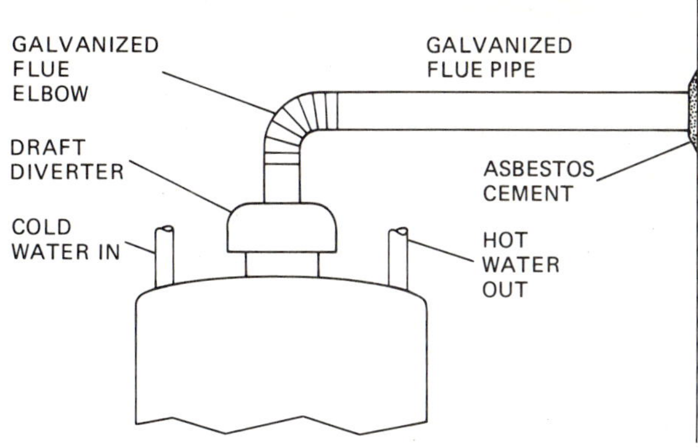

Figure 9-9A. Chimney connection

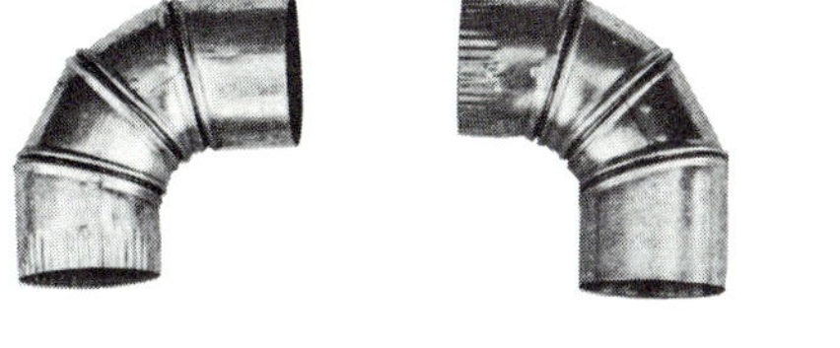

Figure 9-9B. Galvanized stovepipe

remainder of the connection is a run of galvanized stovepipe to the chimney. A drawing of the finished connection is given in Figure 9-9*a*, and pictures of galvanized stovepipe in Figure 9-9*b*.

Notice that one end of the stovepipe has a series of little dents all the way around. This is called the crimped end. Two sections of stovepipe are connected simply by placing the crimped end of one into the plain end of another. Figure 9-10 shows this connection.

Figure 9-10. Galvanized stovepipe connection

After the gas line and the flue are connected, the tank should be filled with water. All water and gas connections should be tested for leaks. Water leaks will be easy to spot, but the gas joints should be tested using soapy water. If a leak is present, a group of bubbles will identify it. Any leaks in the water or gas lines should be corrected immediately by tightening the joint. *Never* use a match or open flame to test for gas leaks!

• SOLAR WATER HEATERS •

Recent advances in solar water heating, that is, using the sun to heat water, have made it economical to supply the hot water for a building in that manner. Solar hot-water heaters have been in use in many southern states for 40 years or more. The early solar hot-water heaters were large heating boxes made of wood or metal, painted black on the inside and covered with glass. The inside of the boxes contained many connected pipes, and the boxes were mounted on the southern roof of the house. An early solar hot-water heater is shown in Figure 9-11.

The connection of a solar hot-water heater is different from the others we have discussed. As long as the sun is out, the solar heater will supply hot water, but to have hot water all night long, this hot water must be stored in some way. Water heated during the sunny hours is usually stored in a large tank located in the attic of the building. Figure 9-12 is a drawing of a typical solar hot-water system hookup.

Further information on solar heating or hot-water systems will be covered in Chapter 12.

• HEAT-EXCHANGE HOT-WATER HEATERS •

Many houses have a separate water-heating system used to heat the house during the winter. These

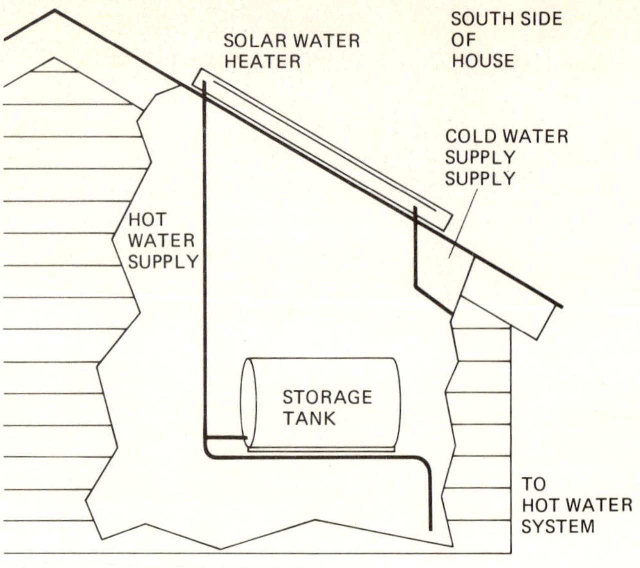

Figure 9-12. Placing the solar water heater

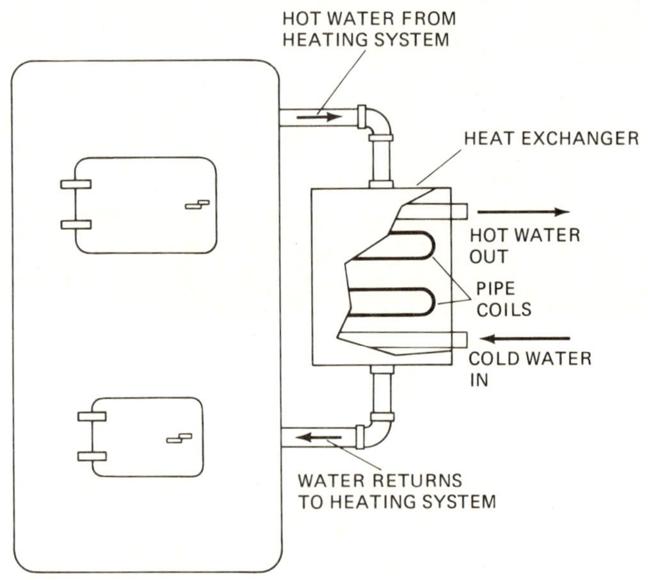

Figure 9-13. Heat exchanger

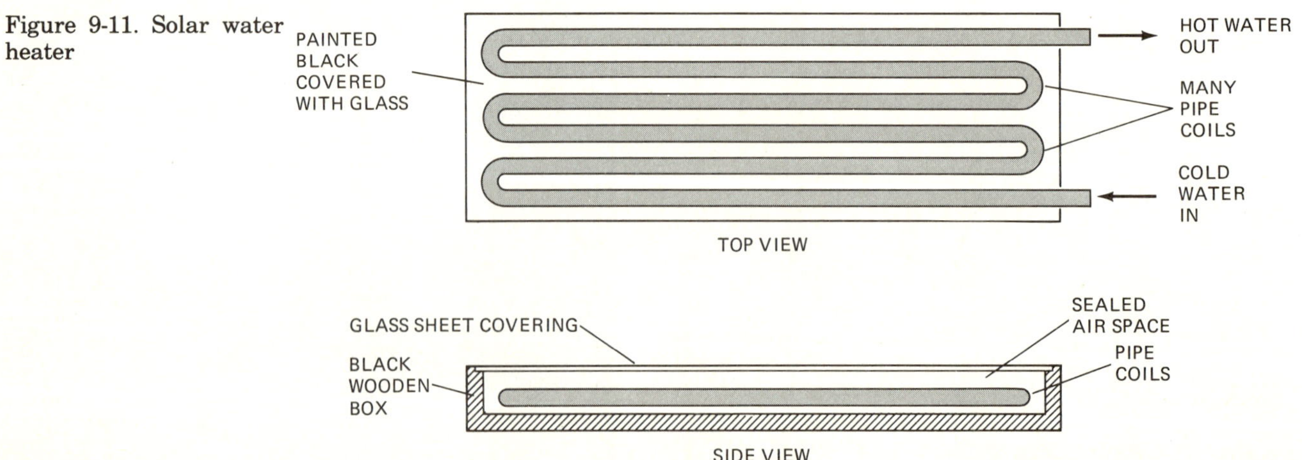

Figure 9-11. Solar water heater

hot-water or steam heating systems will be discussed in detail in Chapter 12. The heat-exchange hot-water heater uses the water or steam of the house heating system to heat water for use in the bath or kitchen.

The tank of the heat exchanger is filled with the hot water from the heating system's boiler. The cutaway view in Figure 9-13 shows the coil of pipes inside the tank. These pipes carry the water for personal use through the exchanger and out to a storage tank. This type of exchanger is necessary, since the water in the boiler is contaminated with rust and other matter.

• QUESTIONS •

1. What controls the temperature of hot water in an automatic hot-water heater?
2. The electric hot-water heater can be installed almost anywhere in the house. (T or F)
3. The rust trap is important in the installation of the gas hot-water heater because it collects rust and other foreign matter. (T or F)
4. Why are safety valves installed in the hot-water system?
5. The heat-exchange water heater is used with a hot-water or steam heating system. (T or F)

• SPRINKLER SYSTEMS •

This section will briefly discuss two different systems, the fire-extinguisher system and the lawn-sprinkler system. Though the basic principle of the two systems is the same, to spray water over an area, their operation is very different.

The Fire-Extinguisher System

This system is generally installed in business, manufacturing, and hotel or motel buildings. Its sole purpose is to provide a fire-extinguishing spray of water over the interior of the building. The pipes are installed near the ceiling of the building to allow good coverage of the area. Since this system must be ready to operate at any time in case of a fire, the system must always be pressurized and ready.

The control system and the sprinkler valves for this system are very different. There are certain problems which occur when a system like this must be turned on by heat or fire. The first one, the problem of opening the sprinkler valve when fire or too much heat occurs, is easily overcome.

The faucet of this sprinkler system, the sprinkler head, has this feature built in. The secret of the fire-extinguishing sprinkler head is a small bit of metal which will melt at a set temperature. Just as the melting temperature of solder can be set, this metal can be manufactured to melt at a set temperature. One of these sprinkler heads is shown in Figure 9-14.

Notice the small piece of metal in the center of the sprinkler. It is called a fusible plug. This is the metal which will melt when the heat in the building reaches about 160°F [70°C]. When this metal melts, the springs bring the arms together and the valve begins to spray water. You can see that this would turn on that *one* valve.

But what about the rest of the valves in the sprinkler system? One spray of water may not put out the fire. The other sprinkler heads in the system will also be turned on when the fusible plug melts. The sprinkler system is designed so that each sprinkler will have enough pressure to spray water even if all are opened by heat.

Though a sprinkler system may be unsightly, much equipment, many buildings, and some lives have been saved by its installation. It is an ever-watchful fire fighter and does its job well.

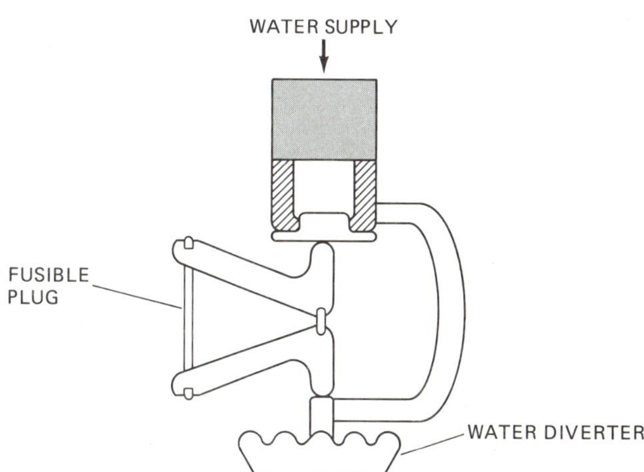

Figure 9-14. Sprinkler-system head

The Lawn-Sprinkler System

A lawn-sprinkler system is controlled by a manually operated valve rather than an automatic valve. A person can water the entire lawn at one time by opening a single valve. Many lawns and gardens in the southern and western sections of the United States are watered in this manner. The northern states have some sprinkler systems, but if proper drainage precautions are not taken, undrained water can freeze during the cold winters and split the pipes.

Lawn-sprinkler-system pipes are buried underground, and the sprinklers are connected to the system in two ways. The type of connection depends on the type of sprinkler being connected. Two types of sprinklers are shown in Figure 9-15.

CIRCULAR SPRAY HEAD

WATER BIRD

Figure 9-15. Lawn-sprinkler-system heads

You should know that these two types are not the only ones available, but they are typical of the two different installations. The pop-up or round-head sprinkler is designed to be installed level with the ground. As the water pressure arrives at the sprinkler, the center pops up and the water spray covers a set area. The second type (the Water Bird, a trademark) is installed above the ground and normally sprinkles a larger area. Average ground areas covered by each of these sprinklers are shown in Figure 9-16.

· AUTOMATIC DISHWASHERS ·

Over the past 20 years, the automatic dishwasher has become a major appliance in many kitchens. Many homes built today have a dishwasher installed. There are two different types of dishwashers;

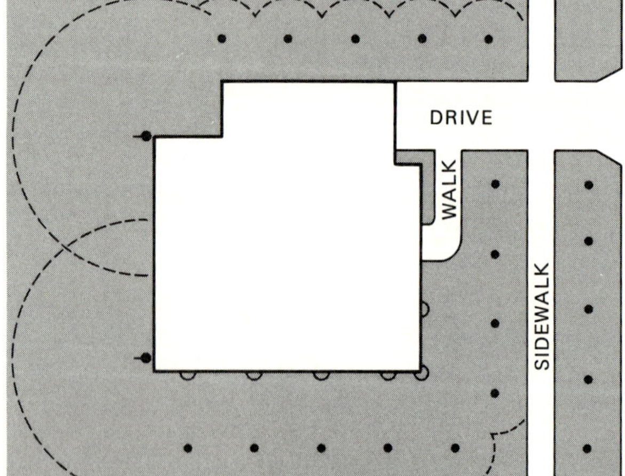

• CIRCULAR SPRINKLER, 10–12′ DIAMETER
⏶ SEMICIRCULAR SPRINKLER, 10–12′ DIAMETER
•– WATER BIRD SPRINKLER 20–25′ DIAMETER

Figure 9-16. Lawn-sprinkler-system plan

one type is installed by the plumber, while the other type is portable and requires no installation. Figure 9-17 is a picture of an automatic dishwasher.

The automatic dishwasher is generally installed next to the kitchen sink and becomes part of the kitchen cabinets. It is manufactured to the correct height so that it can be placed under the kitchen counter. The dishwasher requires two connections by the plumber, one to the hot-water line for water supply and one to the sewer system for discharge or waste, and an electrical connection by the electrician. The plumbing connections are shown in Figure 9-18.

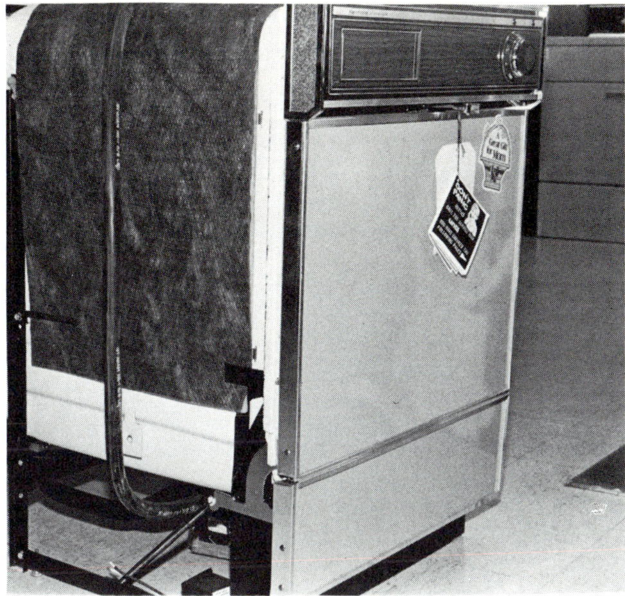

Figure 9-17. Automatic dishwasher

As you can see, the hot-water connection and the waste connection use the same outlet as the sink. When the dishwasher is installed away from the sink, separate water and waste lines must be run.

• QUESTIONS •

1. A fire-extinguisher system uses a fusible plug which melts when a certain temperature is reached. (T or F)
2. All the sprinkler heads in a lawn-sprinkler system will be turned on when the fusible plug melts. (T or F)
3. The main valve of the fire extinguisher keeps pressure on the system at all times. (T or F)
4. The automatic dishwasher can use the same connections as the kitchen sink. (T or F)
5. The lawn-sprinkler system can freeze in the winter in northern states if proper precautions are not taken. (T or F)

• SUMP PUMPS •

The sump pump is a major sewer-system accessory. Many times, the sewer outlet or the septic tank will be above the level of the laundry tub or washer in the basement. Since the waste from these fixtures will not run uphill, some type of pump is needed to move the waste water up to the sewer level so that it can drain away.

There are three types of pumps used for this uphill movement of waste in the sewer system: the centrifugal pump, the water ejector, and the air-displacement ejector. We will deal with the most commonly used type, the centrifugal pump. Pictures of two types of centrifugal sump pumps are shown in Figure 9-19.

In operation, each of these pumps is placed in a drainage pit, called a sump. The sump is a watertight hole which receives waste water that cannot drain into the main sewer. When the waste water in the

Figure 9-18. Automatic dishwasher installation

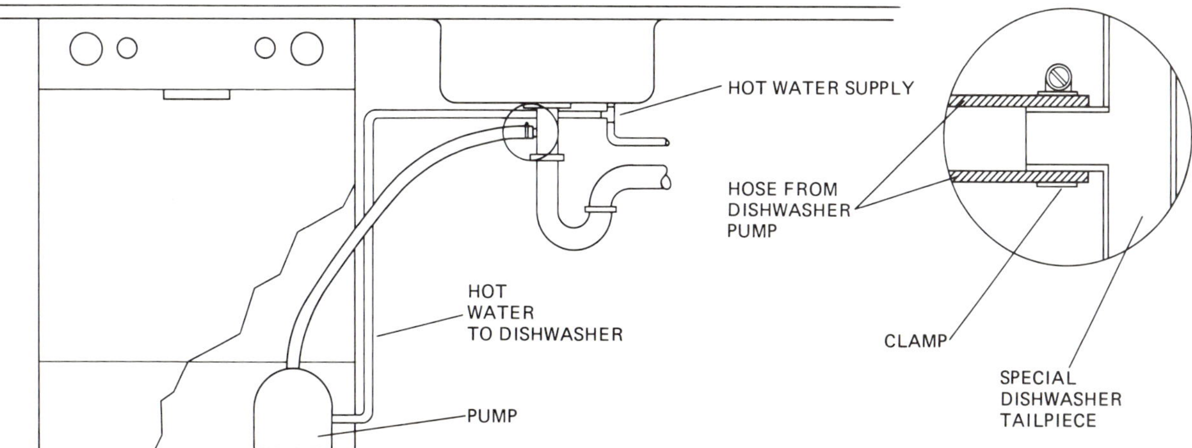

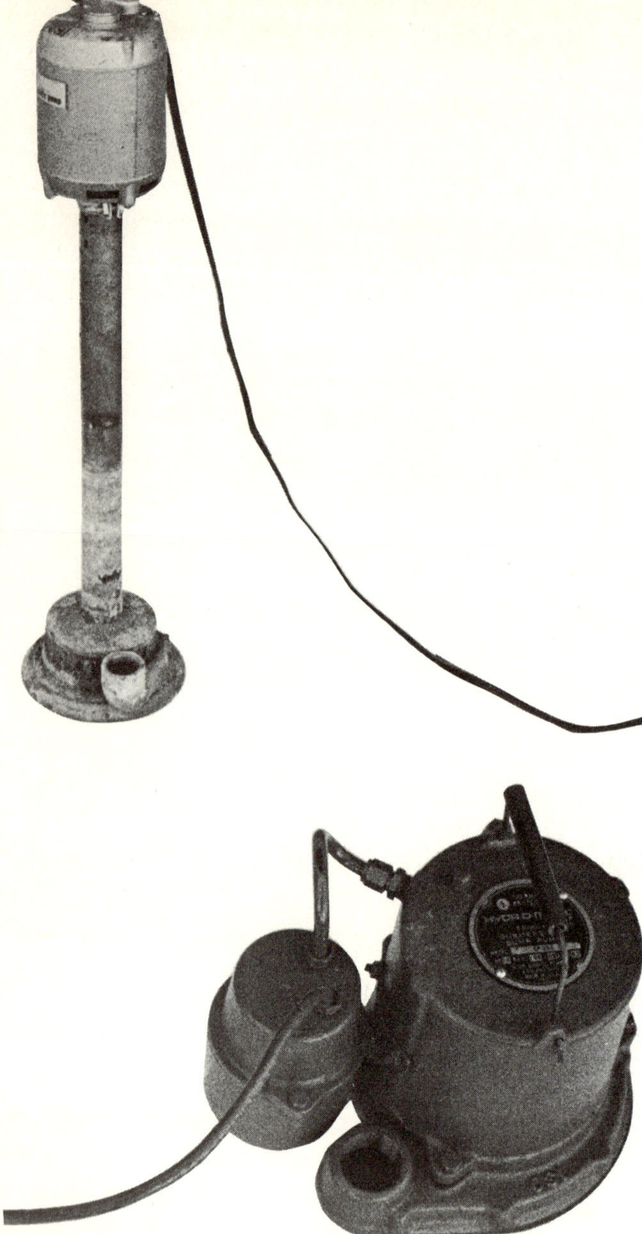

Figure 9-19. Centrifugal sump pumps

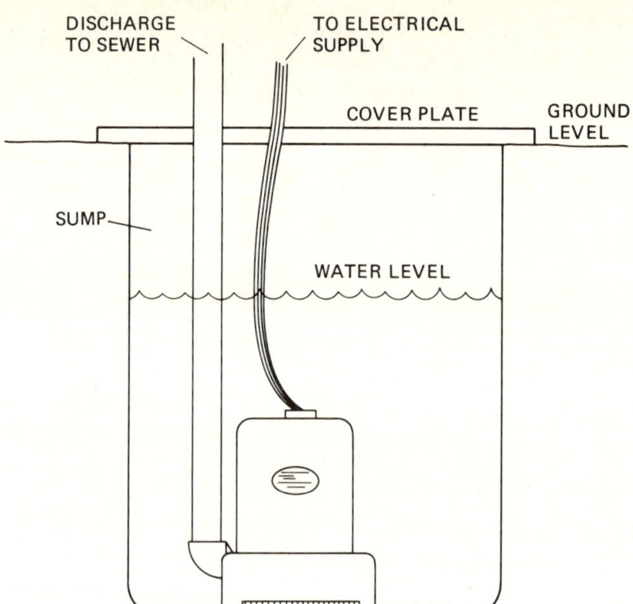

Figure 9-20. Submersible sump pump

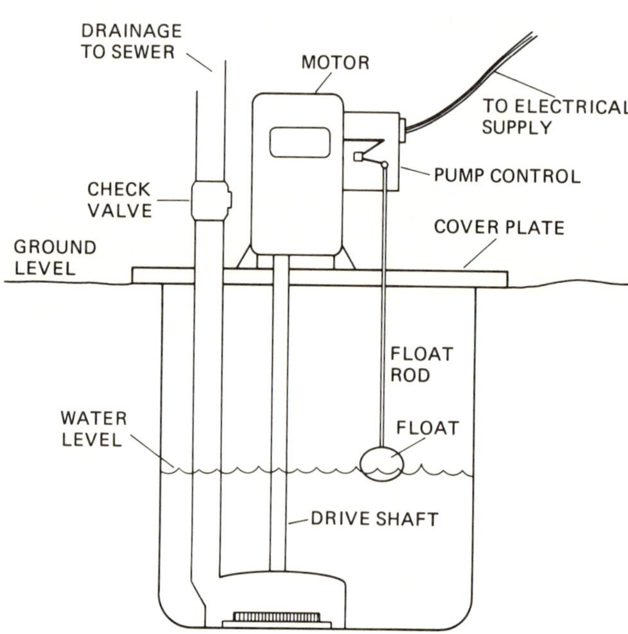

Figure 9-21. Vertical sump pump

sump rises to a set level, the pump is automatically turned on and the waste water is pumped out of the sump and into the main sewer. A cutaway view of a centrifugal sump pump and its connections is given in Figure 9-20.

The submersible pump is designed to work completely under water. The electric motor and the electric wires are insulated from the water, and the motor is turned on and off by a pressure switch. When the water rises far enough above the pump to close the switch, the pump is turned on and the water is pumped away. As the water level falls, the switch is turned off and the pump stops.

The vertical sump pump works on a different principle. The electric motor is at the top of a long shaft and the centrifugal pump is at the bottom. This sump pump is placed so that the centrifugal pump portion is at or near the bottom of the sump and the electric motor is above the top of the sump pit. A connected vertical sump pump is shown in Figure 9-21.

This pump is turned on and off by a mechanical switch, operated by the float rod. As the water rises in the sump pit, the float moves upward. When the water reaches the set height, the float rod turns on the switch and the pump begins removing water from the pit. As the water level goes down, the float rod turns the switch off and the pump stops.

The outflow pipe of both the submerged and the vertical sump pump should have a check valve installed close to the top of the sump pit. This check valve prevents the waste water from draining back into the pit. Figure 9-21 shows the location of the check valve.

Many local plumbing codes require that the sump pit be covered and vented. The main reason for this is the same as the reason for having traps on all the fixtures: to keep sewer gas out of the building. A second reason is to keep someone from falling into the pit.

• QUESTIONS •

1. What are the three types of sump pumps which are used in sewage systems?
2. What is the most common type of sump pump?
3. What are the two most common sump pumps?
4. What are sump pumps used for?

• SUMMARY •

This chapter introduced you to some of the different plumbing accessories on the market today. The hot-water system is normally a part of every building, many homes or businesses have an automatic dishwasher, and many buildings require some type of sump pump. As you work with these accessories, either installing or repairing them, you will become more familiar with their parts and uses.

• WORDS PLUMBERS USE •

accessory
thermostat
fusible plug
sump
sump pump
submersible pump
vertical sump pump

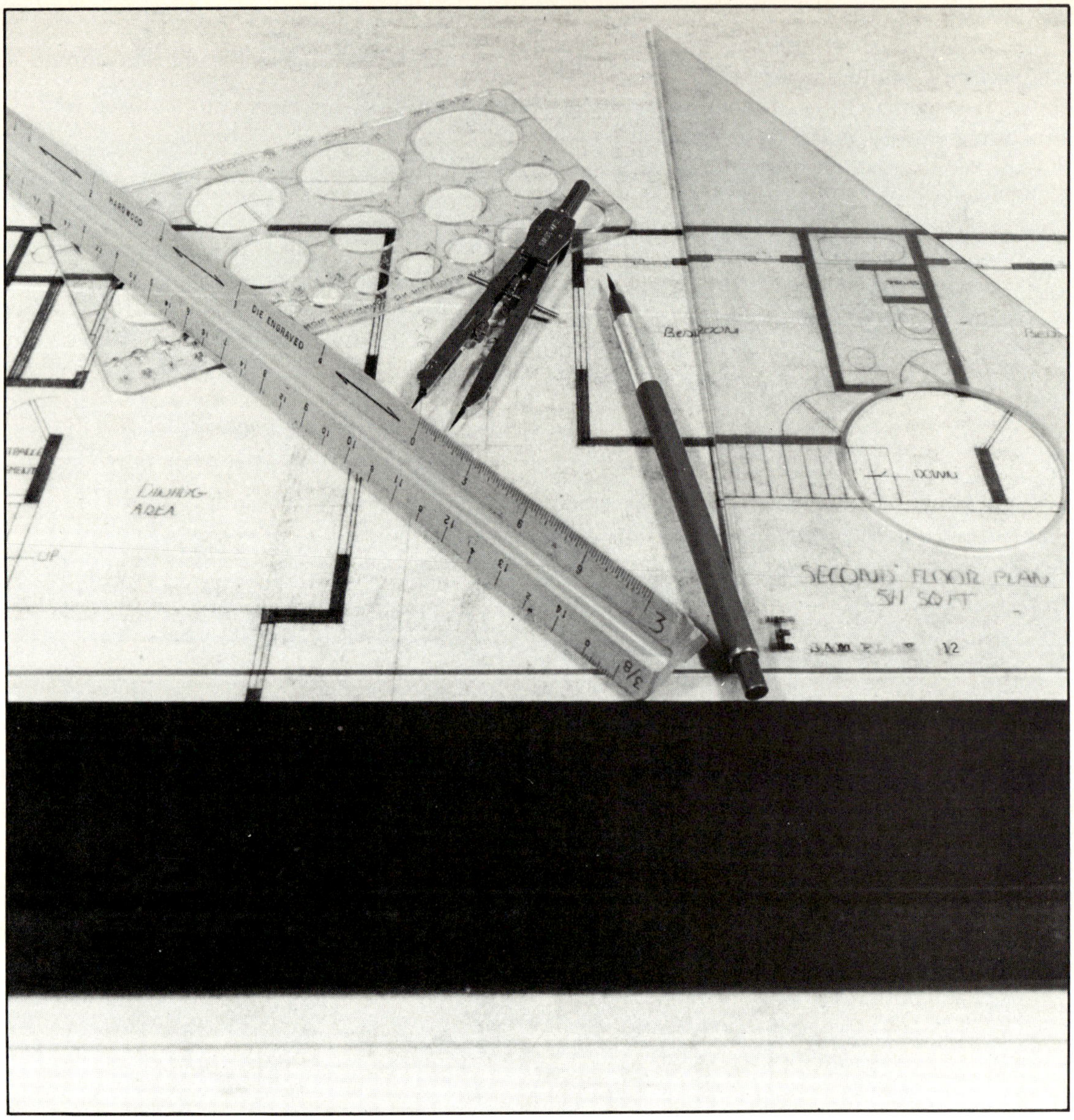

10
DRAFTING FUNDAMENTALS

While plumbers generally deal with pipes, faucets, fixtures, and other hardware with which they are familiar, they are often faced with the problem of repairing or connecting an unfamiliar piece of equipment. It is for this reason that they must have a working knowledge of drafting, the craft of making technical drawings of equipment. A basic knowledge of this craft allows the plumber to disassemble, identify, and repair broken or worn parts, and reassemble an unfamiliar piece of equipment. Plumbers must also have some knowledge of technical drawings to keep up with the many changes which will take place during their lifetime.

At the completion of this chapter, you will be able to:

- Identify the various instruments, scales, and templates used in the preparation of technical drawings.
- Sketch a three-dimensional object in preparation for a technical drawing.
- Outline one-, two-, and three-view technical drawings of a three-dimensional object.
- Be able to identify the three types of pictorial drawings.
- Be able to identify cutaway and exploded drawings.

• DRAFTING TOOLS •

Though the plumber can become proficient in reading a technical drawing by trial and error, it is much easier to spend a little time learning to make them. The first order of business in making technical drawings is to learn the materials and instruments used by the drafter.

Drawing Pencils

You have been using pencils for a long time, probably long before you started in first grade, but did you know the drafter uses 17 different types of pencils? Each of these 17 pencils contains a different grade of lead. These grades range from 6B, very soft, to 9H, very hard. You are probably familiar with the No. 2 or 2B pencil, since that is the most common.

The drafter will use the grade of pencil most suited to the job. When beginning or laying out the drawing, a drafter will use a 4H or 5H pencil to get light lines, some of which will have to be erased. After the layout is complete, an H or 2H pencil will be used to darken the light lines and for the lettering. Figure 10-1 shows the general use and color differences of the various drawing pencils.

In addition to wood-cased pencils, there are a variety of mechanical pencils manufactured. Any mechanical pencils can use many different grades of lead. It is simply a matter of removing one grade and inserting another. A picture of a mechanical drafting pencil and a case of leads is shown in Figure 10-2.

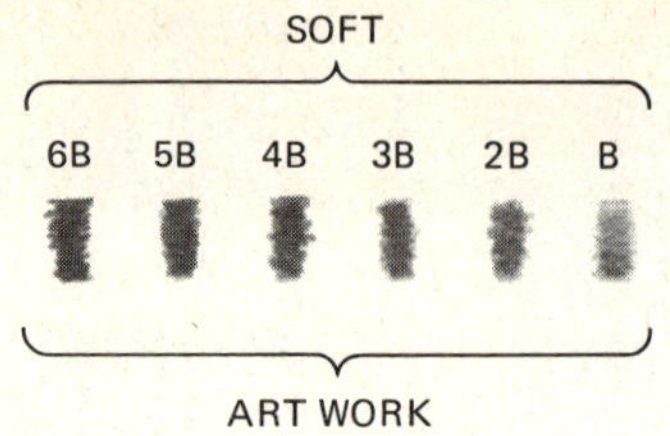

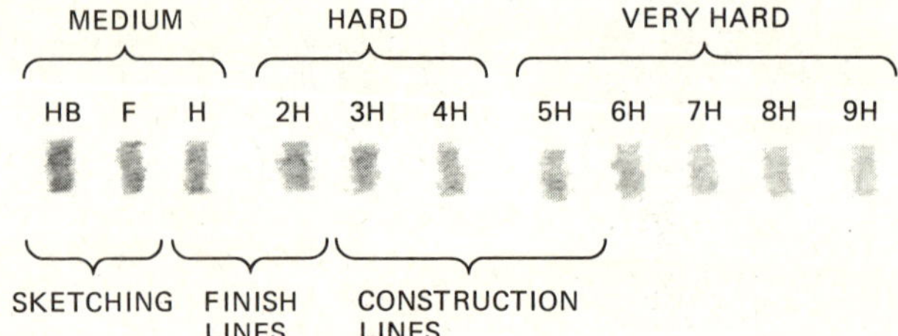

Figure 10-1. Drawing pencil grades and uses.

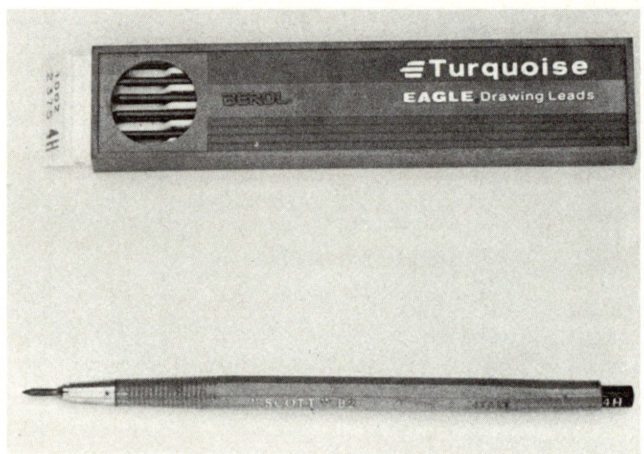

Figure 10-2. Mechanical pencil and lead case

Drawing pencils are sharpened in a different manner from the regular pencil you use in school. A clear, sharp line is required for accurate drawing, and the drawing pencil must always have a clean, sharp point. Most drafters prepare their pencil point using the steps outlined below. Follow through these steps in Figure 10-3.

1. Using a penknife or a drafter's pencil sharpener, cut away the wood so that approximately ⅜ inch of lead is showing.
2. Using a sandpaper pad or a pencil pointer, sharpen the lead to a pointed cone shape.

Be cautious when sharpening a drawing pencil. Do not cut away the end of the pencil which tells the grade of the lead. All pencils look alike.

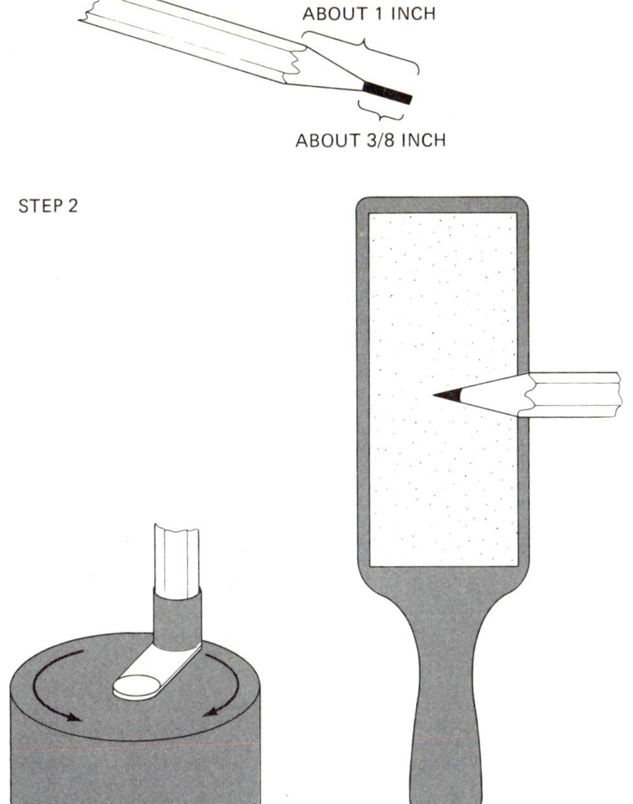

Figure 10-3. Sharpening a drafting pencil

Drawing Boards

The drawing board has a smooth, flat surface and is normally made of a softwood (basswood or soft pine). It may also be covered with a plastic or linoleum sheet. The edges of the drawing board are made of hardwood or metal to make a "true" or straight edge for the T-square.

Figure 10-4. Drawing board and table

Industrial drafting is normally done on a drawing table. The top of this table is used as the drawing board and has the capability of being raised to give the drawing surface some degree of slope. Figure 10-4 has pictures of a drawing board and a drawing table.

T-Square and Triangles

The T-square and triangles are the primary straight-line tools of the drafter. The T-square is used to draw horizontal lines and to hold the triangles in position. The T-square is generally made of wood and consists of a head and blade. These two parts are joined at a 90-degree angle, and many times, as shown in Figure 10-5, the blade has clear plastic on the edges.

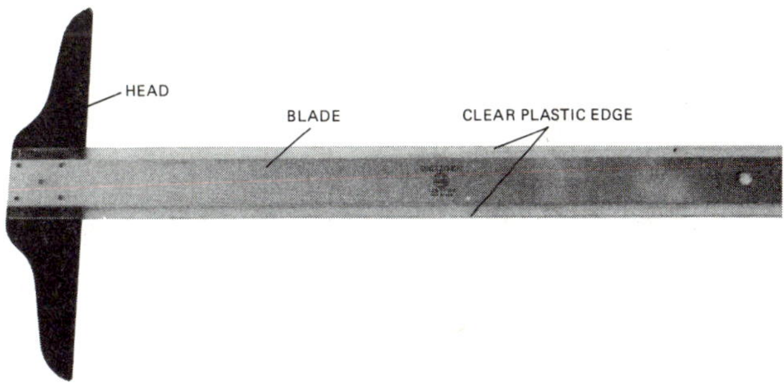

Figure 10-5. T-square

The triangles are almost always made of clear plastic, though some may be tinted (red, green, etc.). The two main triangles used are the 30-60-90-degree triangle, commonly called the 30-60; and the 45-45-90-degree triangle, commonly called the 45. These are shown in Figure 10-6.

The 30-60 and 45 triangles can be purchased in

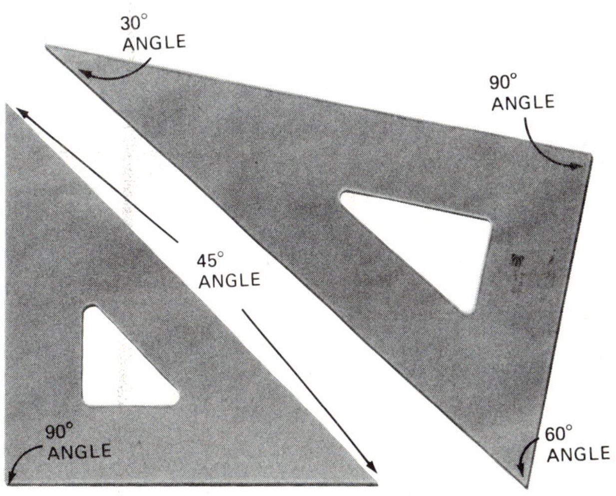

Figure 10-6. Drafting triangles

many sizes. The size of the triangle does not affect the angles, only the distance between them. Normally a 6- or 8-inch 45 and an 8- or 10-inch 30-60 are adequate for most situations.

Protractors

While many different angles can be made using the triangles, some odd angles require the use of a protractor. This is a clear plastic drawing aid with a center point and a full circle (360 degrees) marked off on the edge. This aid can be used to mark off any angle from one as small as 1 degree to one as large as 179 degrees. Figure 10-7 shows a typical protractor.

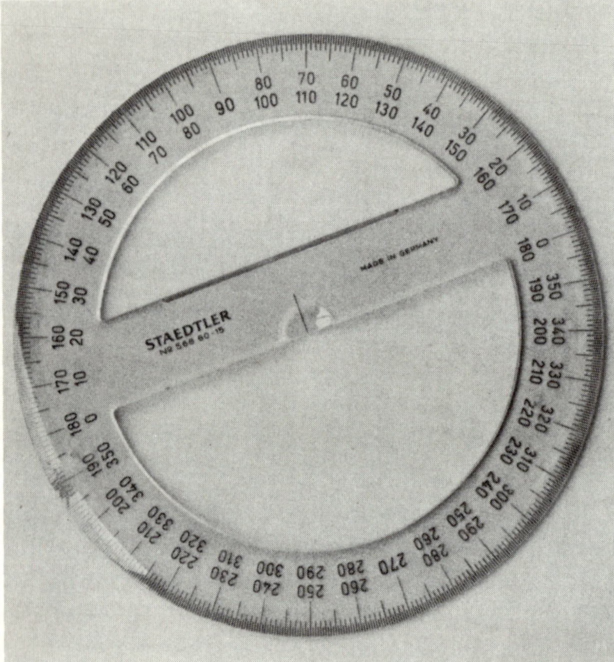

Figure 10-7. Protractor

Scaled Drawings

Many of you think of a scale as something used to find the weight of an object. This is not so in technical drawing. Many times, the object we wish to draw is much too large to draw full size, a hot-water heater, for instance; others are too small to have good detail, like the ball valve in a single-lever faucet. This means that a scale or ratio must be made so that the plumber can determine the true size of the object. If we say we are drawing something full size, we mean that for each inch on the object, there will be 1 inch on the drawing. If we say half size, each inch on the object will be ½ inch on the drawing. In that way, we could reduce the size of the drawing for a large object by reducing the scale of the drawing: ¼″ = 1″. The first figure, ¼″, in this case refers to the size on the drawing, the second, 1″, is the full-sized object.

In the same manner, we could make small parts look larger on the drawing by increasing the scale of the drawing: 1″ = ½″. Notice that the first number in the scale always refers to the drawing, while the number after the equals sign always refers to the real object.

Measuring Scales

You will have to learn another type of scale also, the measuring scale. This scale is the drafter's ruler. Just as you used a ruler in measuring pipe, so you will also use this scale in measuring drawings. There are different scales available for different jobs. There is an engineer's scale, an architect's scale, a drafter's scale, and a metric scale. We shall review the architect's scale, since the plumber must become familiar with it.

The architect's scale is normally a triangular scale. It has six measuring surfaces. Each of the measuring surfaces is divided into a different scale. One has full inches, another ½-inch divisions, another ¼-inch divisions, and so on until the smallest division, ⅛ inch, is reached. Each scale is marked, and ¼ marked on the edge of the scale means that each division (¼ inch) equals 1 inch or 1 foot, depending on the size of the real object being drawn. An architect's scale is shown in Figure 10-8.

In the 1980s the United States will be using the European system of metric measurement. This system uses as length measurements the meter, the centimeter (cm; 1/100 of a meter), and the millimeter (mm; 1/1000 of a meter). A metric drawing scale is normally divided into 30 centimeters and 300 millimeters. This metric scale can be used to set up a scaled drawing in the same manner as an architect's scale, 1 mm = 1 meter.

Figure 10-8. Architect's scale

Compass and Dividers

Two main instruments used by the drafter are the compass and the dividers. The compass is used to make accurately sized circles and arcs (a part of a circle), while the dividers are used to transfer a measurement accurately from one place to another or to mark off equal sections of a line.

The compass and the dividers look very much alike. The difference is that the compass has a steel point on one leg and a pencil or pen point on the other, while the dividers have a steel point on each leg. Two different types of compasses and two different types of dividers are shown in Figure 10-9.

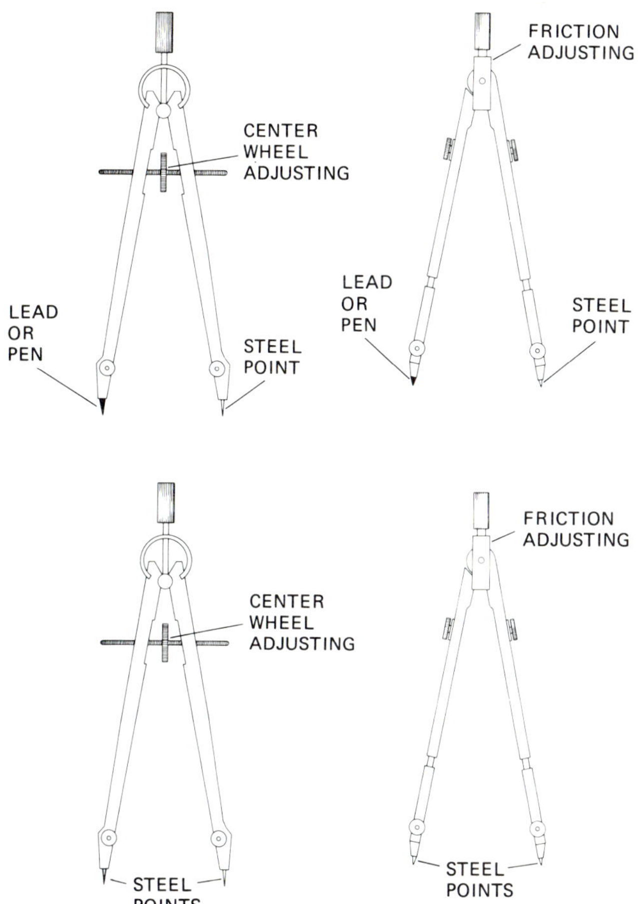

Figure 10-9. Compass and dividers

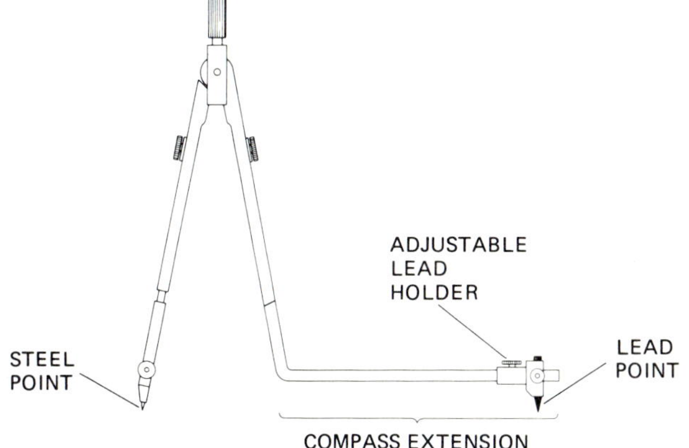

Figure 10-10. Compass with extension

At some time, the drafter may have to draw a circle or an arc which has a radius larger than the spread of the compass. A compass extension is available to lengthen the radius of the circle drawn by the compass. The compass extension is shown in Figure 10-10.

Irregular Curves and Templates

Many of the items drawn by a drafter have irregular lines; that is, not all the lines are straight nor are they parts of a circle. One of the best examples of this is the roof flashing used at the top of the vent for a sewer stack. The pattern for this flashing is a series of irregular lines. The irregular curve, part of the drafter's equipment, is used to make these lines.

The irregular curve, sometimes called the french curve, is made of the same type of plastic as the triangles. Like the triangles, they may be tinted, and they are also made in many different sizes. Two irregular curves are shown in Figure 10-11.

Figure 10-11. Irregular curves

In addition to drawing irregular curves, the drafter may have many items on one drawing that are alike, circles to identify holes for screws or bolts, for instance. To make this task easier, the drafter can use a circle template. A circle template is shown in Figure 10-12. Notice that the circles go from $\frac{1}{16}$ inch to 3 inches in diameter. Figure 10-12 also shows other symbols available in templates for drafters.

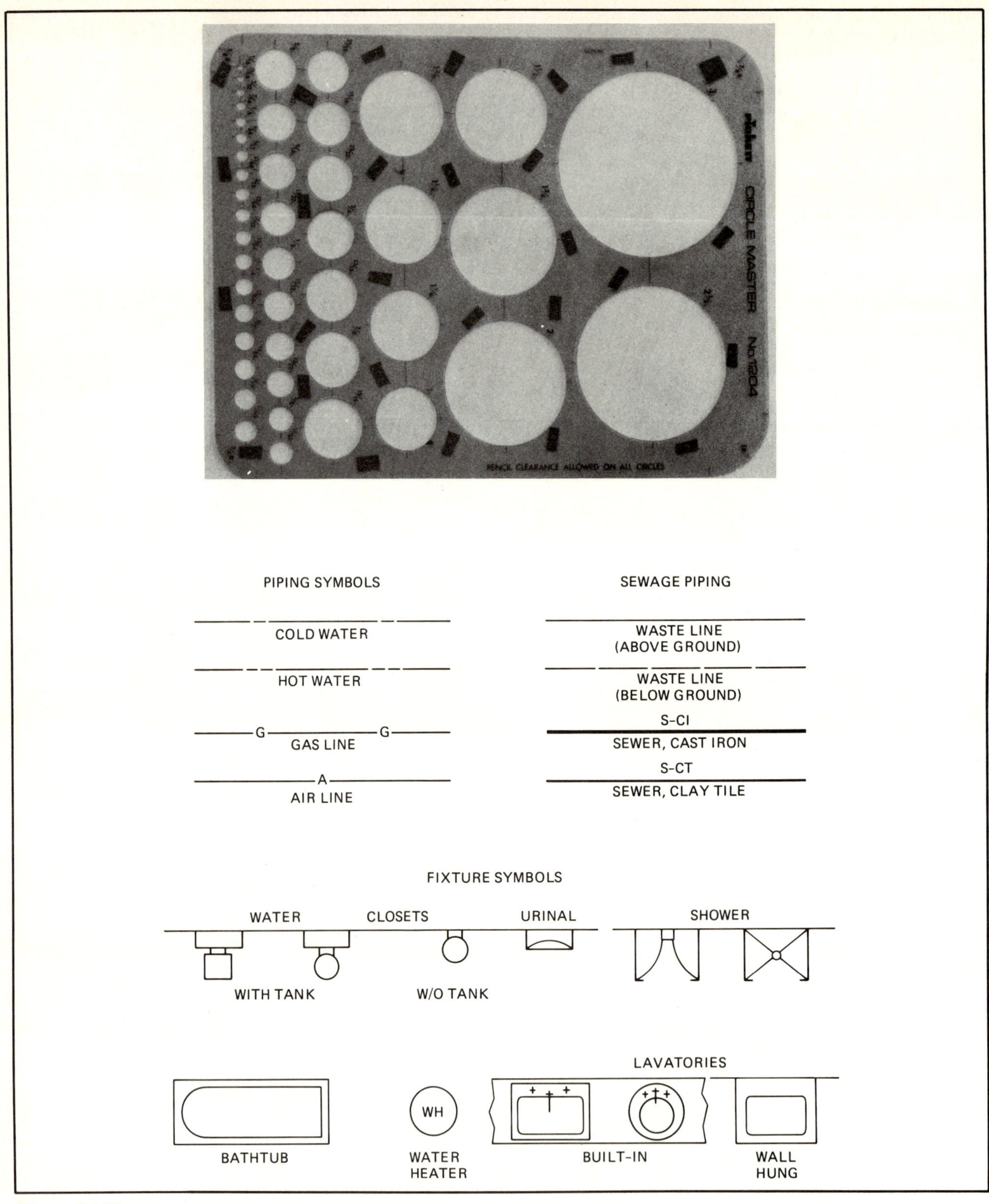

Figure 10-12.

• LINE IDENTIFICATION •

One day you, as the plumber, will have to look at a technical drawing and identify the different parts of an object. You will have to know the difference between the common types of lines used in that drawing. There are nine common types of lines used in a technical drawing. These nine common lines are shown in Figure 10-13 along with their meanings and uses.

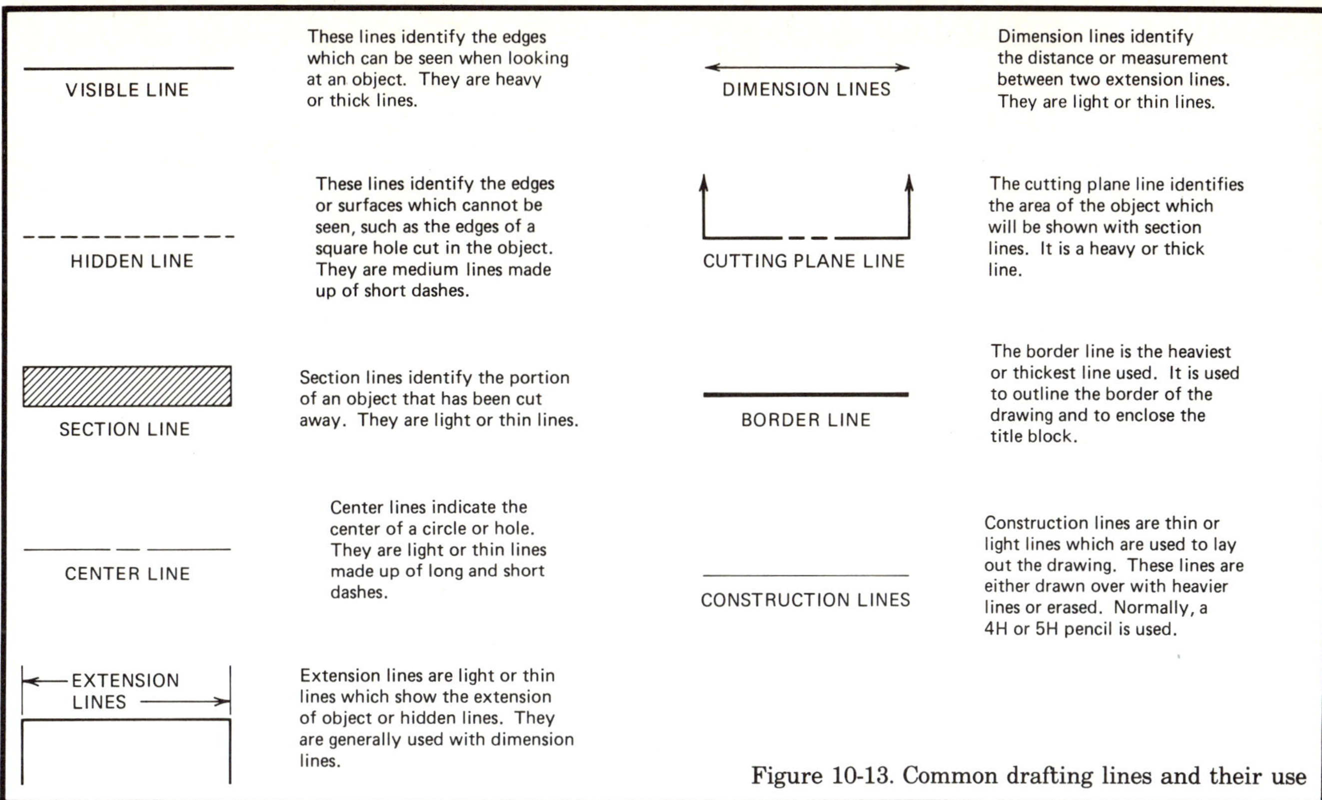

Figure 10-13. Common drafting lines and their use

• QUESTIONS •

Figure 10-14 is a typical simple technical drawing. Identify the different types of lines which are indicated by these letters.

1. Line A
2. Line B
3. Line C
4. Line D
5. Line E
6. The group of lines in F
7. Line G.

The tools we have learned about in the first part of this chapter are not used only by the drafter. They are also used by the architect, the builder, and the other tradespeople who work in the construction business. They are used in drawing the blueprints or technical drawings of a building or a part of a building. Blueprints and blueprint reading will be discussed in Chapter 11.

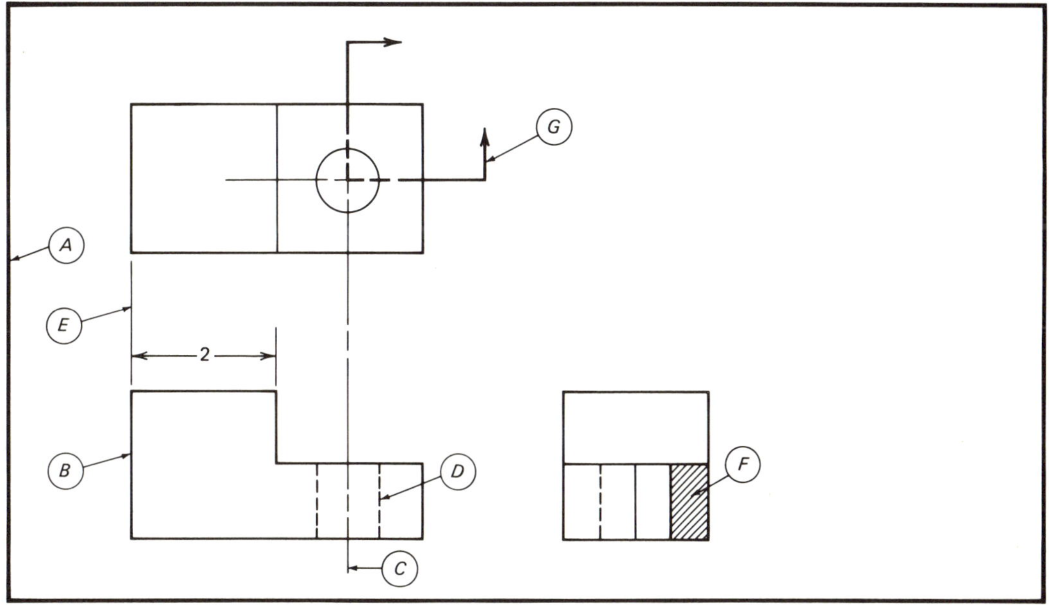

Figure 10-14. Simple technical drawing

• SKETCHES •

A sketch is a freehand drawing of an object. The sketch is generally made to show the drafter the size and shape of an object before the technical drawing is made. A sketch can also be used to show an architect or a plumber how the owner of a building wants the building, or some part of the building (such as the kitchen), to look.

The sketch is a rough drawing of the object to be drawn, but some things must be kept in mind when the sketch is made. While the sketch can be made quickly, it must be easy to read. The lines should be neat; invisible lines, center lines, and object lines should be identifiable; and the sketch should be in proportion. That is, if it is twice as wide as it is tall, the sketch should show that.

The sketch should also show the dimensions of the object as well as the width or diameter and the depth of holes. Any measurement must be transferred to thé sketch. "A picture is worth a thousand words," and so is a complete sketch. Figure 10-15 shows a two-view sketch of a pair of pliers.

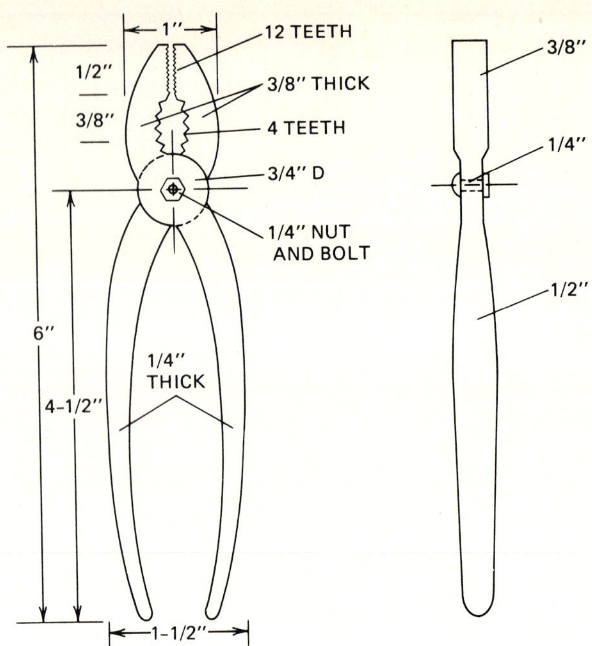

Figure 10-15. Two-view sketch of pliers

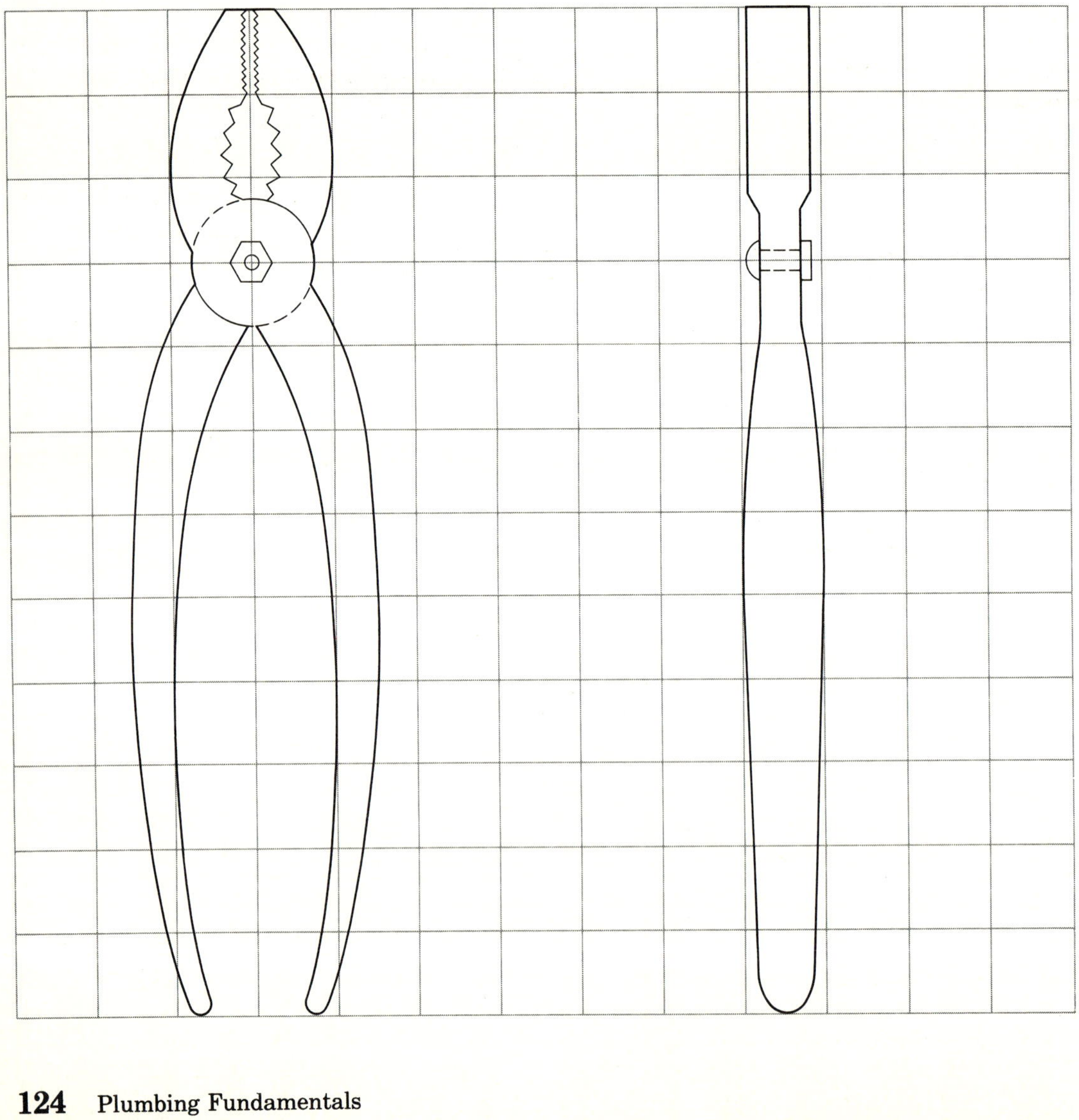

Figure 10-16. Sketch on graph paper

You can see from Figure 10-15 that there are many dimensions and dimension lines on a sketch—enough of them, in fact, to clutter up the drawing and make it hard to figure out. A sketch on graph paper removes the need to list the length and width dimensions on the paper. Figure 10-16 shows the same pliers sketched on ½-inch-square graph paper.

Remember, you may have to make a drawing of something you have sketched. Keep it neat, use the correct lines, and include all dimensions. It is also good to include helpful items, such as the number of teeth on the jaws of the pliers.

• ONE-, TWO-, AND THREE-VIEW DRAWINGS •

Many of the technical drawings you will come in contact with over the coming years will be either two- or three-view drawings. Occasionally you may see a one-view drawing, but a one-view drawing rarely shows the entire picture of an object.

You know that a single die in a pair of dice has six sides because there are six different numbers on each one. If we look at the side of the die with 1 dot, the top will have 2 dots, the left side 3 dots, the right side 4 dots, the back 6 dots, and the bottom 5 dots. The six sides of a die are shown in Figure 10-17.

Each of the sides in Figure 10-17 can be called a different view of the die.

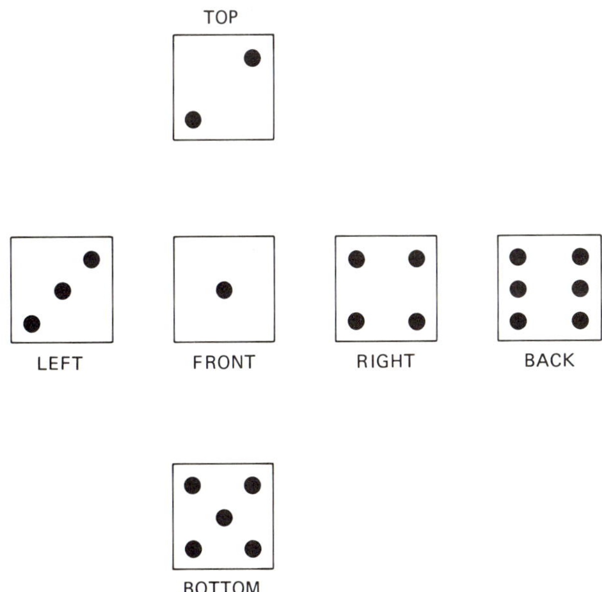

Figure 10-17. Six sides of die stretched out

One-View Drawing

The one- or single-view drawing is a picture of only one side of an object, and any real object has more than one side. If all the sides are exactly alike, as a cube, a one-view drawing will be accurate. Figure 10-18 shows a front view of a 2-inch cube; that is, it is 2 inches high, 2 inches long, and 2 inches deep, or thick.

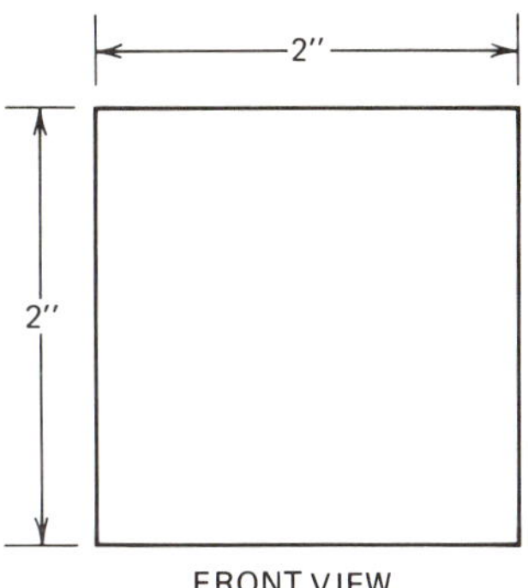

Figure 10-18. Front view of cube

While this drawing is accurate and shows exactly what one side of the cube looks like, it does not really give the person looking at it the feeling that it is a cube. The only thing we really know about the object is that it has at least four edges which are 2 inches long. The dimension which we are missing is the thickness or depth of the object. If, instead of a cube, the object were 2 inches high, 2 inches long, and only 1 inch thick, how would the person looking at the drawing know that?

Two-View Drawings

If we were to show two views of the object discussed above, the person looking at it would immediately know how thick it was. We could use the same view as in Figure 10-18 as the front view and add either the top or the side view. Figure 10-19 shows the front and the right-side view of the 2-inch-high, 2-inch-long, and 1-inch-deep block.

A person looking at those two views can easily tell that the block is only 1 inch deep. Figure 10-20 shows the front view and the top view. These two views also allow the person looking at the drawing to tell that the block is only 1 inch thick.

Now, what if that same block had two holes, one drilled down from the top and one cut in from the right side? Would a two-view drawing show these holes accurately? Figure 10-21 shows the front and the right-side view.

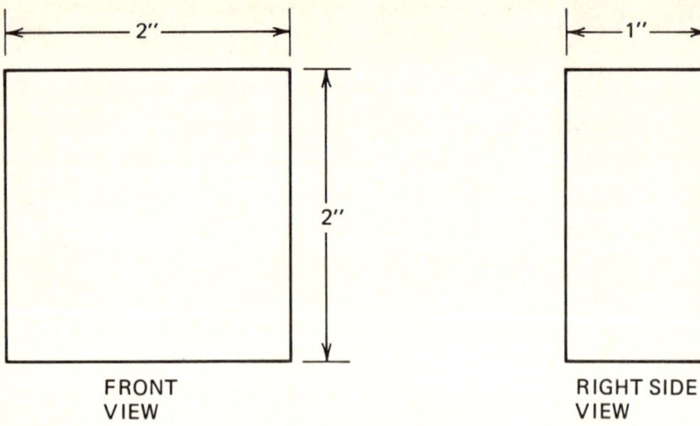

Figure 10-19. Front and right-side view

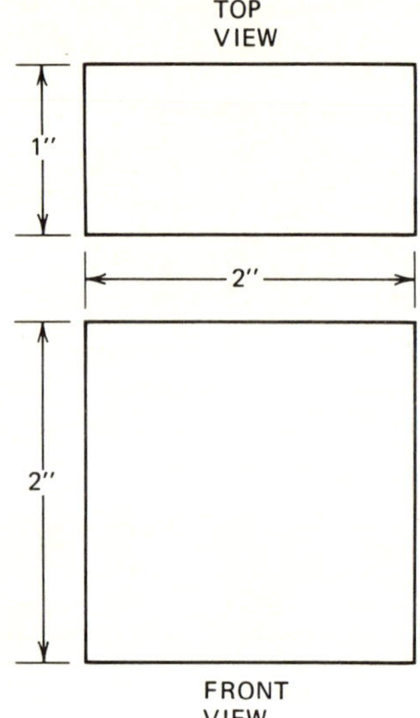

Figure 10-20. Front and top view

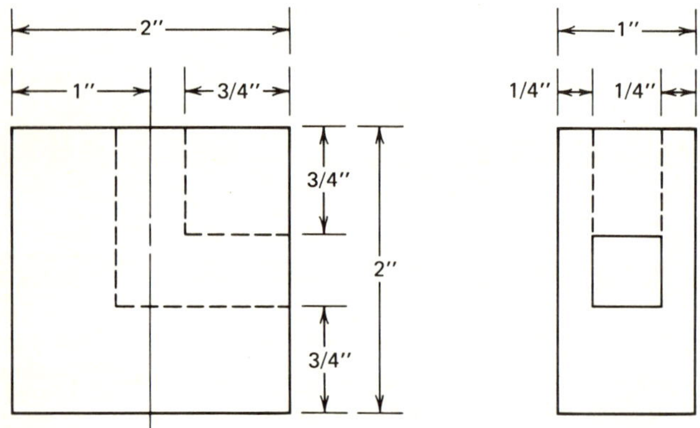

Figure 10-21. Front and right-side view shows only one hole

We cannot tell from these two views what kind of a hole there is in the top of the block. If we were to show only the top and the front, we would not know that there was a square hole in the side. Figure 10-22 shows the top and front views.

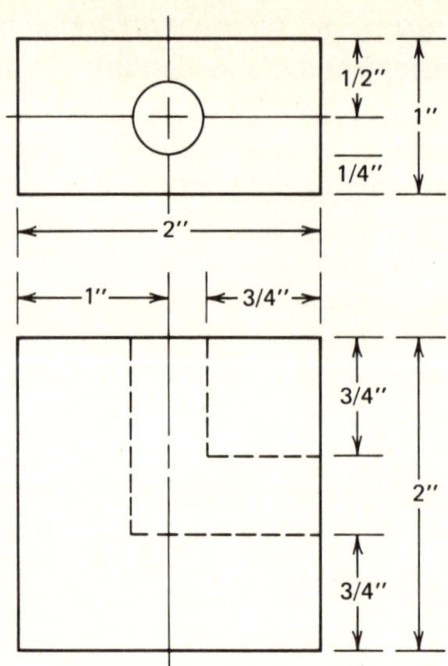

Figure 10-22. Front and top view shows only one hole

Three-View Drawings

So that all possible views and hidden lines of a technical drawing are shown, the three-view drawing is used. These views are normally the front, the right side, and the top view. They are generally drawn in that relationship, and there is no need to label the different views. Figure 10-23 shows a typical three-view drawing of the object in Figure 10-22. (See top of page 127.)

The three-view technical drawing in Figure 10-23 identifies all the differences in the object. It shows the round hole and its size in the top view, the square hole and its size in the right-side view, and the outline of the depth of the holes and the overall size in the front view. Although three-view drawings are widely used in some areas, there is another type of drawing which is used to show the three sides of an object in one picture. These drawings are called pictorial drawings.

• QUESTIONS •

1. The one-view drawing is only accurate if all six sides are the same. (T or F)
2. The three-view drawing normally shows the front, the left side, and the bottom views of an object. (T or F)
3. Which drawing will show all the details of an object?

• MAKING THE TECHNICAL DRAWING OUTLINE •

The first task in making a technical drawing outline is to attach the drawing paper to the drawing

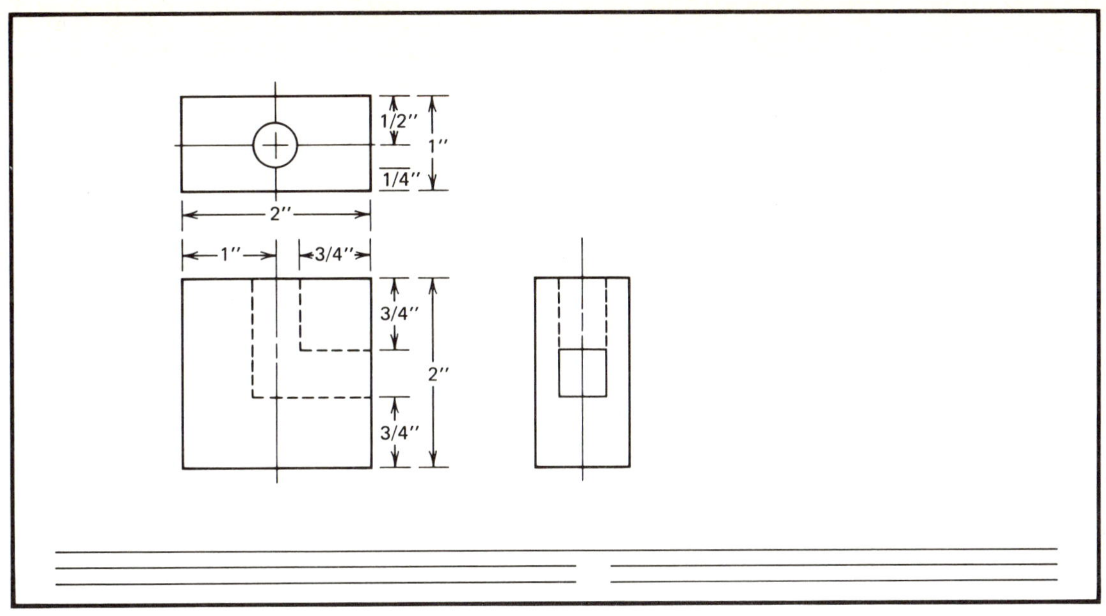

Figure 10-23. A three-view drawing shows both holes

board. The paper should be set so that the edges of the paper are square with the edges of the board. Follow this procedure in Figure 10-24 while going through the steps below.

1. Put the T-square on the drawing board with the head on the metal edge of the board.
2. Place the drawing paper on the board toward the left center of the board.
3. Hold the head of the T-square against the edge of the board and move it up until the blade is in line with the bottom of the paper.
4. Still holding the T-square against the edge, straighten the paper against the T-square blade.
5. Hold the paper in place, release the T-square, and tape the four corners of the paper to the drawing board.

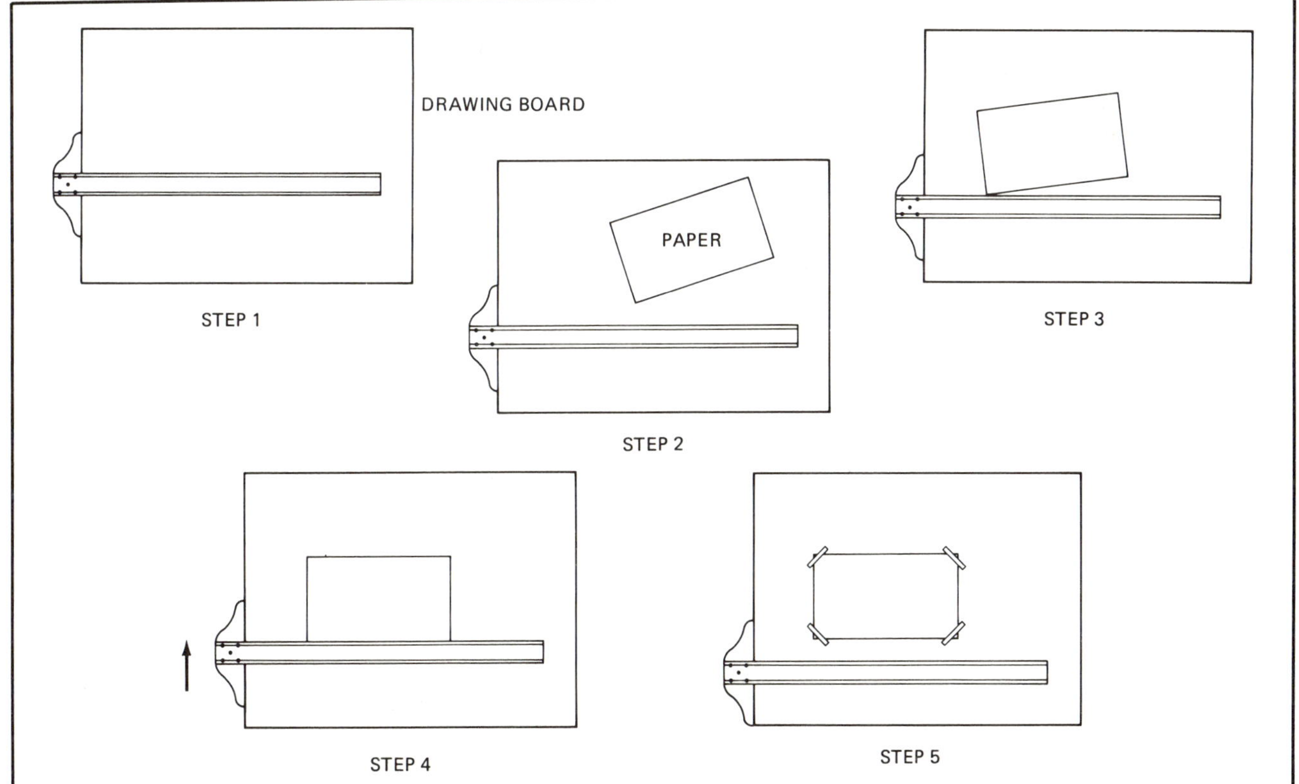

Figure 10-24. Attaching the drawing paper

Horizontal Lines

All lines drawn using only the T-square are horizontal lines. They go from left to right across the paper. Drawing a horizontal line is simply a matter of holding the T-square head against the edge of the drawing board, aligning the upper edge of the blade with the mark showing where the line should be, and then lightly moving the pencil along that edge of the blade. All lines made in this manner will be straight and parallel (Figure 10-25).

Vertical Lines

These lines are at 90 degrees to the horizontal lines made with the T-square, and they are made using the T-square and one of the triangles. The T-square is held to the edge of the board as in drawing the horizontal lines, but then the triangle is placed on the drawing board so that one edge of the 90-degree angle rests on the upper edge of the T-square and the other 90-degree angle edge is aligned with the vertical line mark. Figure 10-26 illustrates how parallel vertical lines are drawn.

The triangles can be used in the same manner to draw parallel lines of 30, 45, and 60 degrees. Other angles can be made by combining two triangles as shown in Figure 10-27.

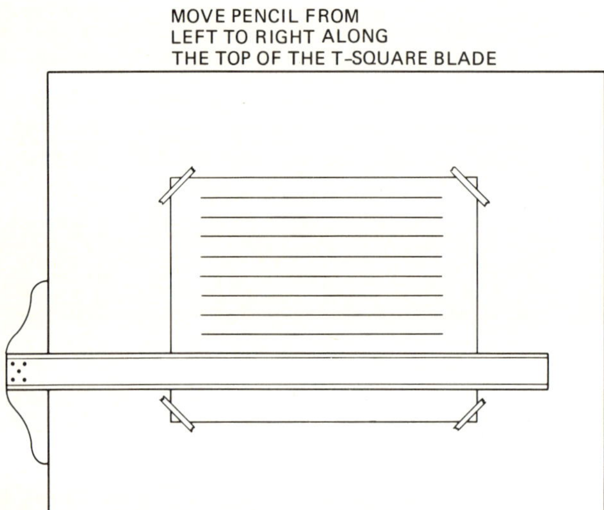

Figure 10-25. Drawing horizontal lines

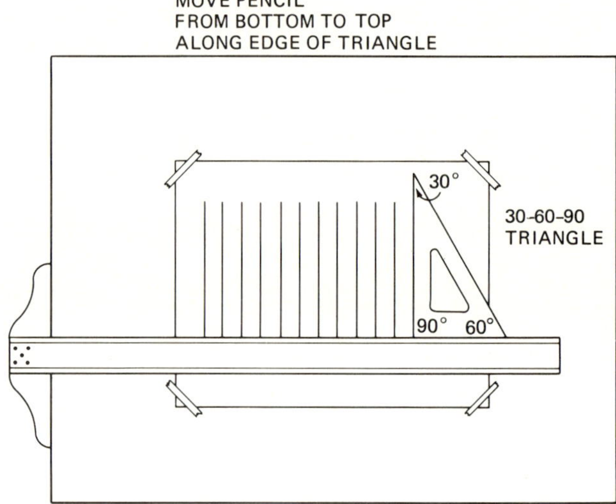

Figure 10-26. Drawing vertical lines

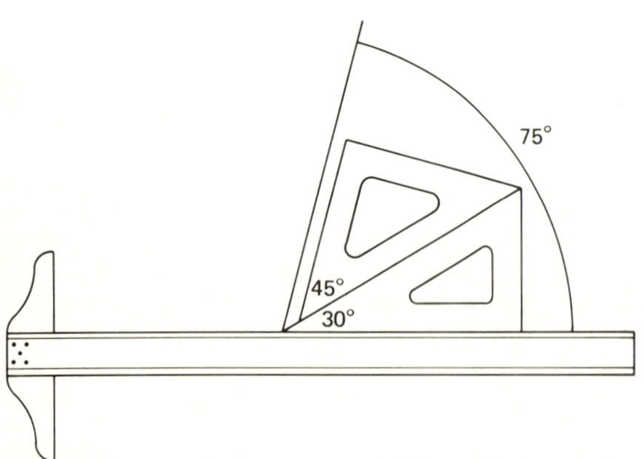

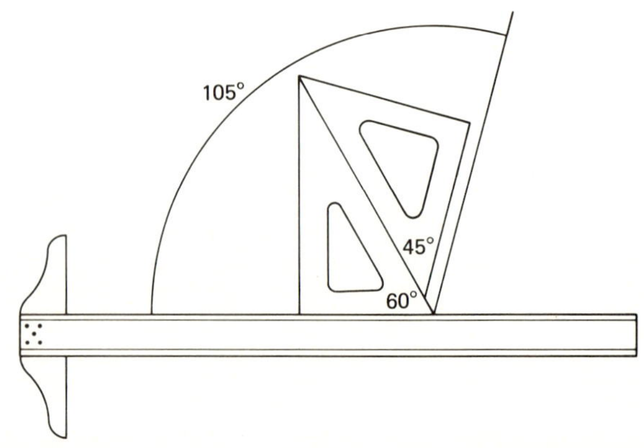

Figure 10-27. Drawing combination lines

Outlining the Drawing

Once the drawing paper is in place, the object outline should be drawn in with light lines. These light lines will be darkened or erased before the drawing is finished. Follow through in Figure 10-28 as you go over the steps below.

1. Using the T-square, draw in a horizontal base line (line 1) near the lower left-hand corner of the drawing.
2. Using the T-square and a 90-degree angle, draw in a vertical base line (line 2) near the left-hand side of line 1.
3. Accurately measure and mark the length of the object on line 1 and the height of the object on line 2.
4. Draw a horizontal line from the mark you made on line 2 and a vertical line up from the mark you made on line 1.

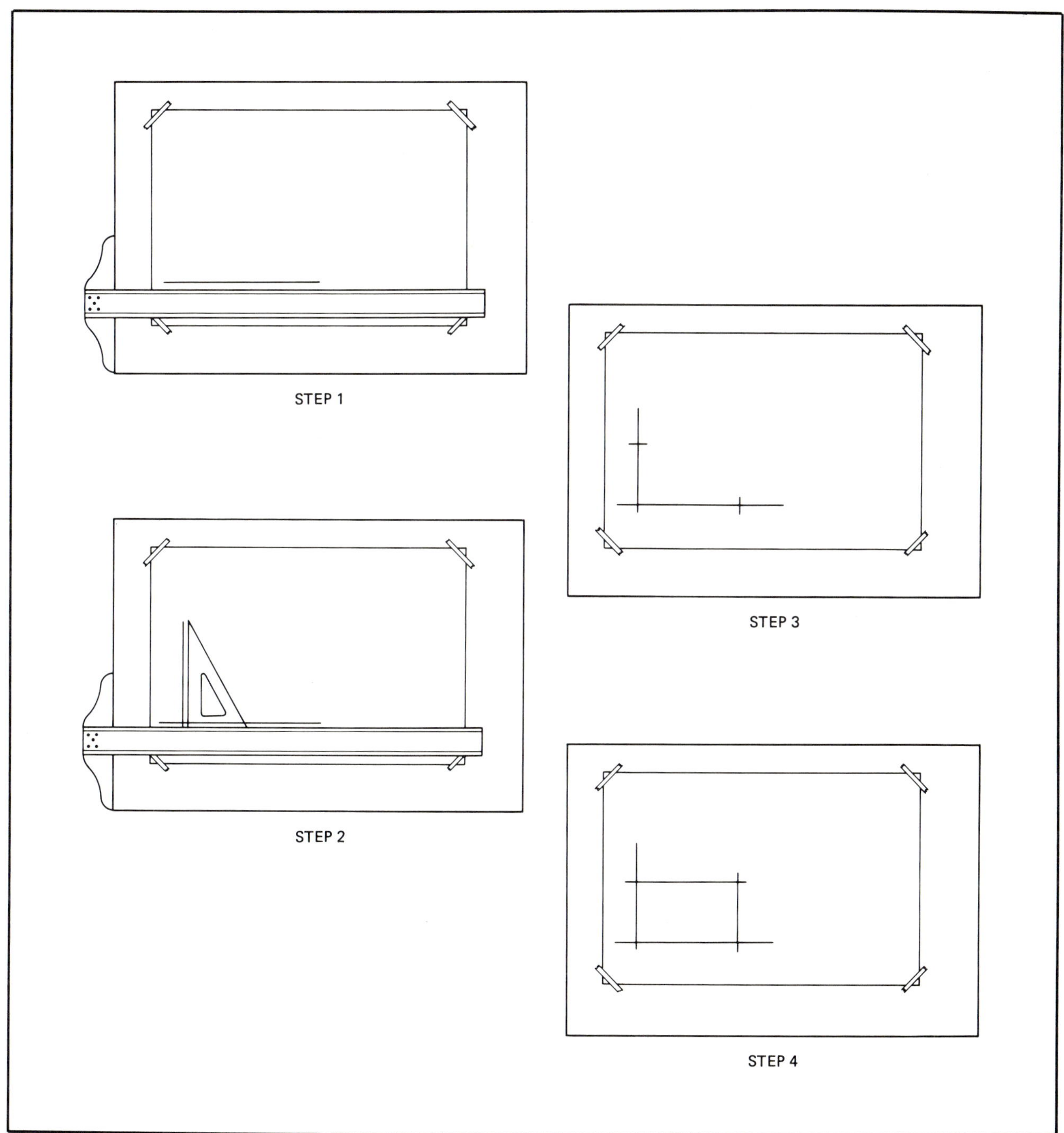

Figure 10-28. Outlining the front view

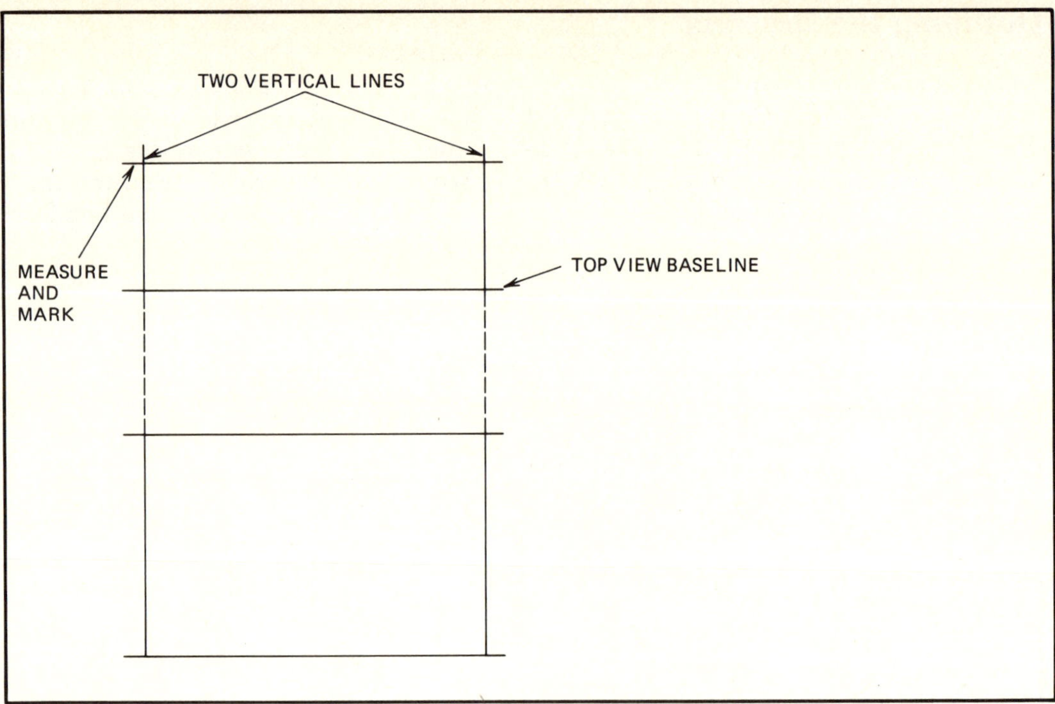

Figure 10-29. Outlining the top view

Following the procedure on page 129 will result in an outline of the front view of the object. Outlining the top view is a matter of extending the two vertical lines and drawing a horizontal base line. The front view is completed by measuring and marking the depth of the object, and drawing a horizontal line through the mark (Figure 10-29).

The right-side view can be done in the same way. First, the two horizontal lines from the front view are projected over to the right side of the drawing paper. The depth can be measured off on the bottom line, and two vertical lines can be drawn to outline this view (Figure 10-30).

There are other methods of finding the depth and marking these lines on the side view. These other methods are known as angular or radial projections. While the drafter uses them, these procedures will not be covered here.

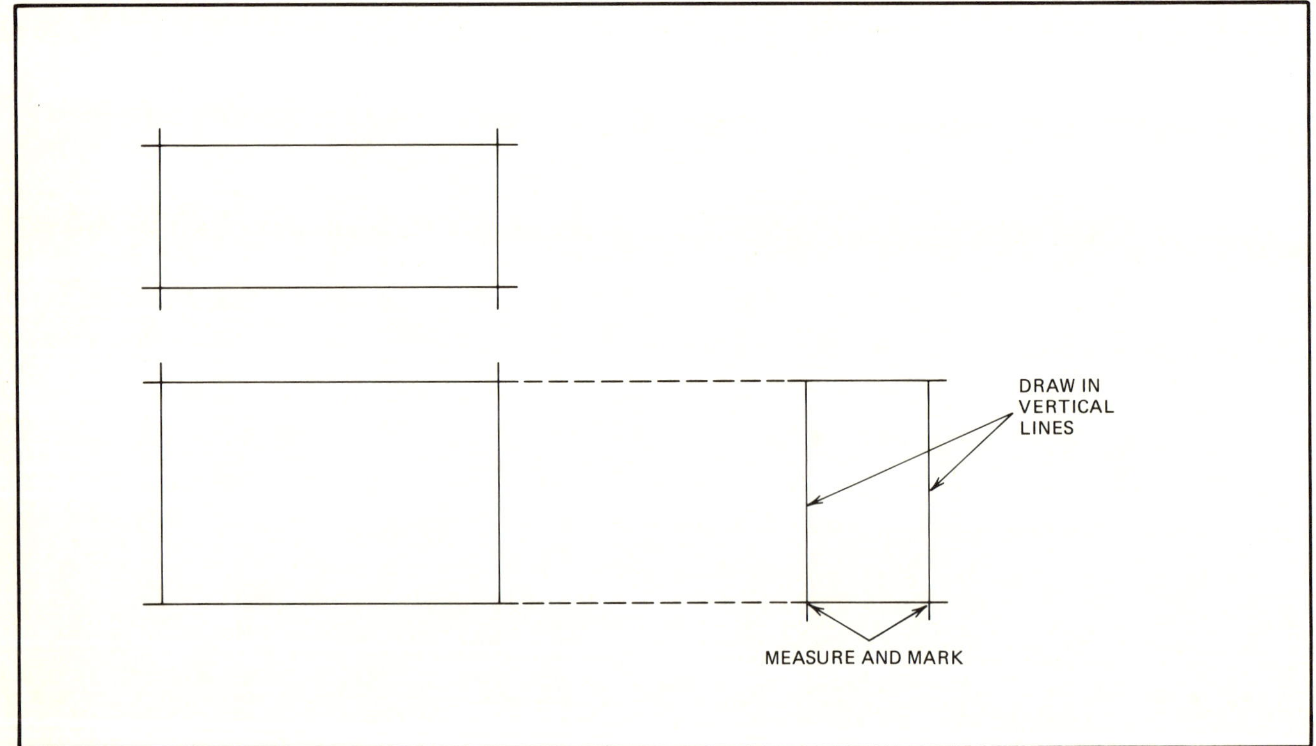

Figure 10-30. Outlining the side view

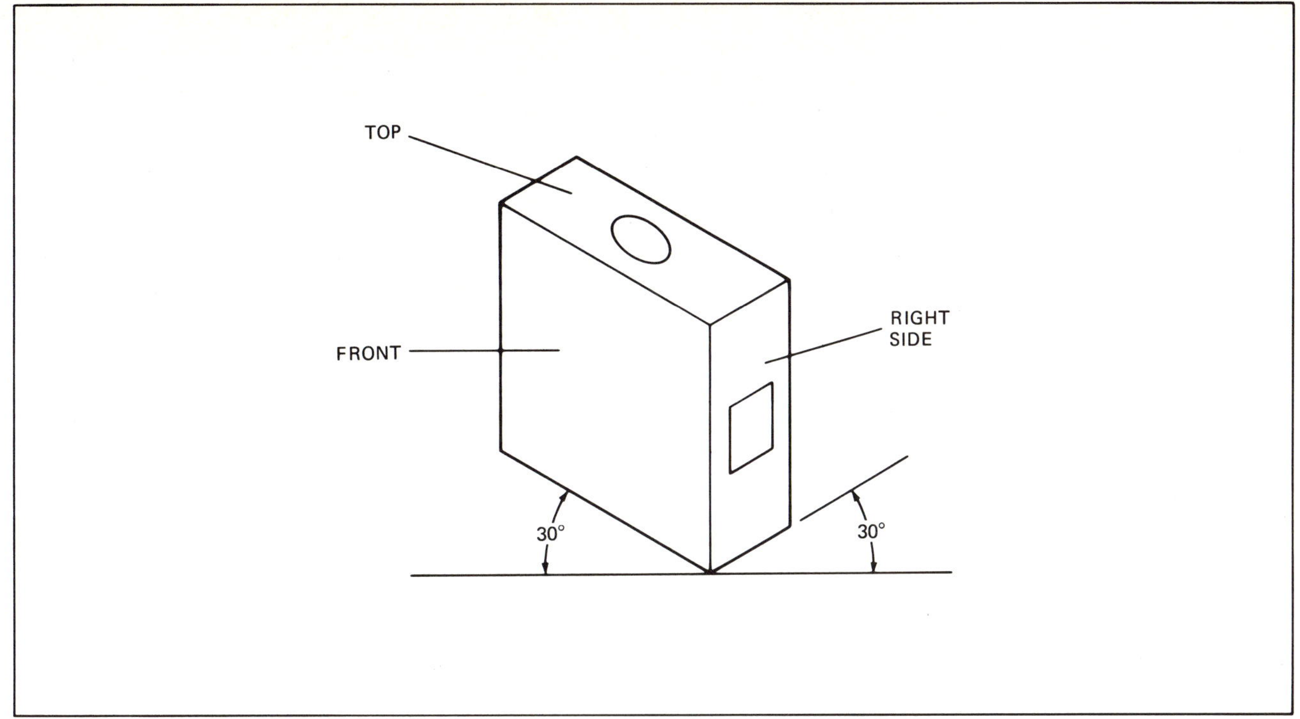

Figure 10-31. Isometric drawing

Pictorial Drawing

There are three main types of pictorial drawings: isometric, oblique, and perspective. The isometric drawing is drawn from three base lines, and will show three sides of the object when drawn correctly (Figure 10-31).

In the oblique drawing, the front view of the object is drawn the same way as for a three-view drawing, then the sides and top are added. There are two oblique drawings, the cavalier and the cabinet (Figure 10-32).

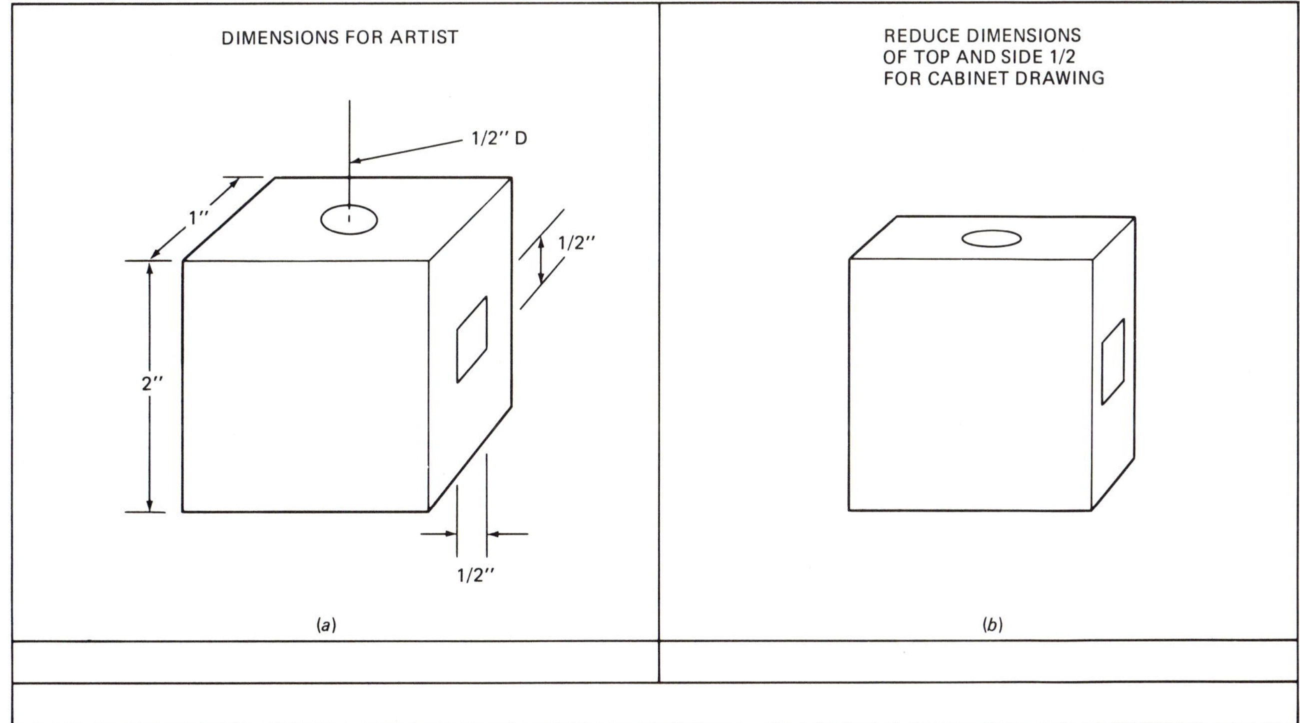

Figure 10-32. Oblique drawings. *(a)* Cavalier; *(b)* Cabinet

Notice the difference between these two. The cavalier drawing shows the top and the right side drawn to full depth, while the cabinet drawing shows them at half size.

Perspective drawings seem to be the most natural looking of these three types. This is because they are based on a vanishing point. If you have ever looked down a long, level, straight road, you may have noticed that the road seems to disappear or come to a point a long way off. This is called the vanishing point.

Pictorial drawings can use one, two, or three vanishing points to make the object appear more natural. Figure 10-33 shows each of these three different kinds of drawings.

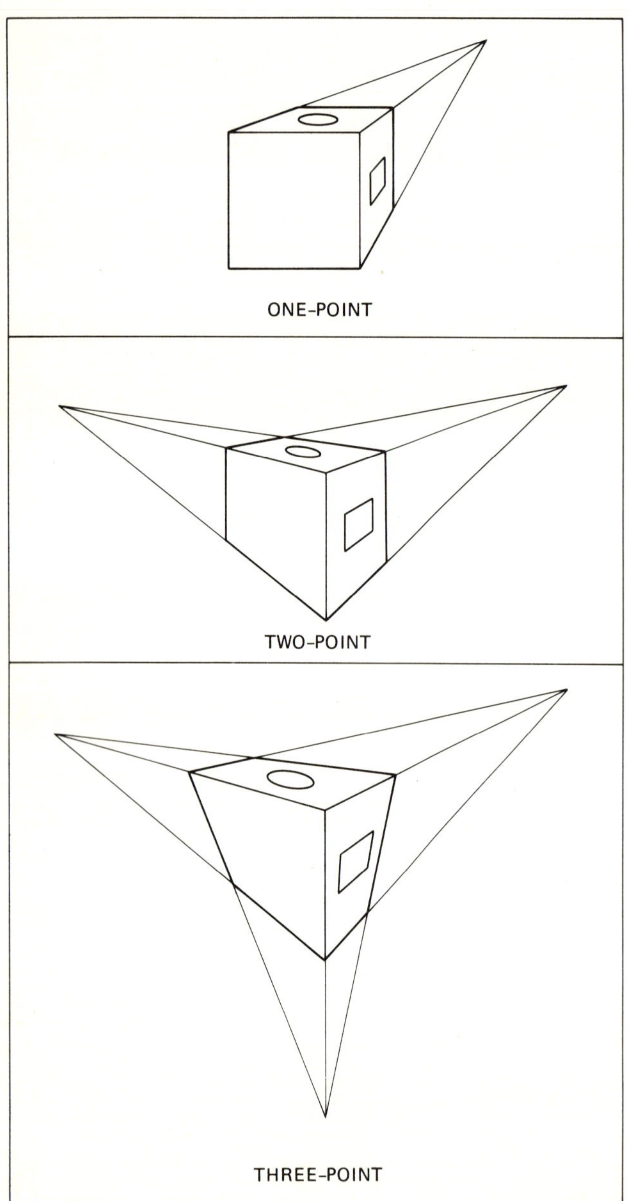

Figure 10-33. One, two, and three vanishing point drawings

• CUTAWAY AND EXPLODED DRAWINGS •

Three-view drawings and perspective drawings show the outside lines and the shape of an object. Three-view drawings also show the interior or hidden lines of an object, but what happens when the inside of that object is filled with *many* parts, like a faucet? The cutaway drawing or the exploded drawing are used in this case.

You may not realize this, but you are already familiar with these two types of drawings. The drawings of the commodes and lavatories pictured in Chapter 5 were the cutaway type, and the drawings of the different types of faucets and their parts were exploded drawings.

Cutaway Drawings

A cutaway drawing shows the object assembled, but the side or top of the object is not drawn. This gives the appearance that it has been "cut away" so that the interior parts of the object can be seen. Figure 10-34 is a cutaway drawing of a gate valve.

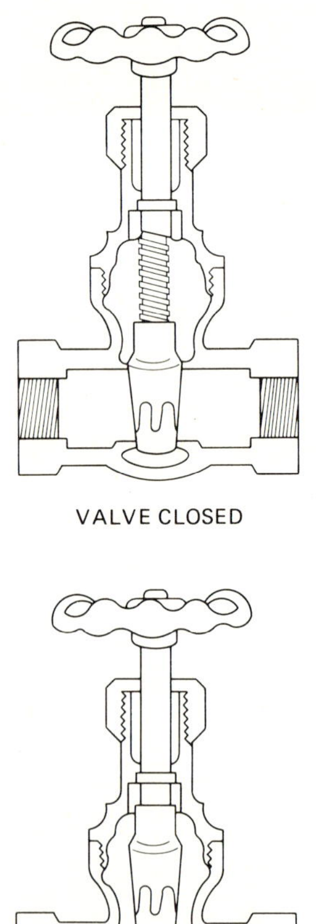

Figure 10-34. Cutaway drawing of gate valve

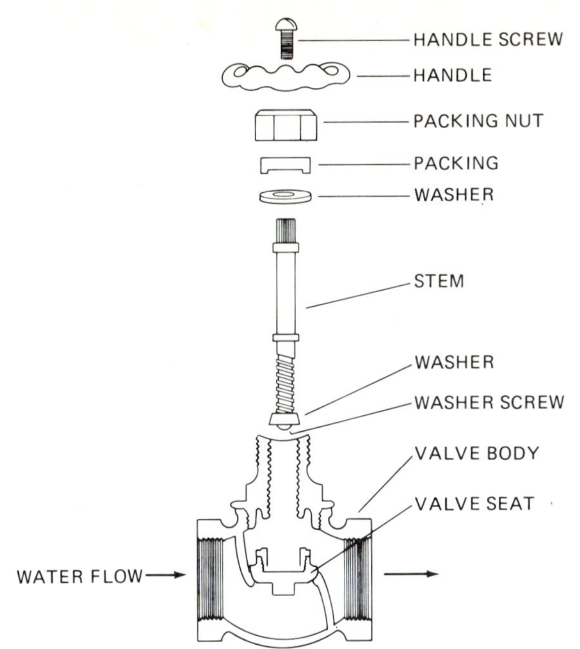

Figure 10-35. Exploded

Exploded Drawings

The exploded drawing shows each part individually. The parts are drawn in the same order as they would be put together, and a center line is usually included to show how the parts are assembled. Figure 10-35 shows an exploded, or expanded, drawing from Chapter 6.

• SUMMARY •

This chapter introduced the various types of technical drawings that you will be using in the plumbing trade. These drawings identify an object and its parts, and knowledge of them will help you install and repair various plumbing fixtures and equipment.

• WORDS PLUMBERS USE •

T-square
triangle
protractor
scale
architect
engineer
compass
dividers
french curve
template
sketch
dimension
one-view drawing
two-view drawing
three-view drawing
pictorial drawing
isometric drawing
oblique drawing
cavalier drawing
cabinet drawing
perspective drawing
cutaway drawing
exploded drawing
elevations

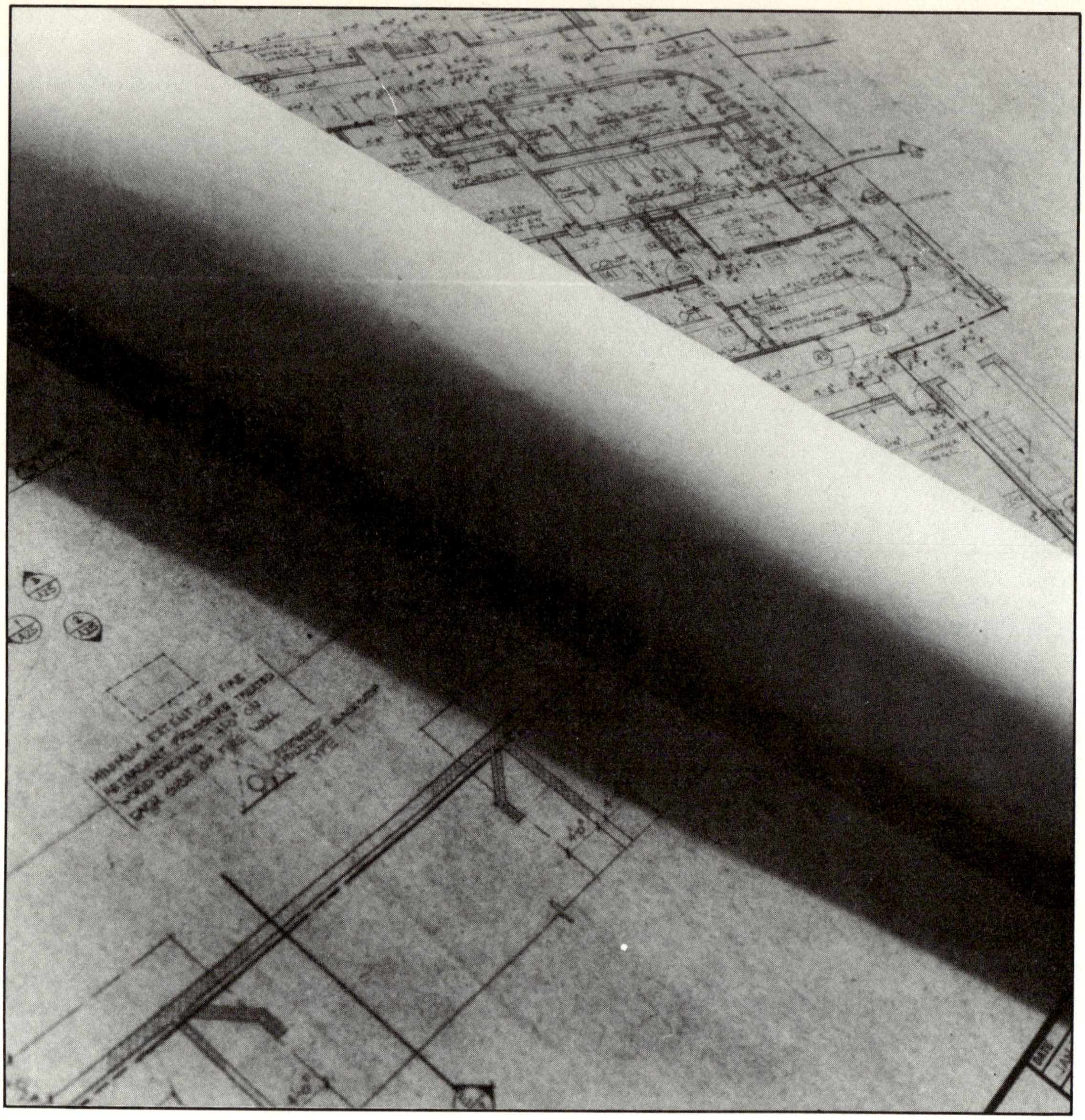

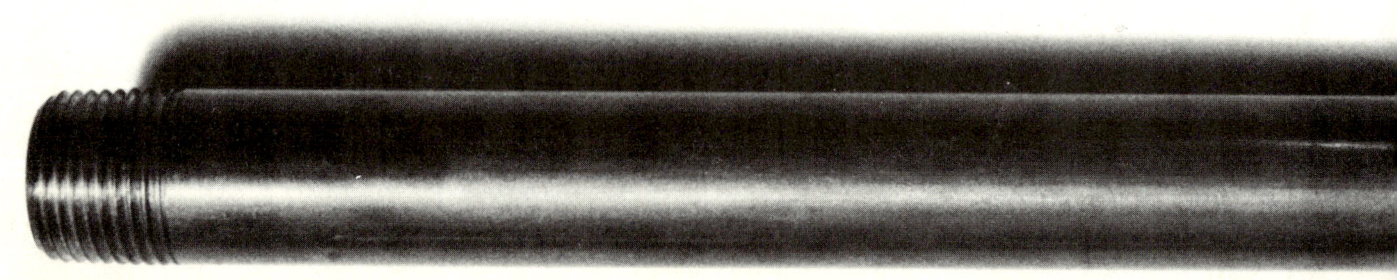

11

READING THE BLUEPRINT

Plumbers may become involved with many small jobs, repairing faucets, fixing leaks, clearing clogged sewer lines, and the like; however, their main stock in trade is constructing new buildings and remodeling old ones. Both new building construction and old building remodeling (renovation) require that plumbers be able to read architectural blueprints.

An architectural blueprint could be called a working drawing for a structure, for example, a house, commercial building, school, or church. It contains all the instructions the carpenters, plumbers, electricians, and other building tradespeople need to build that building. The blueprint will show exactly where the structure will be located (plot plan); what each of the sides of the structure will look like (elevations); the depth, size, and type of construction of the foundation and walls (wall sections); and the size, placement, and number and type of fixtures for each room (floor plan). In addition to the plans above, the blueprint may contain a special section for the plumber, electrician, or other specialized building tradesperson. A specification or written description of the work to be performed may be included in the contract set of documents.

If you, as the plumber, are unable to read a blueprint, identify the symbols, and interpret the location of the plumbing lines and fixtures, you may not have a long career in the plumbing business. To help you learn about blueprints, we will use the blueprint for a typical three-bedroom house.

At the completion of this chapter, you will be able to:

- Identify and interpret common architectural symbols.
- Translate ratio scales from blueprint measurements to full-scale measurements.
- Identify plumbing-component placement from an architectural blueprint.
- Be able to identify and combine the various blueprint plans and complete a rough-sketch plumbing layout and supply estimate.

• PLOT PLAN •

The plot plan, sometimes referred to as the site plan, shows the builder exactly where the building should be located. The plot plan will normally show the street or road on which the building is to be located. It should also show the location of any utilities (electrical, water, sewer) which are to be used. Figure 11-1 is the plot plan for a residence. Look it over carefully.

• QUESTIONS •

1. What is the scale for this plot plan?
2. How large is the plot on which the house is being built?
3. How many feet back from Spring Street is the house located?
4. How long is the house from side to side?
5. The highest point of ground on that lot is 276 feet high. (T or F)
6. How many feet deep is the house (front to back)? (Hint: Use a ruler.)

Now, let us go over some of the questions. The first answer is 110 feet wide by 220 feet deep, not 220 feet wide by 110 feet deep. When we look at a lot, the width is normally that part of the plot which faces the main street.

The house is 71 feet from side to side. The front of the house faces the street, and the left side is 15 feet from the left plot line while the right is 24 feet from the right plot line. If you add 15 and 24 and subtract that from 110, that will leave 71 feet in the middle for the house. To check the answer, add 71 + 15 + 24.

The highest point of ground is identified by that curved line near the rear of the lot. That line is called an elevation or grade line. The figures (203) on that line identify the height of that area, and by finding all these lines, you can determine the grade or slope of the land. In this case, the land slopes down toward the street. The back of the plot if 203 feet and the front is 200, so the land slopes down toward the front.

One thing about this plot plan is of special interest to the plumber. The cold-water service is shown near the center front of the plot. There is no municipal sewer line indicated, and this normally means a septic tank and drain field. On many blueprints, the architect would also indicate where the septic tank and drain field are to be placed (Figure 11-2).

• FOUNDATION PLAN •

The foundation plan outlines the base of the building. It identifies in detail where each outside wall and where some interior walls will be located. While this blueprint is primarily for the carpenter or brick mason, the plumber must also be familiar with it because of the water-supply and sewerage lines.

Figure 11-3 is a foundation plan. There are two foundation plans for this house, since it may be completed with a concrete (slab) floor or with a wood-frame floor. Let us look at Figure 11-3 and find out what the foundation plan for a concrete slab floor tells us.

First of all, notice the scale of this foundation plan. It is different from the plot plan scale. The first item

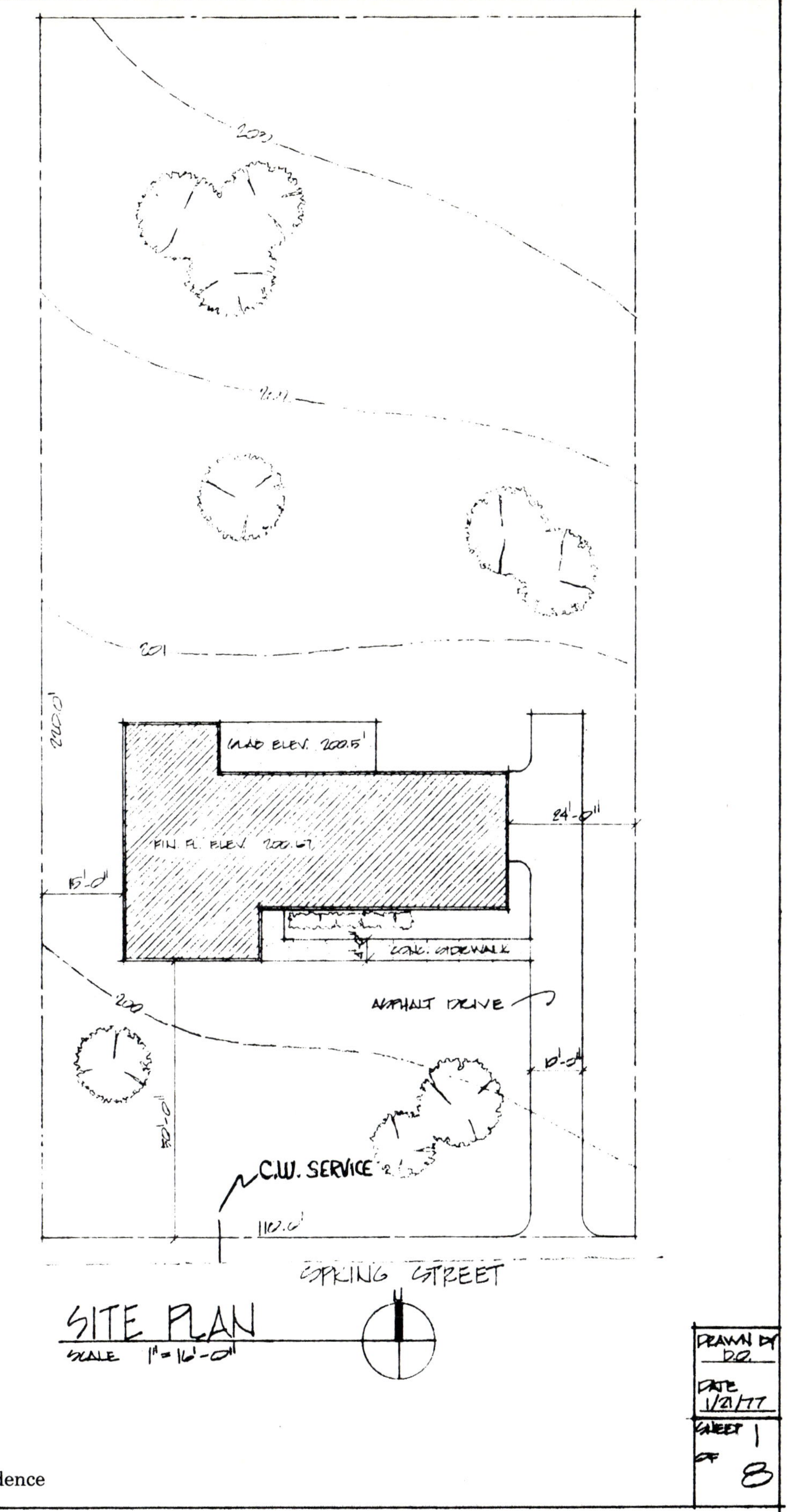

Figure 11-1. Plot plan for a residence

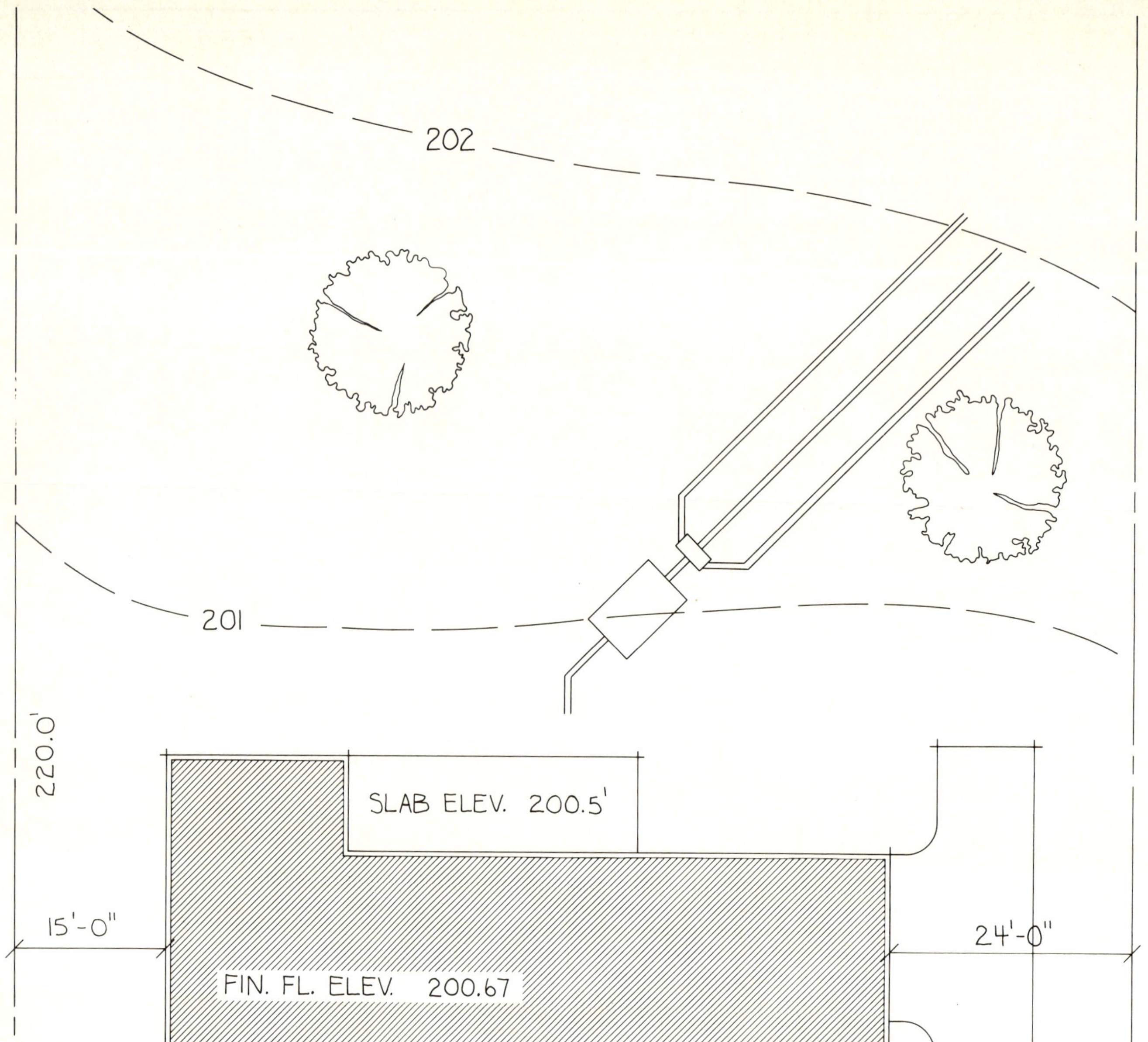

Figure 11-2. Blueprint with septic tank and drain field

identified on a blueprint should be the scale. In this case, ¼ inch on the print will equal 1 foot of the building. Now look at the foundation plan. There are dashed lines on each side of the foundation wall. These lines indicate the width of the footing. The footing is a reinforced concrete (concrete and steel) pad on which the building stands.

In between the dashed lines (footing lines) there are two different sets of lines, one filled in with curved lines and one filled in with straight lines. The inside or curved lines indicate a concrete block wall, and the straight lines indicate a brick outer wall. These materials are identified later in the wall cross section and will be covered at that time.

Look around the outside of the plan. You can see little circles with arrowheads, each with a set of numbers or a letter in the circle. These lines are called cross-section or detail lines and show where cross sections or details will be taken. A separate plan for each cross section will be found in the elevation plan or the detail plan of the blueprint.

The left and right sides of the foundation plan have *cross-section lines* marked A, and show where the long or longitudinal cross section will be shown. The front and back sides of the foundation plan have cross-section lines marked B, and show where the short or transverse cross section will be shown.

Each of the three *detail lines* is identified by a pair of numbers, in this case $\frac{1}{2}$, $\frac{2}{2}$, and $\frac{3}{2}$; and each pair of numbers refers to a different detail drawing. The upper right corner of the foundation plan page

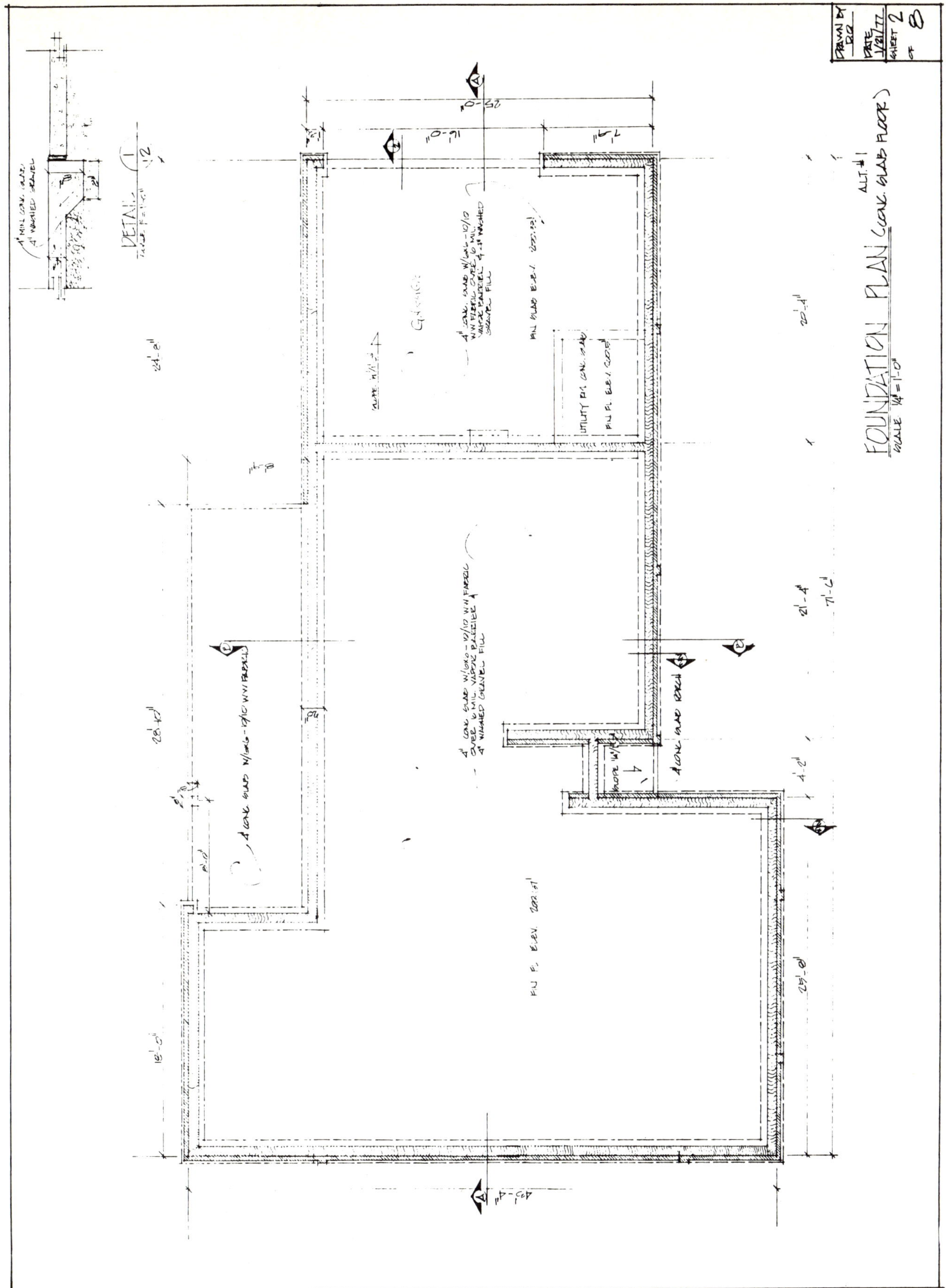

Figure 11-3. Foundation plan for a concrete slab

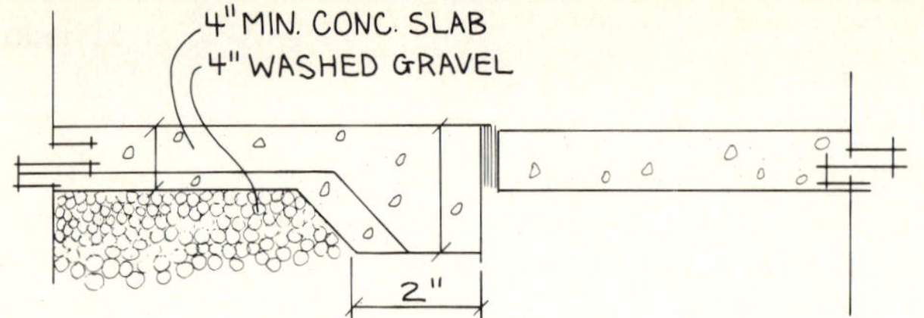

Figure 11-4. Foundation Plan Detail

shows detail drawing $\frac{1}{2}$ (Figure 11-4). This detail drawing is the cross section of the outside edge of the garage floor. Notice that the scale of this detail drawing is different from the scale of the foundation plan.

Sheet 3 on page 146 is the foundation plan for the same building, but in this case, the floor is wood rather than a concrete slab.

Compare Figures 11-3 and Sheet 3. The outside wall appears the same with the addition of vents (for air circulation) and an access door. The cross-section and detail lines are in the same place, but one detail line is not numbered. The unnumbered detail line indicates that the detail here is the same as in the first foundation plan. The major changes in the plan for the wood-floor building are the addition of six concrete and concrete-block piers, the 6 × 10 beams, and the lines which identify the floor joists.

• QUESTIONS •

1. How long is the building?
2. How wide is the building?
3. How wide is the garage door?

• THE FLOOR PLAN •

The floor plan is the most important part of the blueprint. It identifies each room and its size; it shows all walls, closets, doors, and windows; it locates all electric lights, switches, and outlets; and it accurately locates each plumbing fixture. The floor plan shows what the building would look like if the roof and the top third of the wall were cut off. Sheet 4 on page 147 is the floor plan for our building. Look it over carefully and identify the different rooms, components, and built-in fixtures.

Let us look at the different sections of this floor plan in detail. First, the outside wall of the house fits over the outside wall outlined in the foundation plan. The same cross-section and detail lines are shown except for $\frac{1}{2}$, which is only needed in the foundation plan. The outside wall is still made of two different materials, brick on the outside and wood on the inside. In addition, all the windows and doors are located and drawn in place. Figure 11-5 shows the different symbols used for doors and windows.

Locate all the windows on the floor plan. Outside of each window is a small square around a letter. These letters identify the type of window that belongs in that spot. Now locate all the doors in the outside wall. Next to each door there is a circle containing a number. These numbers identify the type of door that will be installed. These letters and numbers are listed along with window and door sizes on another page of the blueprint and are called schedules (Figure 11-6).

It may not seem too important for plumbers to know about door and window schedules, but knowing where the doors and windows are located and

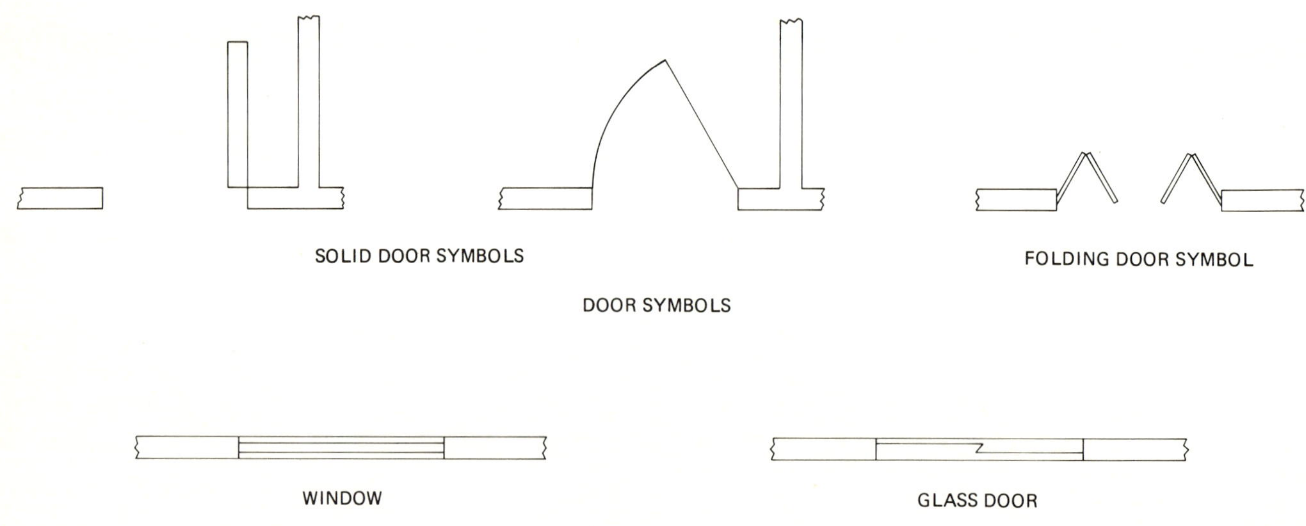

Figure 11-5. Door and window symbols

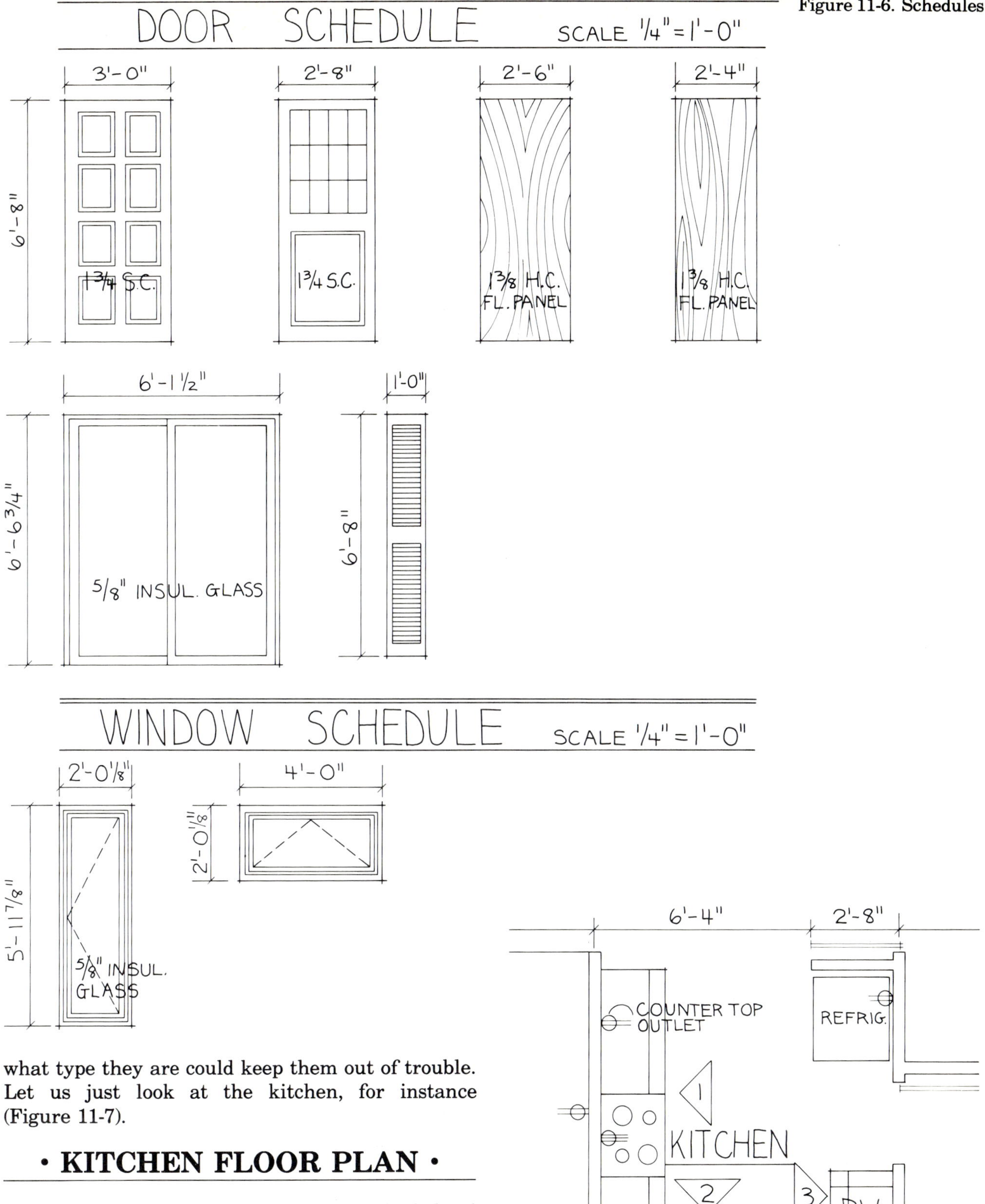

Figure 11-6. Schedules

what type they are could keep them out of trouble. Let us just look at the kitchen, for instance (Figure 11-7).

• KITCHEN FLOOR PLAN •

The kitchen sink is to be installed right behind window B. From what we learned in Chapter 7, we know that a sink this far away from the main vent must have its own vent line. The cheapest and easiest way to run this vent would be straight through the roof in the wall directly behind the sink, but in this case, the window would be in the way. If

Figure 11-7. Kitchen floor plan

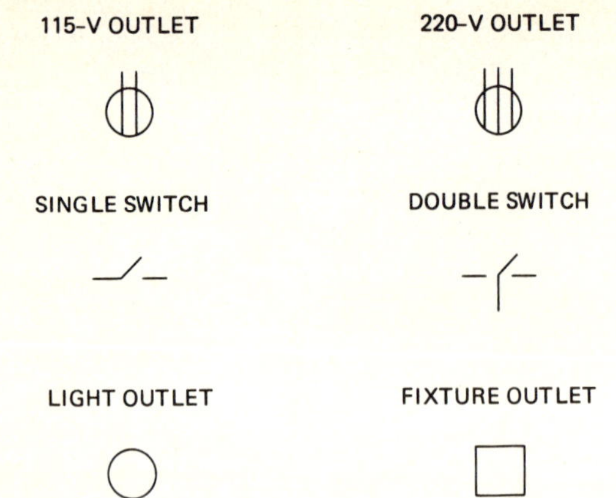

Figure 11-8. Electrical symbols

you had only brought enough materials to run the vent straight up, you would not be able to finish the job without more materials. This loses money and time, and the boss would be upset at that.

The same sort of problem may occur with the electrical wiring. Figure 11-8 shows the basic symbols used to identify electrical equipment in a floor plan.

Look back at Figure 11-7 and see if any of the electrical outlets and lights could interfere with the plumbing lines. They all look as if they are far enough away.

There is a new symbol in the kitchen, a numbered triangle. This triangle tells you that there is another drawing of the wall off the triangle's point. These drawings, normally found on another page of the blueprint, are called elevations; in this case they would be kitchen elevations. Figure 11-9 shows the three kitchen elevations for this floor plan.

You have already seen the blueprint symbol for a kitchen sink, but there are other plumbing symbols which you know. Figure 11-10 shows some of these other symbols.

• BATHROOM FLOOR PLAN •

The floor plan on Sheet 4 (page 147) shows two bathrooms, one entered from the hall and one entered from the master bedroom. They are called bath 1 and bath 2. Figure 11-11 shows this area of the blueprint at a scale of ¼″ = 1′–0″.

You should be able to identify the bathtubs, commodes, and lavatories in each bathroom. These are the plans which the plumber will use to install the water and sewer pipes for each bathroom. Remember the scale, ¼″ = 1′–0″. If the architect draws the commode plan ¼ inch off, the plumber will install the commode 1′–0″ off, and the architect is responsible. But if the plumber does not measure the blueprint correctly and installs the commode 1′–0″ off, the commode will have to be moved at the plumber's expense. Accurate measurement of a floor plan is very important.

Look at the wall between the two bathrooms. That wall is 6 inches thick while all the rest of the walls in the house are 4 inches thick. A 6-inch wall is called a plumbing wall because the bell end of 4-inch cast-iron sewer pipe will not fit a 4-inch wall.

The floor plan for the kitchen contained numbered triangles to identify the different elevations. A bathroom will normally have only one elevation, the one which shows the lavatory and cabinets. These elevations are normally drawn on the same page as the kitchen elevations. Figure 11-12 shows the elevations for bath 1 and bath 2.

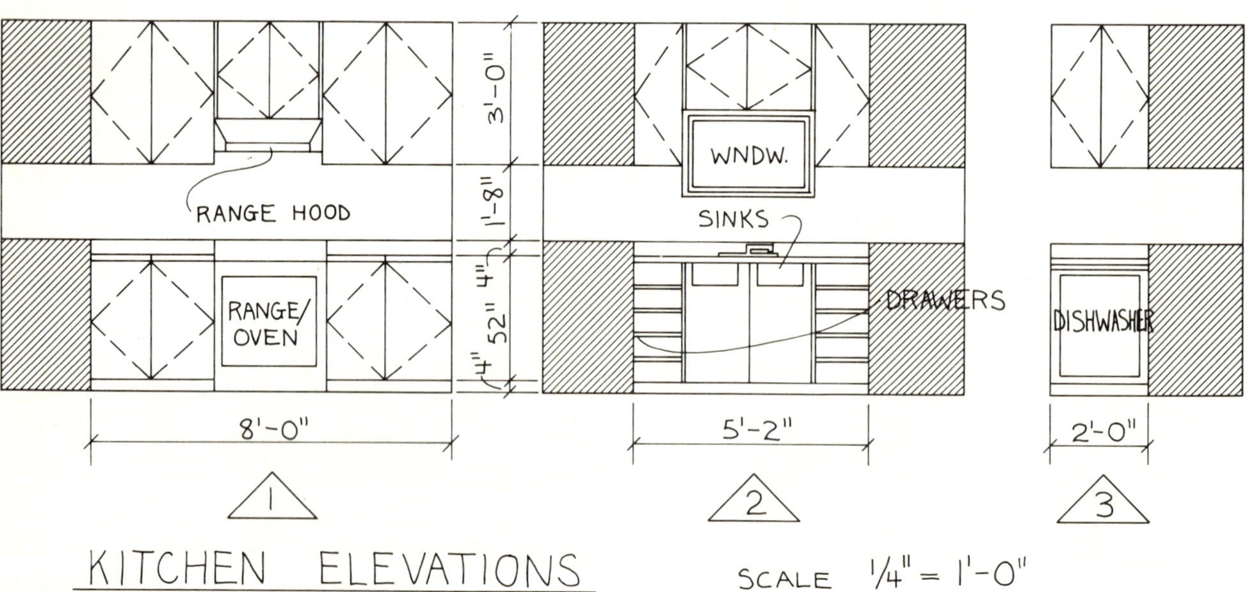

Figure 11-9. Kitchen elevations

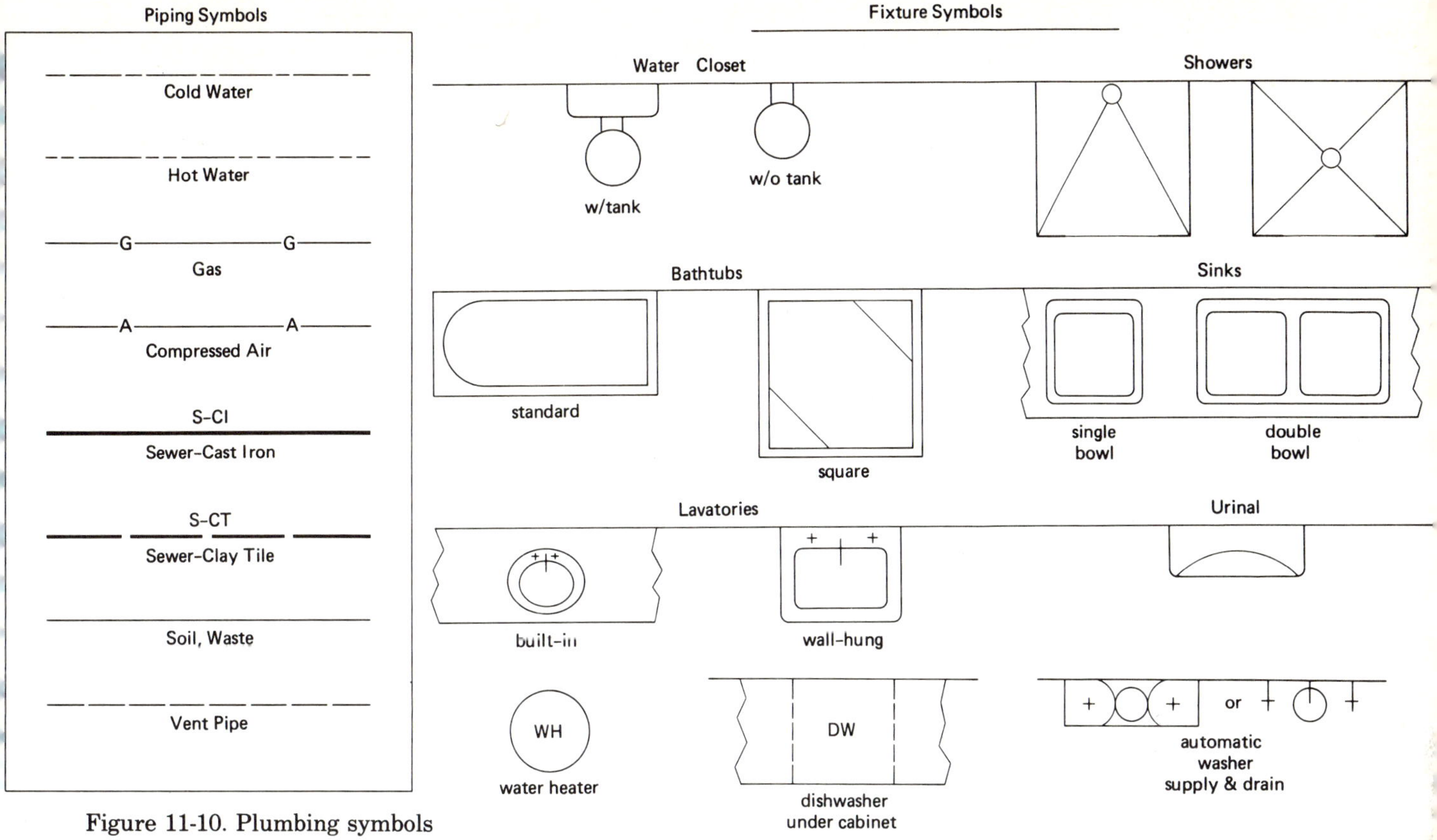
Piping Symbols
Cold Water
Hot Water
G
Gas
A
Compressed Air
S-CI
Sewer-Cast Iron
S-CT
Sewer-Clay Tile
Soil, Waste
Vent Pipe
Fixture Symbols
Water Closet
w/tank
w/o tank
Showers
Bathtubs
standard
square
Sinks
single bowl
double bowl
Lavatories
built-in
wall-hung
Urinal
WH
water heater
DW
dishwasher under cabinet
or
automatic washer supply & drain

Figure 11-10. Plumbing symbols

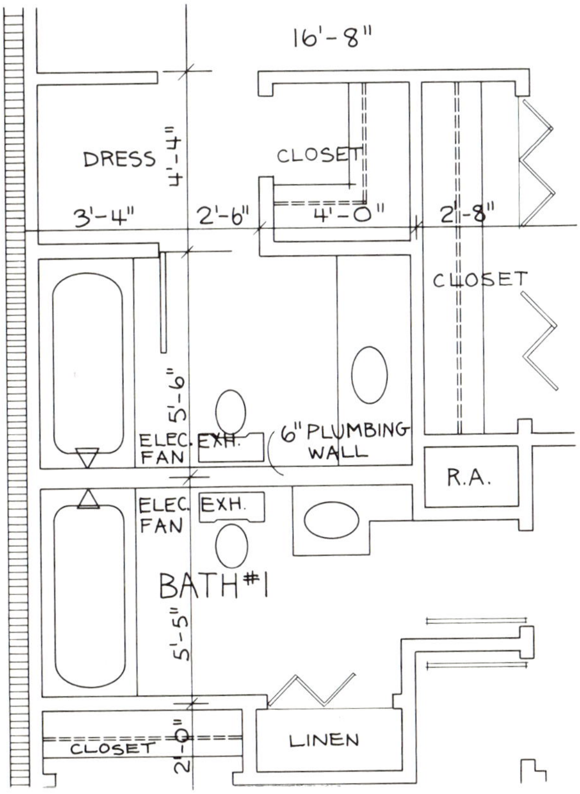
16'-8"
DRESS
4'-4"
CLOSET
3'-4"
2'-6"
4'-0"
2'-8"
CLOSET
5'-6"
ELEC. FAN
EXH.
6" PLUMBING WALL
R.A.
ELEC. FAN
EXH.
BATH #1
5'-5"
CLOSET
2'-0"
LINEN

Figure 11-11. Bathroom detail

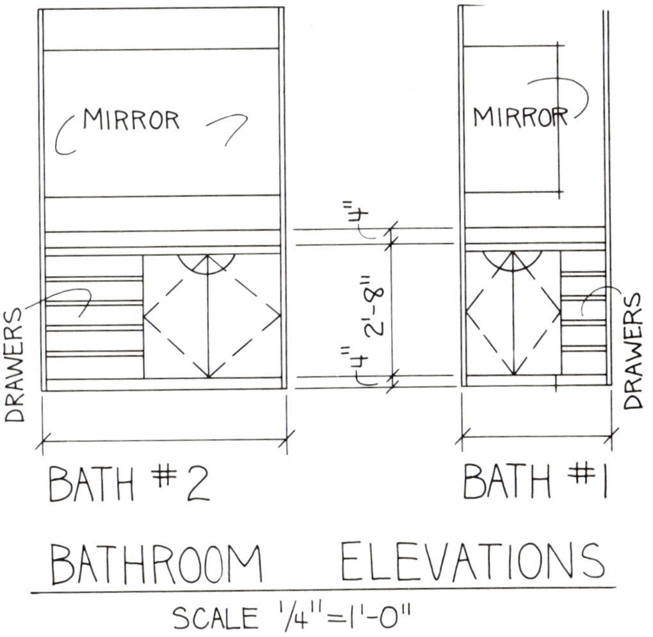
MIRROR
MIRROR
4"
2'-8"
4"
DRAWERS
DRAWERS
BATH #2
BATH #1
BATHROOM ELEVATIONS
SCALE 1/4"=1'-0"

Figure 11-12. Bathroom elevations

These two bath elevations show the plumber that both lavatories will be cabinet-mounted and at the standard height off the floor.

• UTILITY-ROOM FLOOR PLAN •

The utility-room floor plan is also of concern to the plumber. This room normally contains the hot-water heater, the automatic clothes washer, the clothes dryer, and many times, a laundry tray. Figure 11-13 shows the utility-room floor plan for the building. Identify each of the fixtures and appliances and what types of water supply or sewer lines would be required.

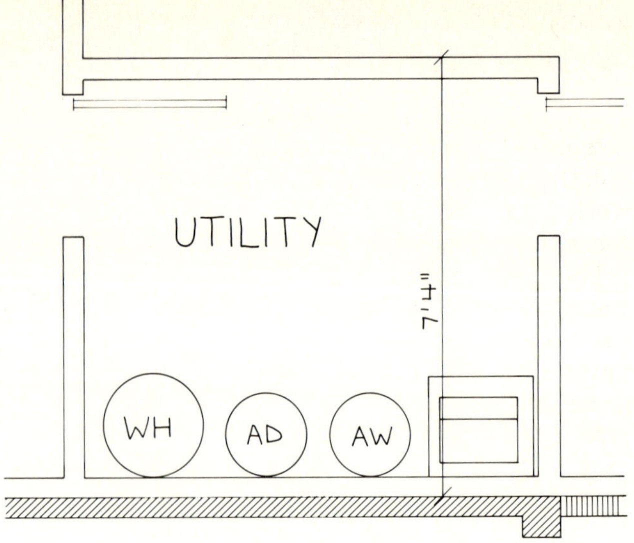

Figure 11-13. Utility-room floor plan

• DETAIL PLANS •

Both the foundation plan and the floor plan contained wall section lines indicated by numbers in a heavy black arrow. We discussed the section $\frac{1}{2}$, since it only applied to the foundation plan. Figure 11-14 shows the detail page of this blueprint.

Notice first that there are two sets of drawings for both $\frac{2}{2}$ and $\frac{3}{2}$. One is for the concrete slab floor and other is for the wood floor.

Now check the scale of the detail drawing; 1 inch on the drawing equals 1 foot on the building. Let

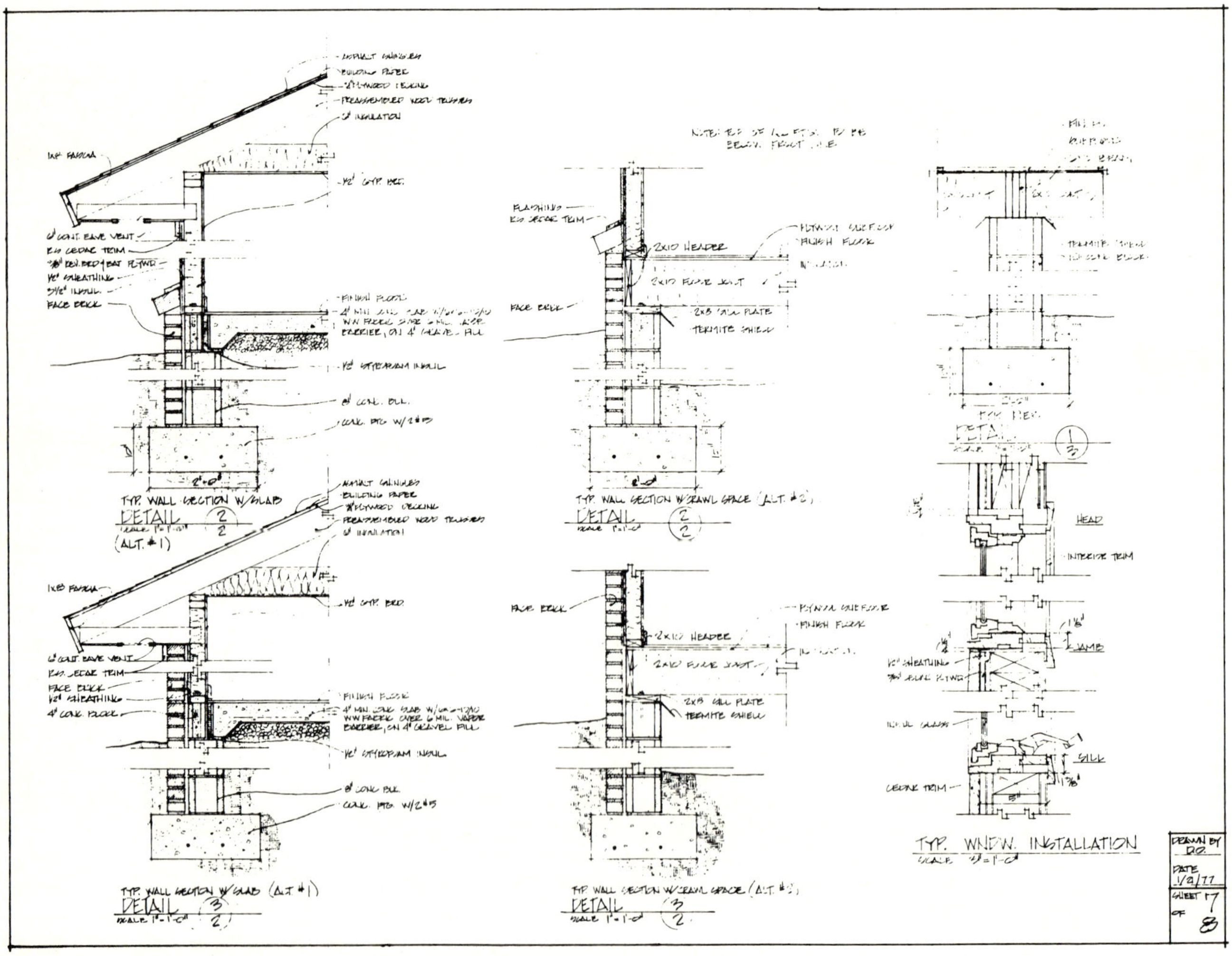

Figure 11-14. Detail

us start at the bottom of the detail drawing for the slab floor numbered $\frac{2}{2}$. The concrete footing is shown; it is 2 feet wide and 10 inches high. The two black dots near the middle indicate the steel reinforcing rod. On top of the footing, there is an 8-inch concrete block wall, and next to that, on the outside of the concrete blocks, a brick wall, one brick wide.

The wall section is important to plumbers for at least two very good reasons. First, they must know the thickness of the footing so they can run the sewer, water, or natural-gas lines deep enough to be below the footing. Second, they must know what type of materials make up the outside wall in case they have to install hose bibs on the outside of the house.

• ELEVATIONS •

Just like the object of any technical drawing, a house has six sides, a right and left side, a front and back side, and a top and bottom. So far, we have just been discussing the bottom, the foundation and floor plan side. There is a difference between a technical drawing and a set of blueprints; the typical set of blueprints will contain drawings of all the sides and the roof of the house as well as the foundation and floor plan. Sheets 5 and 6 show the elevations of the building we are studying.

Sheet 5 shows what the building should look like from the front (front elevation), the left side (left-side elevation), and the rear (rear elevation). In addition to these three views, this page contains the room finish schedule. This is of concern to the carpenter, the painter, the decorator, and the owner, but not to the plumber.

The front elevation will generally show the floor and footing elevation. In the case of this building, the floor is 200.67 feet and the *top* of the footing is 196.5 feet. You must use these two figures as well as the depth of the footing $\left(10 \text{ inches from detail drawing } \frac{2}{3}\right)$ to determine how deep the house sewer, water, natural gas, or other service line must be buried. In this case, it is about 5 feet from the floor of the house to the bottom of the foundation (Figure 11-14).

Sheet 6 shows the right-side elevation. In addition it shows the longitudinal section AA (remember the foundation and floor plans?) and the transverse section BB as well as the door and window schedule. The side and cutaway views of the building and the room finish, door, and window schedules are not as important to the plumber as the foundation and floor plans.

The last page of this set of blueprints is Sheet 8. It contains the roof framing plan, two HVAC (heating, ventilating, and air conditioning) plans, one for the slab floor and one for the wood floor (crawl space), the kitchen and bathroom elevations, and the truss construction drawing. The plumber is not too concerned with the roof framing and truss construction plan, and we have already looked at the kitchen and bathroom elevations. However, plumbers may be required to install the heater and heating ductwork, and even if they are not, they should know about the HVAC plan to be sure that the water and sewer lines do not conflict with anything on that plan.

Chapter 12 of this book deals with HVAC systems and their installation. As a beginning to that chapter, let us look over these two plans, starting with the plan for a wood floor (crawl space) installation. Figure 11-15 is the HVAC plan for the building with a wood floor and a crawl space.

• THE HEATING SYSTEM •

We have been dealing with a variety of scales on each plan, and this one is also different, ⅛″ = 1′–0″. Notice first that the garage is not heated, nor is the utility room. The main heat duct runs along the center of the building, and each heater register is indicated by a small rectangle and a size. In this plan, there are seven 6 × 10-inch and six 6 × 12-inch registers. Each of these registers is connected to the main heat duct of a feeder line. This feeder line is normally a galvanized sheet-metal duct pipe which connects the register to the main line. This duct pipe may be square, rectangular, or round. These different duct pipes are shown in Figure 11-16 on page 152.

Main Heat Duct

The main heat duct is made up of different-sized ducts. The larger ducts are close to the furnace, and the size of the main duct decreases (gets smaller) as the feeder lines remove air. This is a volumetric principle and will be explained in Chapter 12.

Placement of the registers and ducts is planned after the foundation and floor plans are completed. The ducts, both main and feeder, are generally planned so they do not interfere with either the structural or electrical equipment or the plumbing of the building. The heating unit and the blower are indicated by the crossed squares. This equipment is located in a closet next to bath 1.

Figure 11-17 shows the HVAC plan for the building with a concrete slab floor. If you compare these two HVAC plans, you will notice that the only major change is that all the ductwork for the slab floor plan is in the attic rather than in the crawl space.

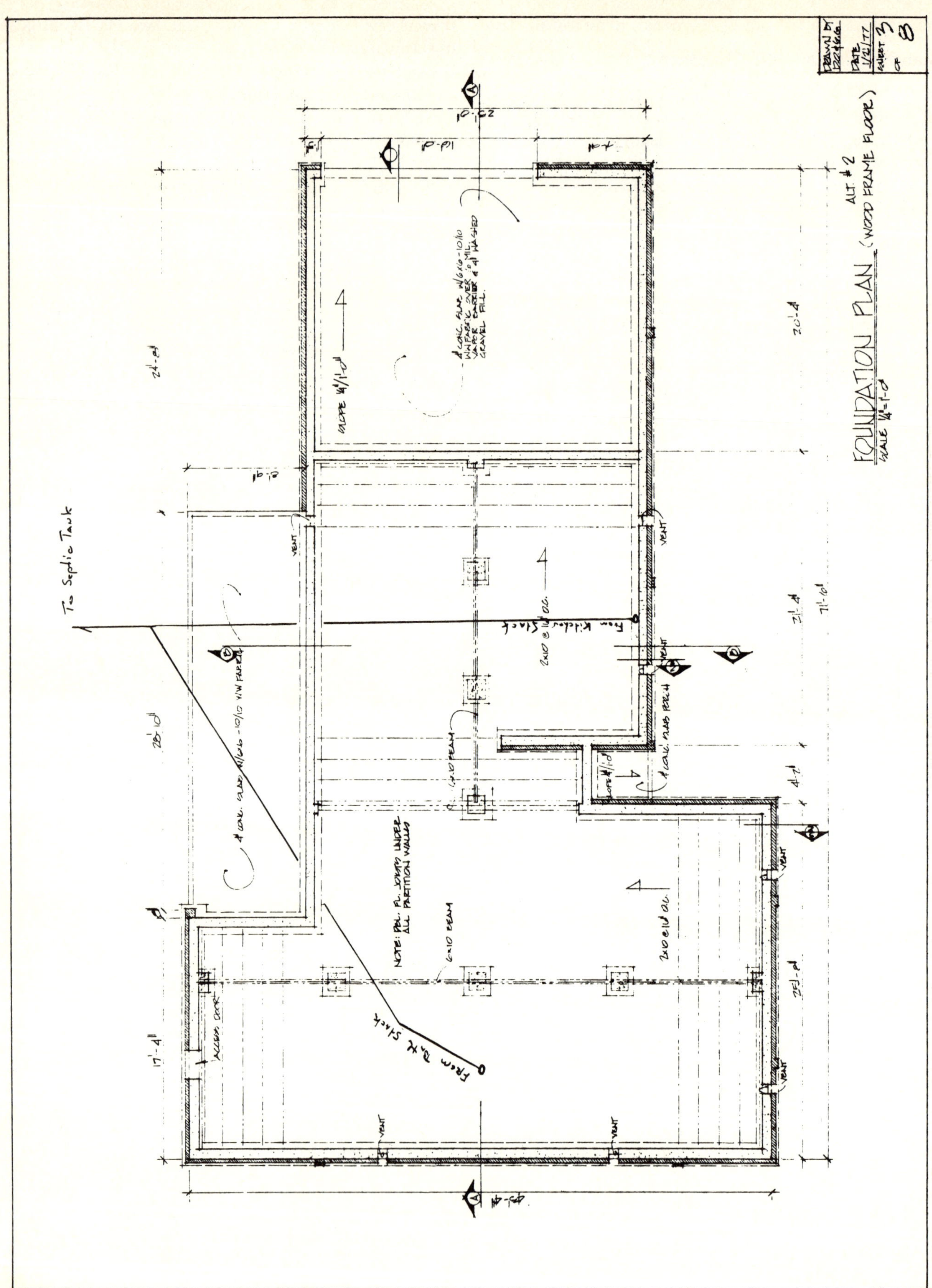

Sheet 3. Foundation plan for a wooden floor

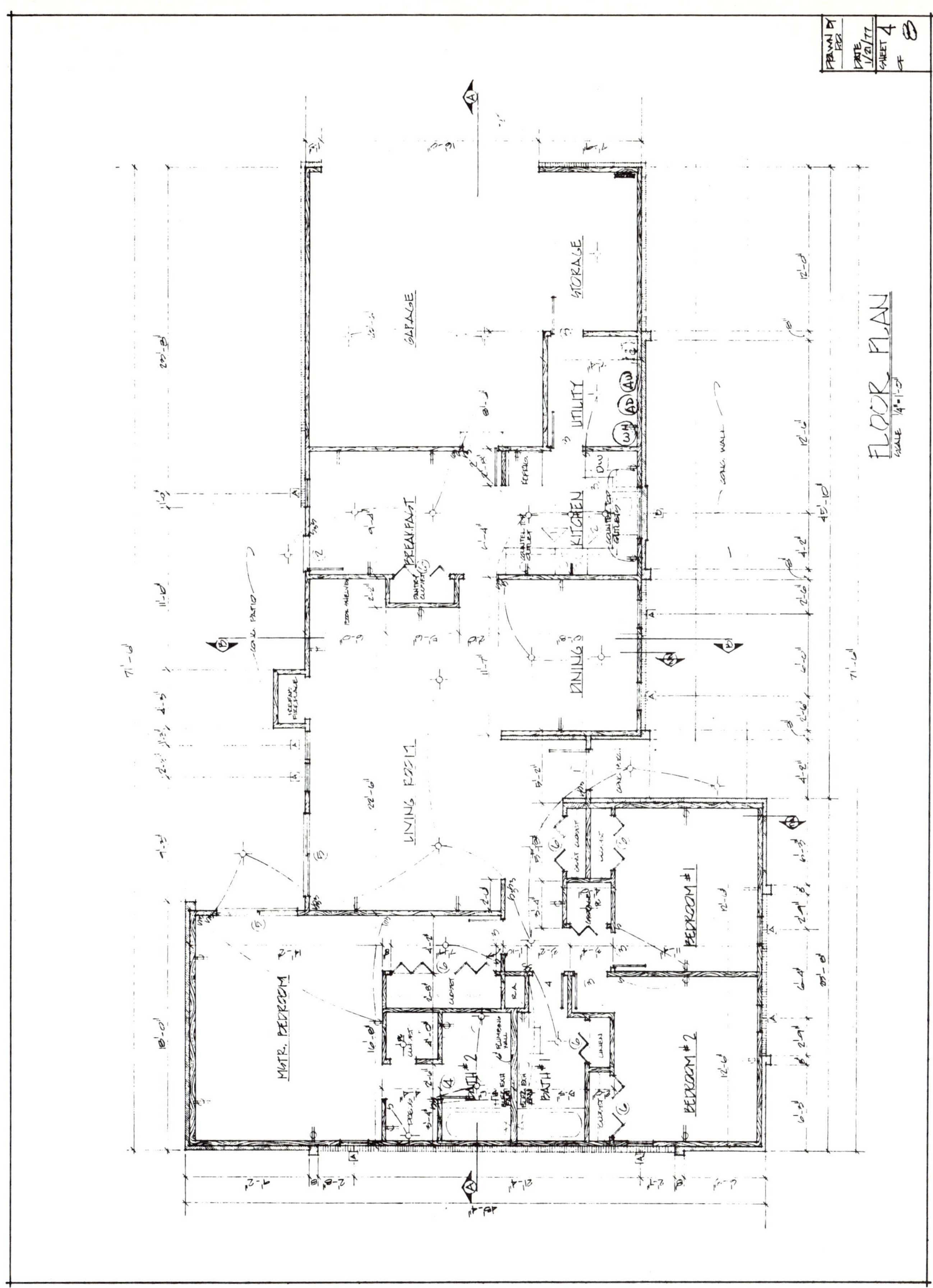

Sheet 4. Floor plan

Sheet 5. Elevations (front)

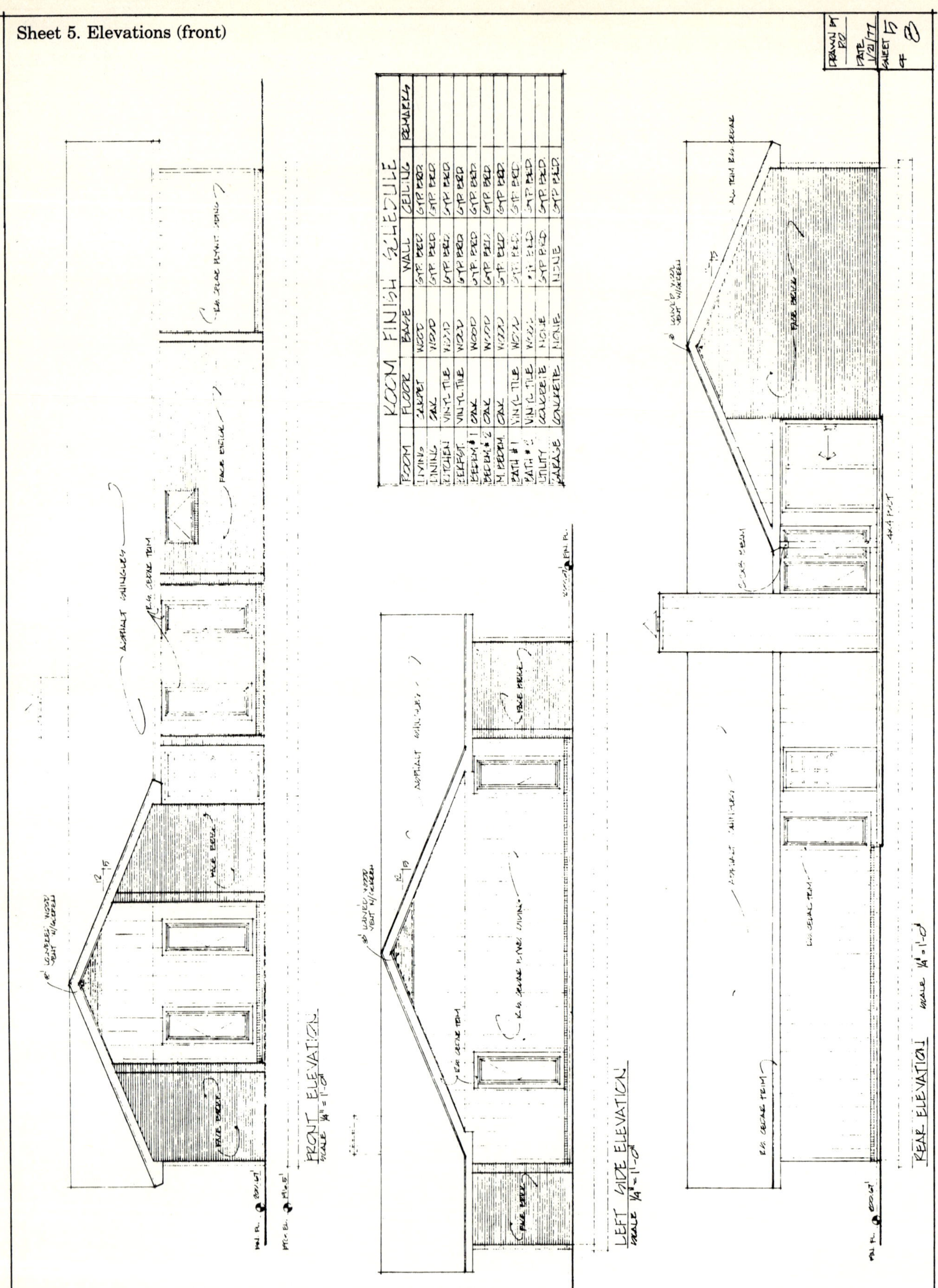

Sheet 6. Elevations (Left)

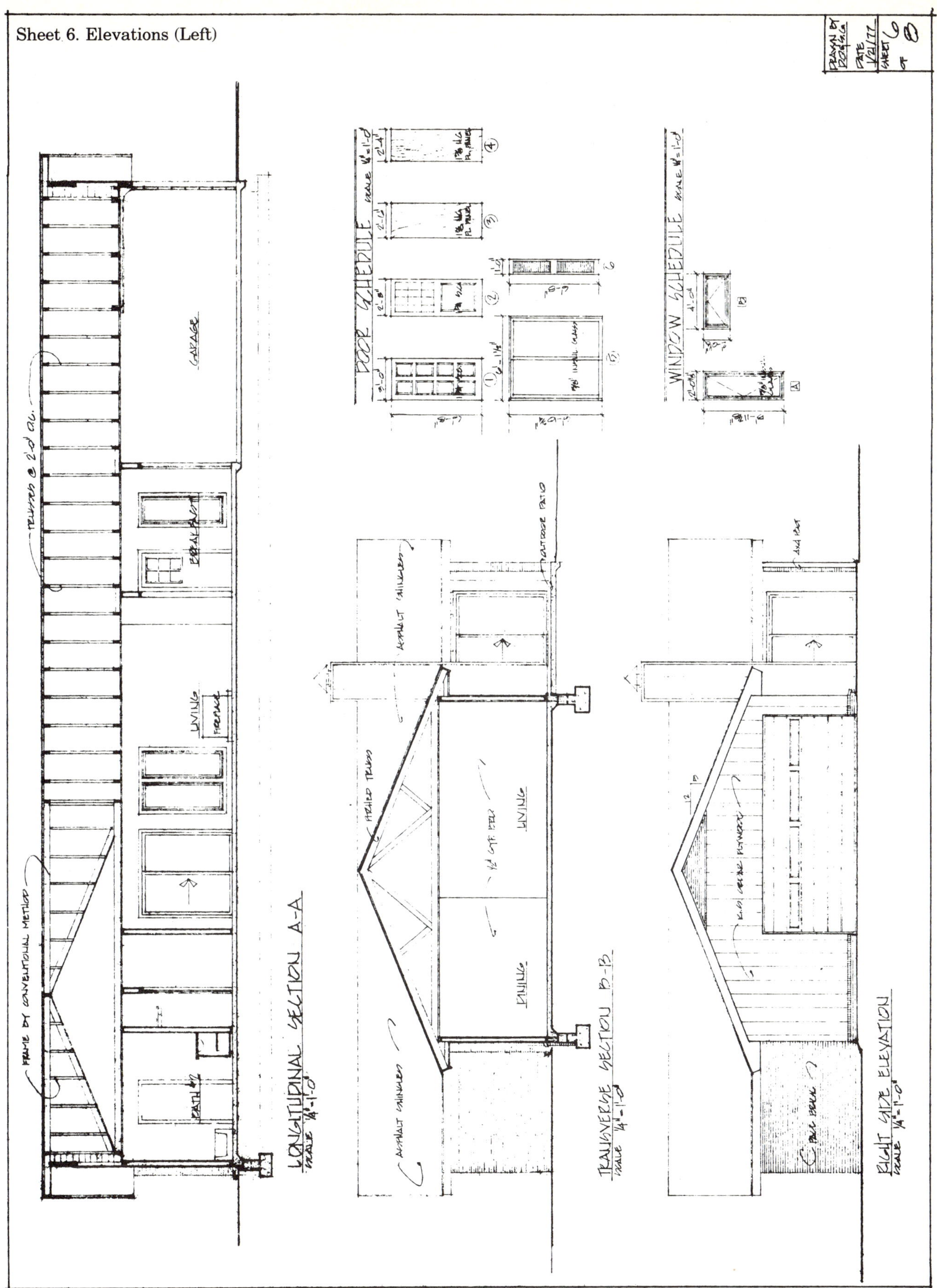

Sheet 8. (1 line)

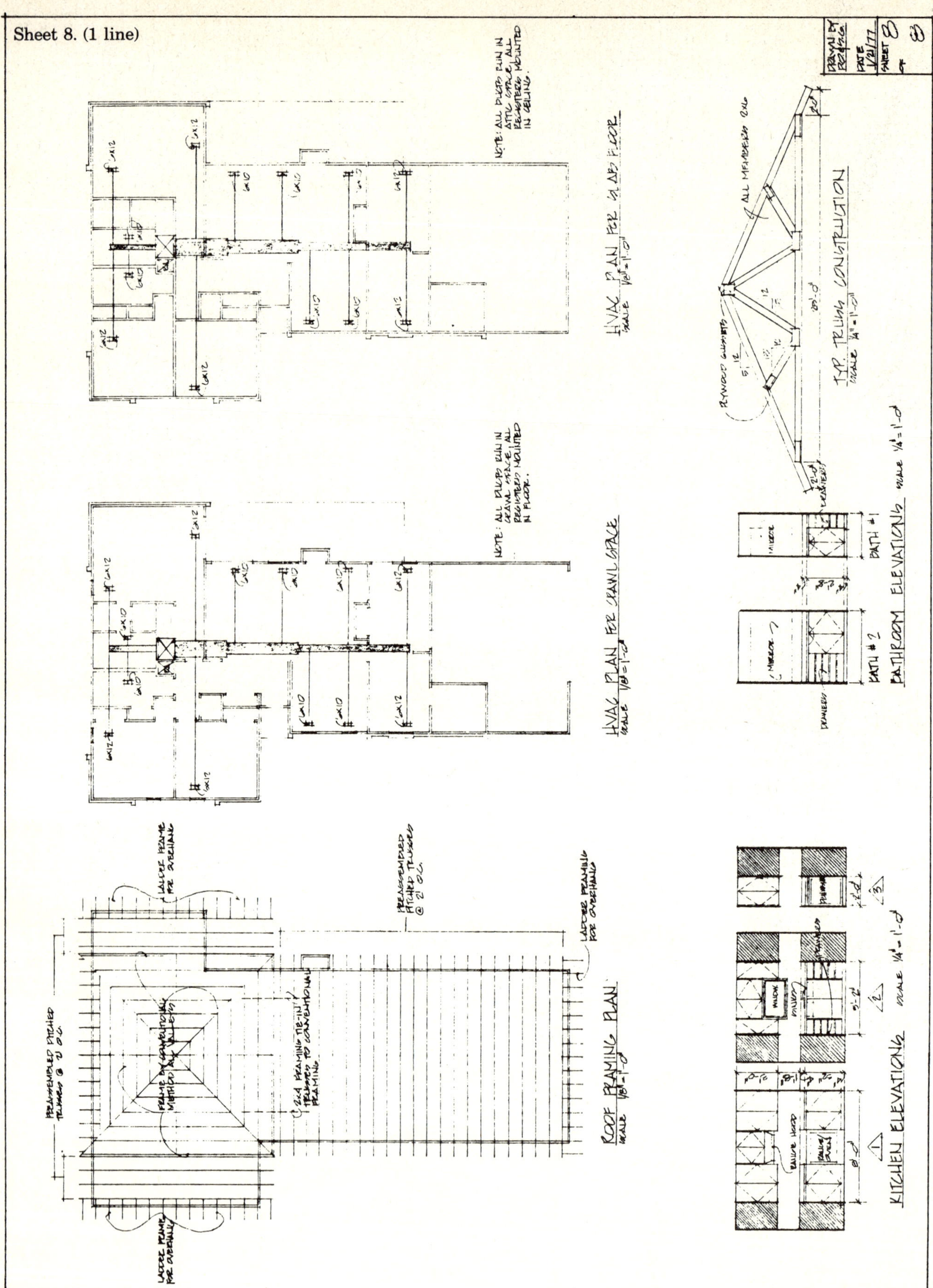

Figure 11-15. HVAC Plan for wooden floor

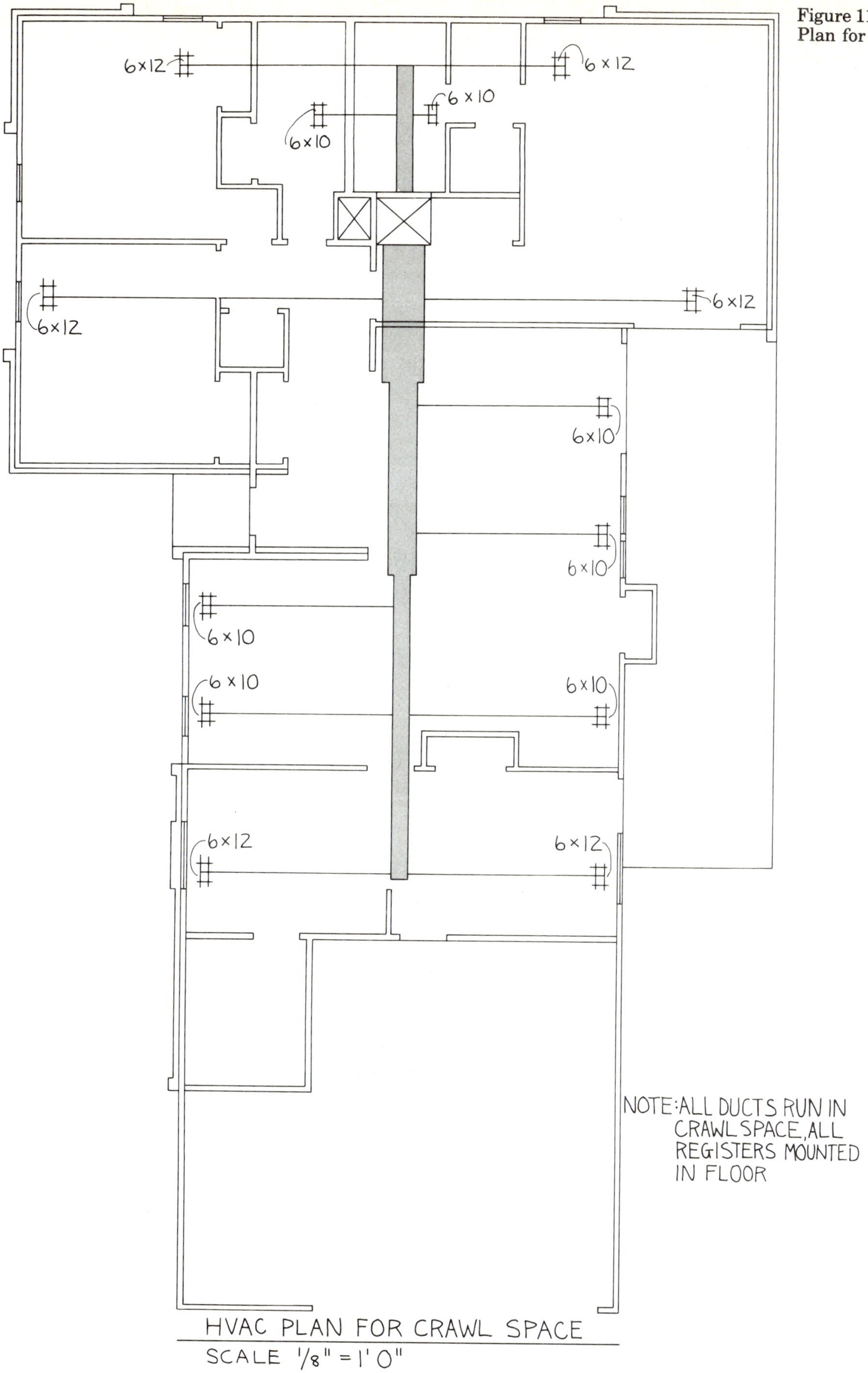

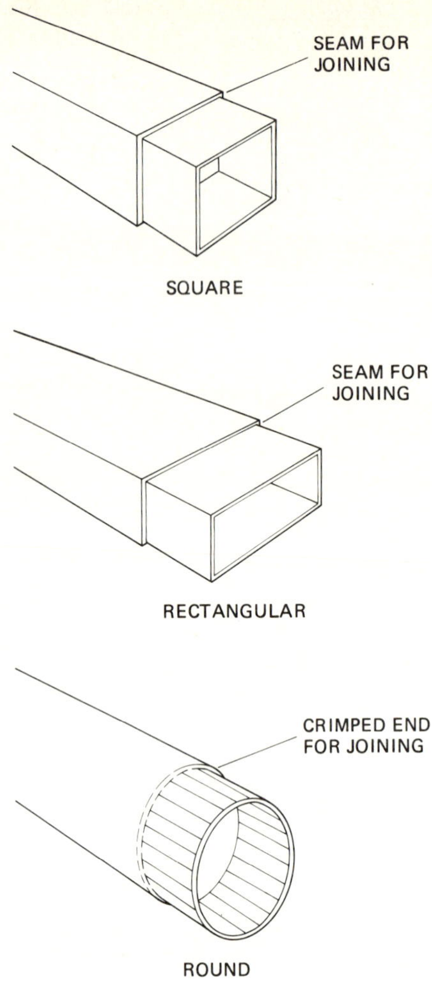

Figure 11-16. Heating plan

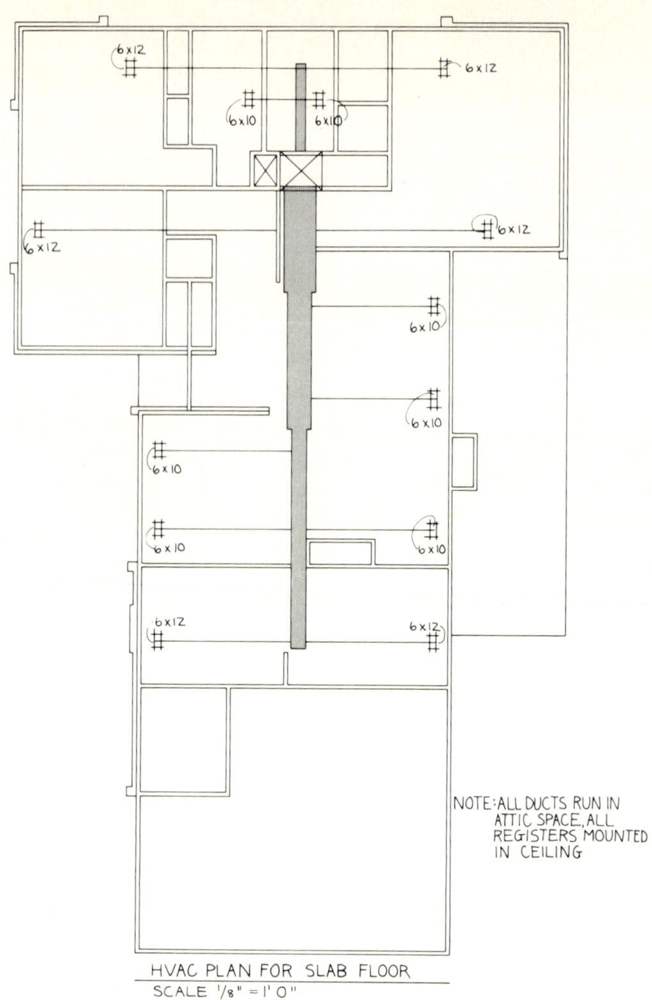

Figure 11-17. HVAC Plan for concrete slab

This heating arrangement would seldom interfere with the plumbing installation.

• PLUMBING LAYOUTS •

Thus far, the information on blueprints and blueprint reading has made you aware of the location of all the plumbing fixtures and accessories necessary to complete this building. Figure 11-18 shows a completed bathroom plumbing layout and the different symbols used for sewer, vent, and hot- and cold-water pipes.

Notice that the bathtub and lavatory each have hot- and cold-water supply, a sewer or waste pipe, and a vent pipe. The commode has a cold-water supply; and because it is next to the sewer stack vent, only a sewer or waste pipe is used. Now let us look at the two bathrooms in the floor plan of our blueprint (Figure 11-19).

The two commodes and the two bathtubs are back to back; that is, they have a common 6-inch plumbing wall between them. The two lavatories are on different walls, but are not very far away. If the stack were placed between the two commodes and waste and vent lines were run to the tub and lavatories, all fixtures could be connected. Figure 11-20 shows this plumbing layout.

The hot- and cold-water layout could be completed the same way. The supply could be brought in at any place in the common wall and the lines run to each of the fixtures as required (Figure 11-21).

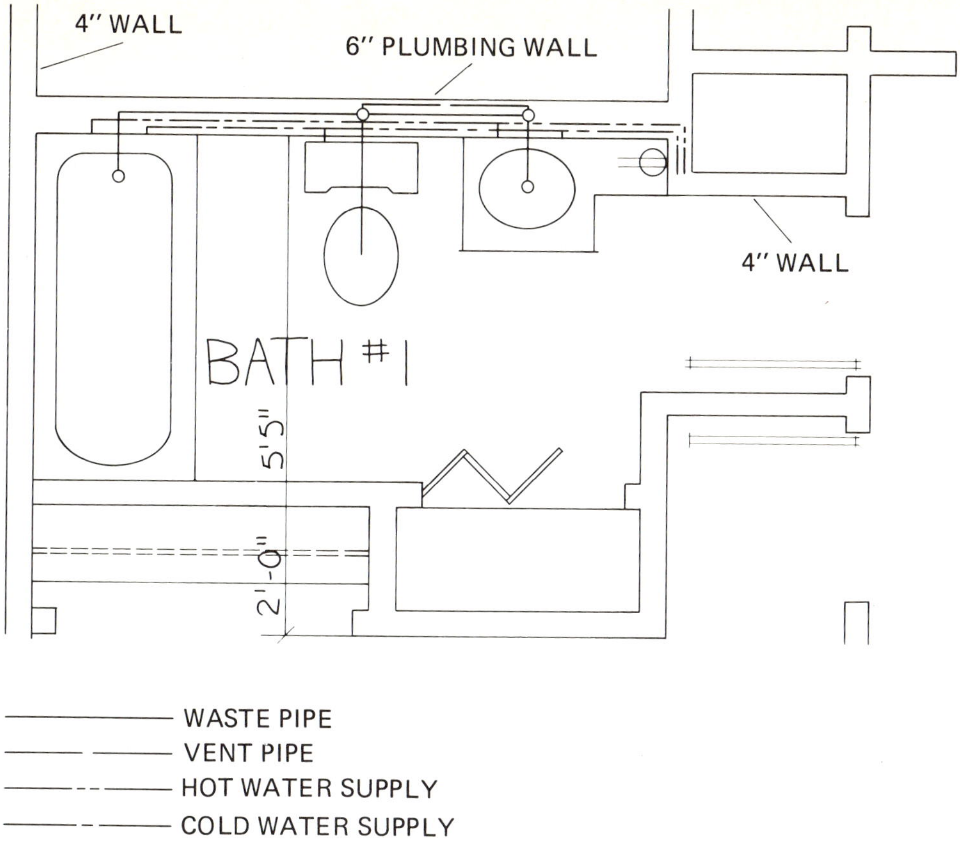

4″ WALL
6″ PLUMBING WALL
4″ WALL
BATH #1
5'5"
2'-0"
WASTE PIPE
VENT PIPE
HOT WATER SUPPLY
COLD WATER SUPPLY

Figure 11-18. Plumbing layout

16'-8"
DRESS
4'-4"
CLOSET
3'-4"
2'-6"
4'-0"
2'-8"
CLOSET
5'-6"
ELEC. FAN
EXH.
6" PLUMBING WALL
R.A.
ELEC. FAN
EXH.
BATH #1
5'-5"
CLOSET
2'-0"
LINEN

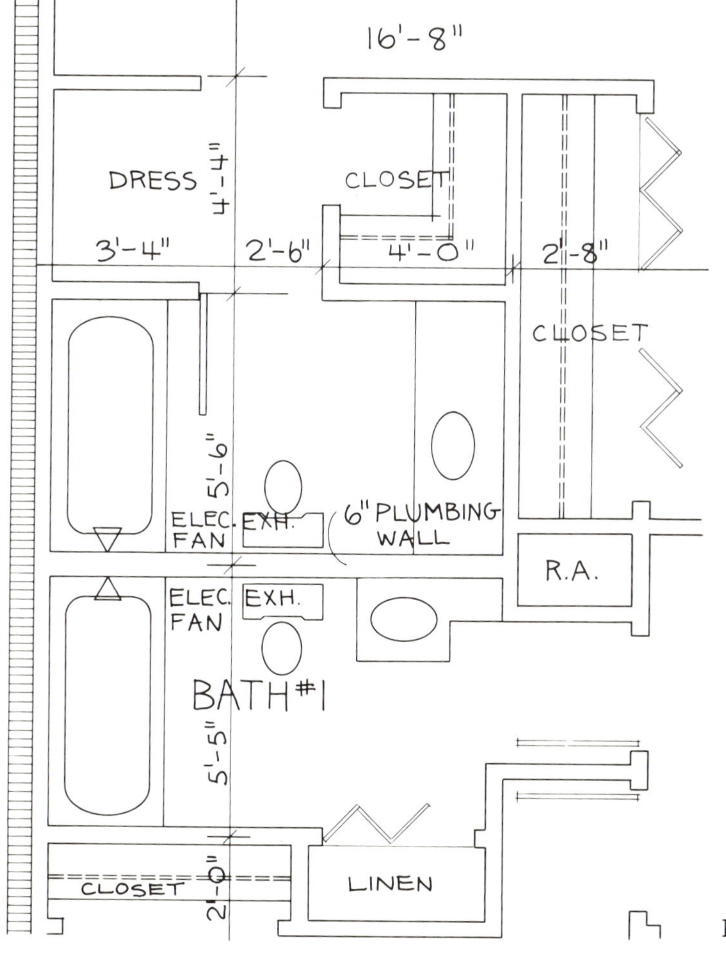

Figure 11-19. (1 line)

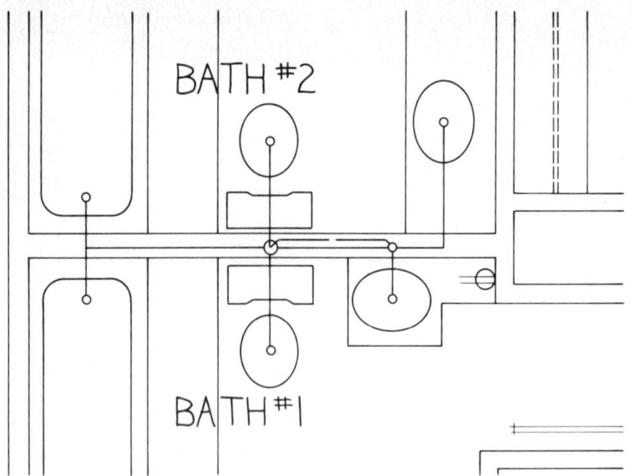

Figure 11-20. (1 line)

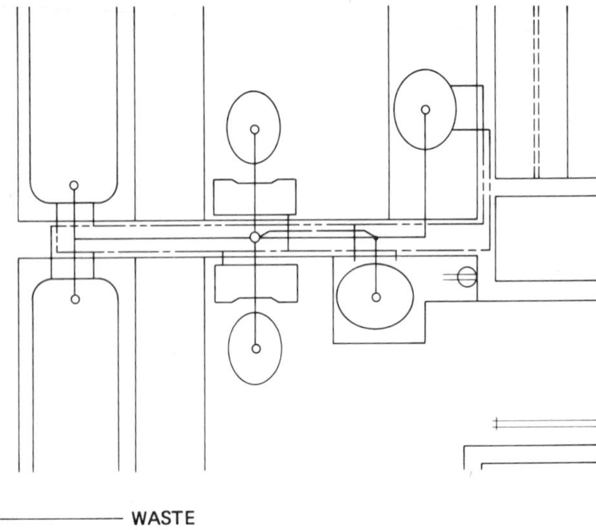

Figure 11-21. Hot and cold water layout

The kitchen and utility-room layouts can be done in the same way. These two rooms do not have a common wall, but the fixtures and accessories are all fairly close together. The kitchen sink, automatic washer, and laundry tray requ're a sewer or waste line, vent line, and hot- and cold-water lines, while the water heater only requires hot- and cold-water lines. Figure 11-22 shows the waste and vent layout using the standard symbols.

The hot- and cold-water lines can be planned and drawn in using the standard symbols. Figure 11-23 shows the completed plumbing layout for the kitchen and the utility room.

Now that we have the plumbing layouts for each of the rooms of the house, we can draw in the main sewer lines and the main hot- and cold-water lines. This could be done on the foundation plan or on the floor plan. We will use the foundation plan, since it is less cluttered and is easier to use.

The main sewer line must run from the sink stack (next to the kitchen sink) and from the bathroom stack (between the commodes) out to the septic tank. The first task is to locate these three points on the foundation plan. Then the main sewer lines are drawn connecting these points in the most efficient manner (Figure 11-24).

Next, the hot- and cold-water lines must be drawn. Again, locate the cold-water supply (from the plot plan) and the water-supply points from Figures 11-26 and 11-28 on the foundation plan and identify them. Now, connect the three cold-water points and the two hot-water points. The foundation plan would then look about like Figure 11-25.

We used the foundation plan for the concrete slab

Figure 11-22. Waste and vent layout

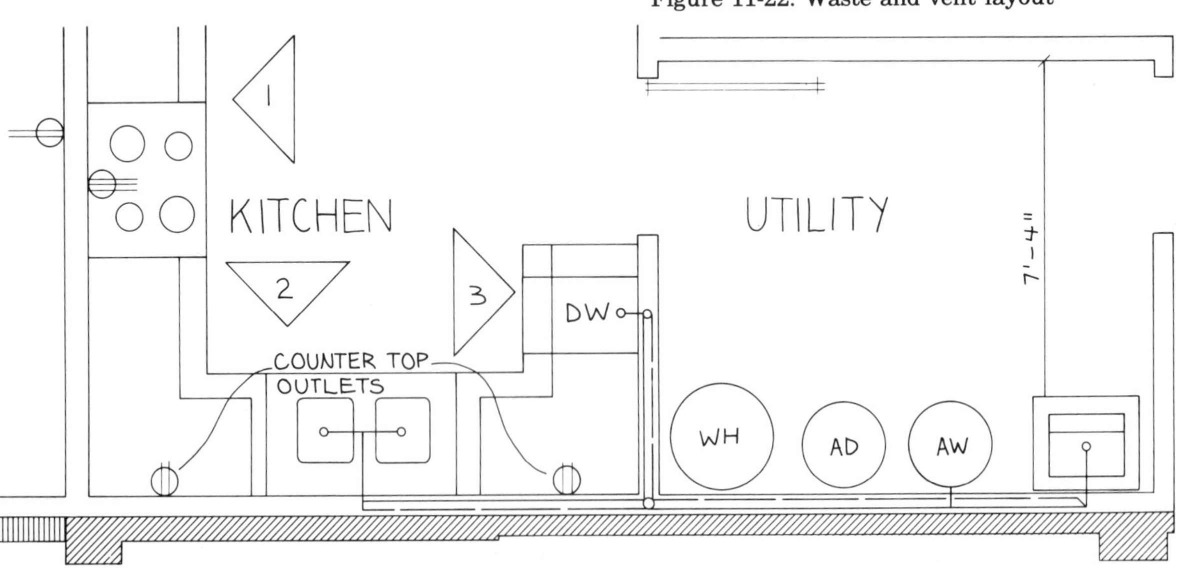

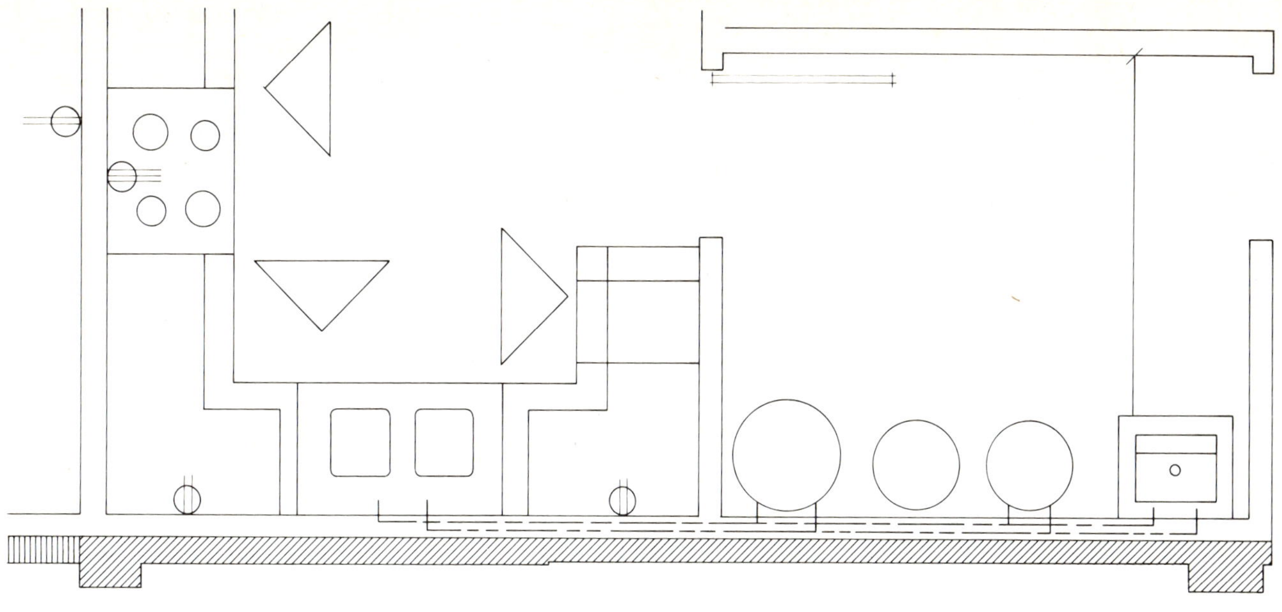

HOT WATER SUPPLY   COLD WATER SUPPLY

Figure 11-23. Kitchen/utility room plumbing layout

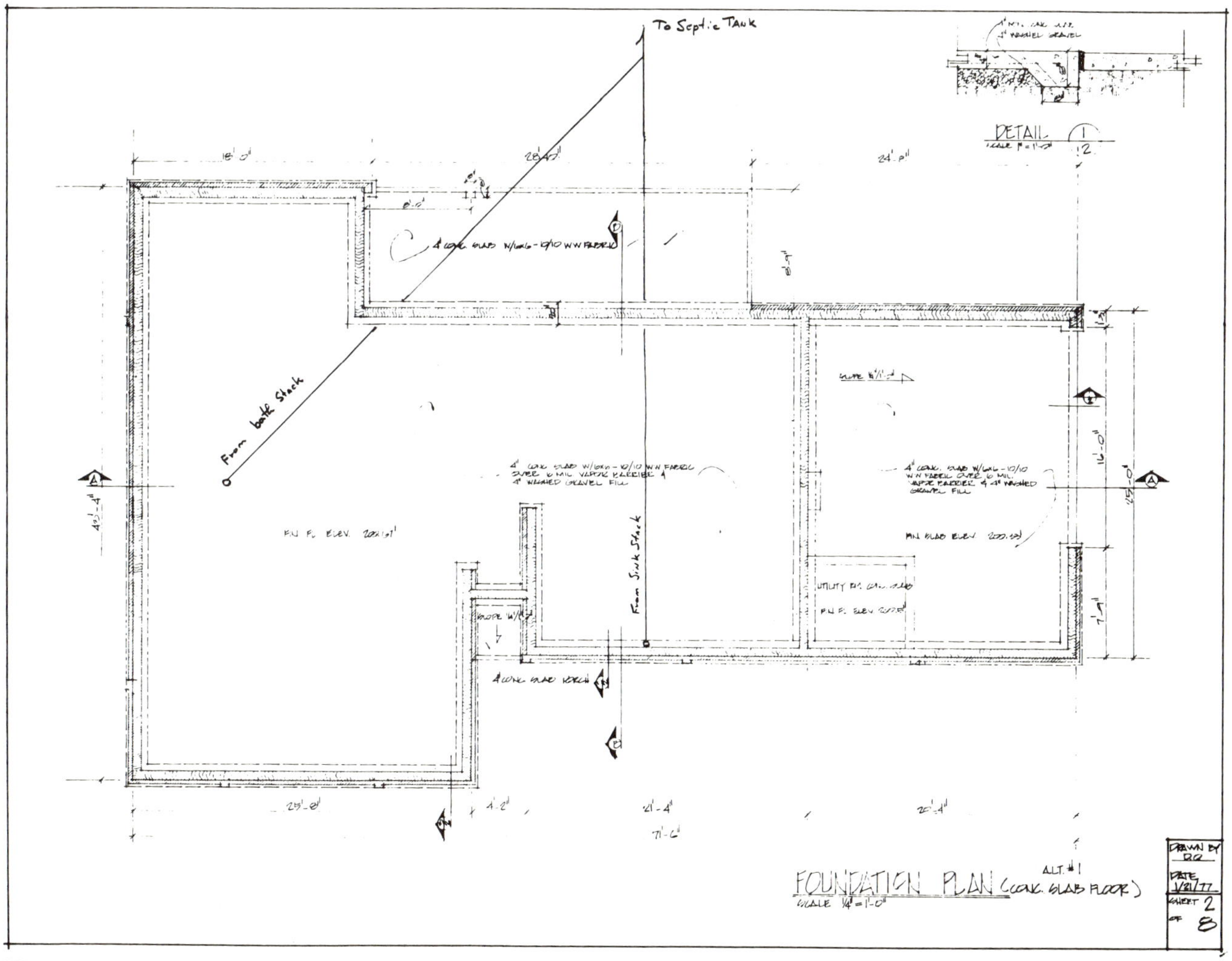

Figure 11-24. Foundation plan with main sewer lines

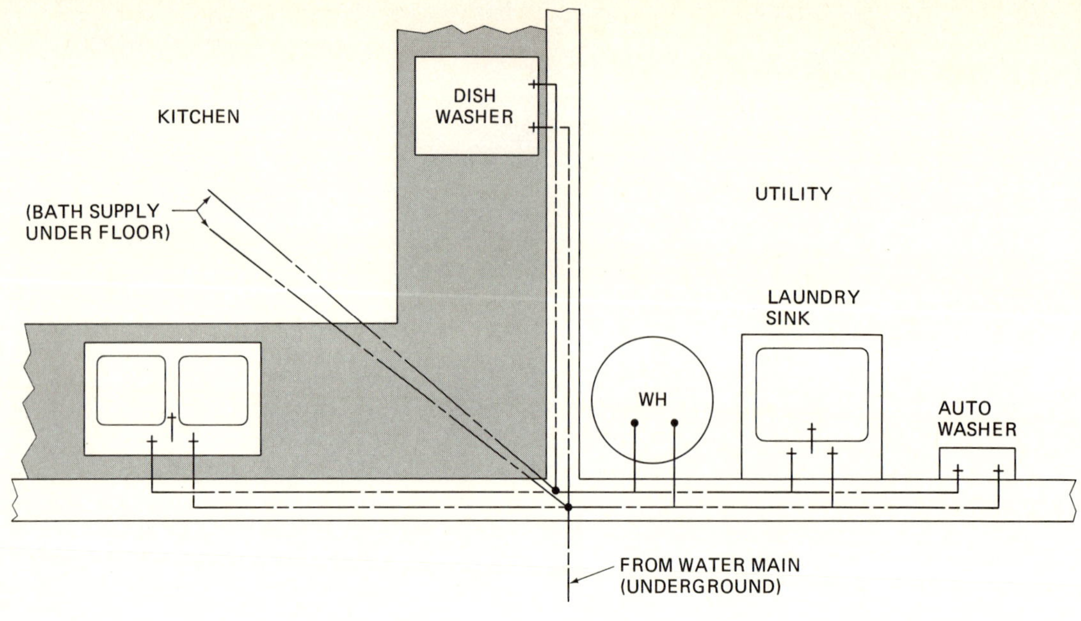

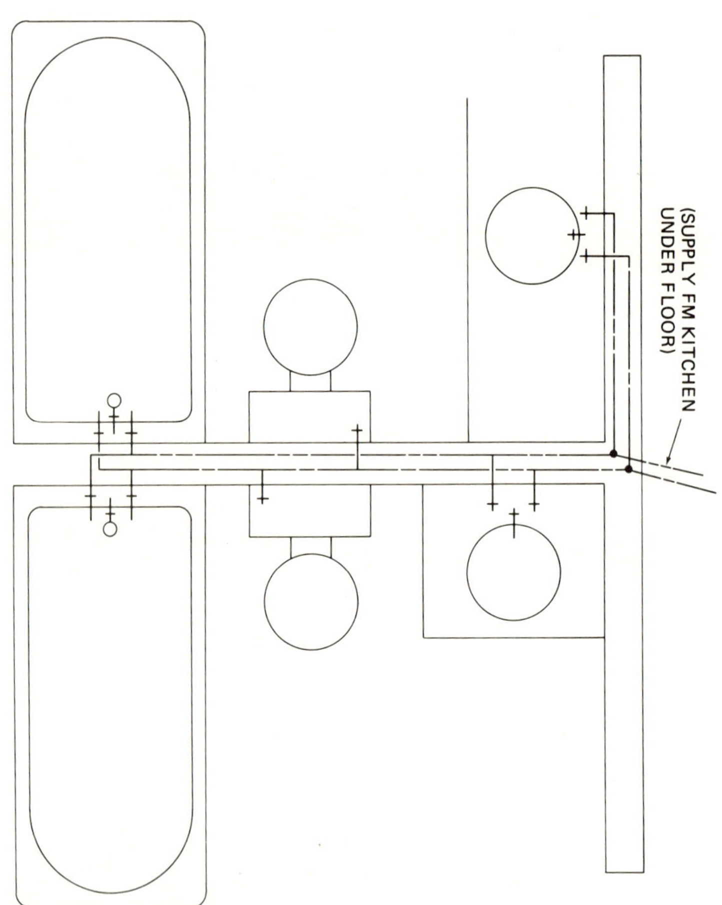

Figure 11-25. (a & b) Foundation plan with hot and cold water layout

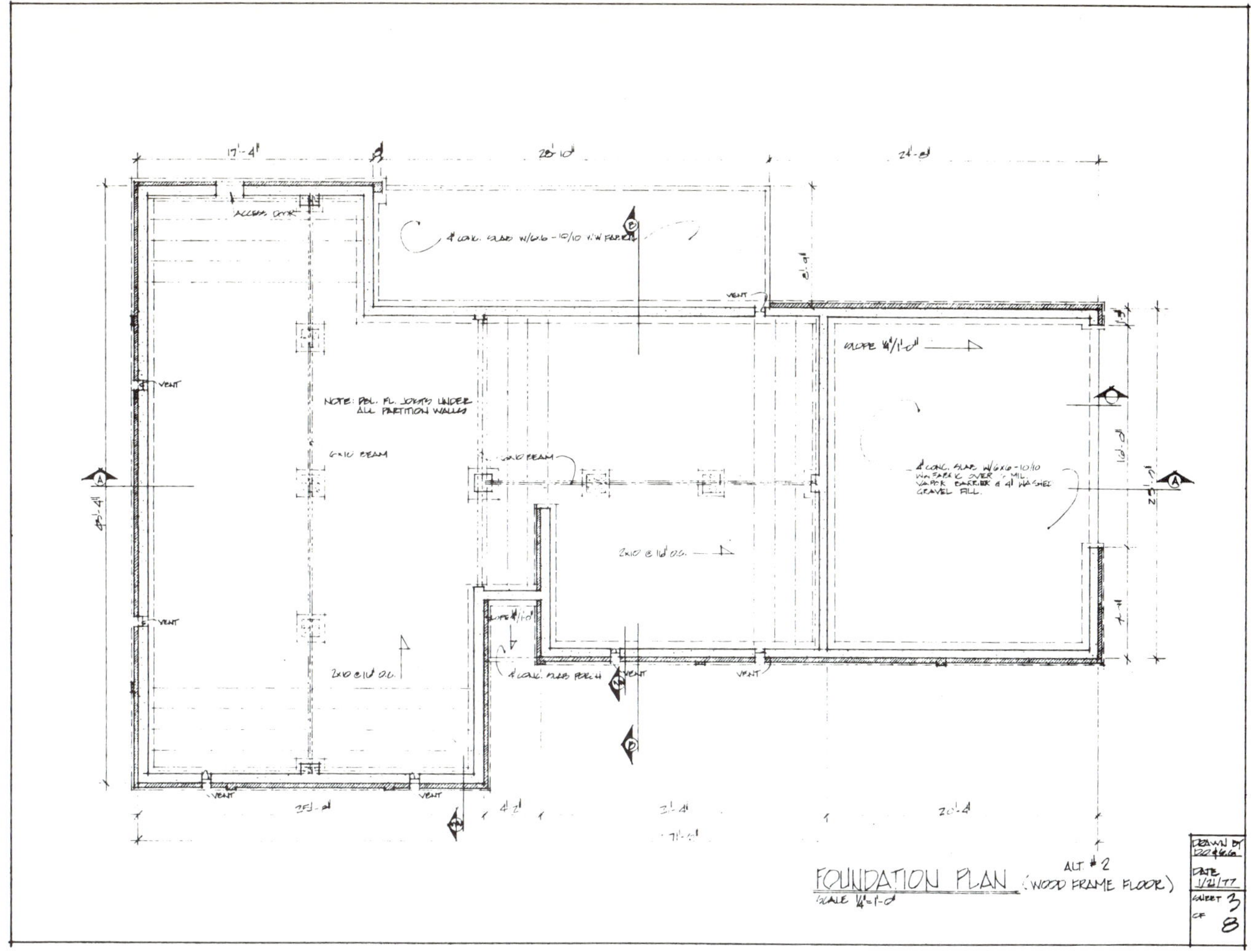

Figure 11-26. (a & b) Foundation plan with wooden floor

floor in the explanation above. If this building were to have a wood floor and crawl space, that foundation plan would be used. Figure 11-26 shows the difference between these two layouts.

• SUMMARY •

This chapter introduced you to blueprints and blueprint reading. Each of the different plans and details of an architectural blueprint was explained from a plumber's point of view. There are many different symbols on a blueprint which identify plumbing fixtures and appliances, as well as many for members of other building trades. The plumber should be aware of all symbols on the blueprint to eliminate errors and conflicts.

Use of the blueprint to make a plumbing layout can save the plumber a lot of time. It can also help in cost estimating a particular building and in ordering and furnishing proper materials. Examining and understanding a blueprint is a necessary tool for the plumber.

• WORDS PLUMBERS USE •

blueprint
plot plan
elevation line
grade line
foundation plan
cross-section line
detail line
floor plan
schedules
interfere
detail plan
elevation
longitudinal section
transverse section

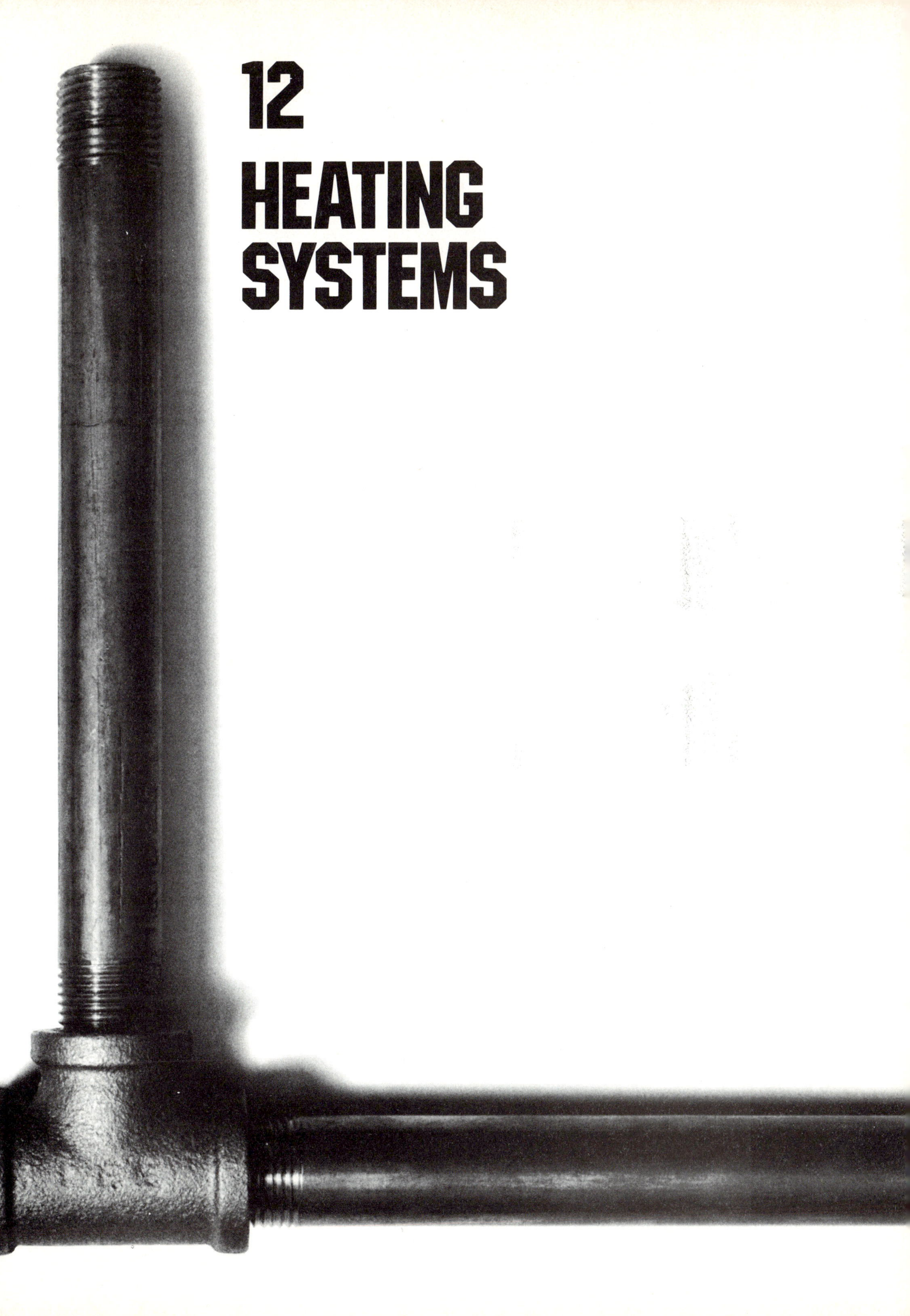

12 HEATING SYSTEMS

Early history tells us that the places in which people lived were heated by open-pit fires or by fireplaces. You have probably seen these types of building heating devices in the movies or in books or magazines which you have read. While these early heating systems worked, they were not very efficient. On a cold evening, the heat from an open-pit fire or a fireplace only heated a portion of the room, and much of the heated air was used in the combustion of the fire. Most of the warmed air went up the chimney with the smoke.

Present-day homes are heated primarily by one of three types of closed heating system, hot-water, steam, or warm-air heating. If there is a fireplace in the house, it is usually used for decorative or other purposes rather than for heating.

In many areas of the United States, the plumber also works with the heating system of a building. At some time in your future as a plumber, you may be called upon to repair or install one of these systems. This chapter will introduce the hot-water, steam, and warm-air heating systems.

At the completion of this chapter, you will be able to:

- Identify three common heating systems.
- Name the components of each system.
- Describe each system's heat-transfer function.

• THE HOT-WATER HEATING SYSTEM •

The hot-water heating system is a relatively low temperature heating system. The water used to transfer the heat from the place it is heated, the boiler, to the place where it heats the room air, the radiator, is at a temperature between 120°F [50°C] and 210°F [99°C]. We all should know that water boils at 212°F [100°C], so the water used within the hot-water heating system is below the boiling point at all times.

The heat-transfer function of the hot-water system is simple. Water is heated in a large boiler and moves through a series of pipes to one or more radiators, where the heat from the water warms the air in the room. Then the water, which has cooled off while in the radiator, is returned to the boiler to be reheated and go through the cycle again. Figure 12-1 shows a simple gravity-flow hot-water heating system.

Hot-Water Heating-System Flow

There are two types of hot-water heating systems. One (the gravity system) is an upward-flow gravity

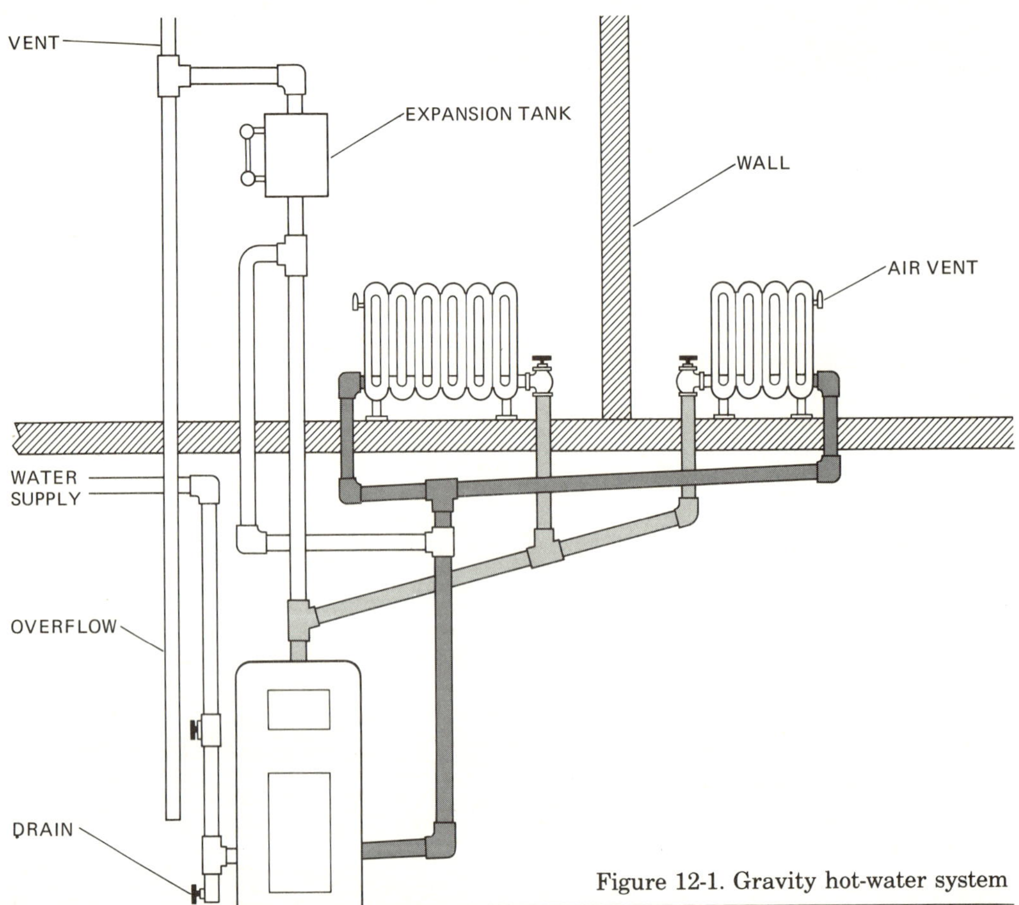

Figure 12-1. Gravity hot-water system

Figure 12-2. Forced-circulation hot-water system

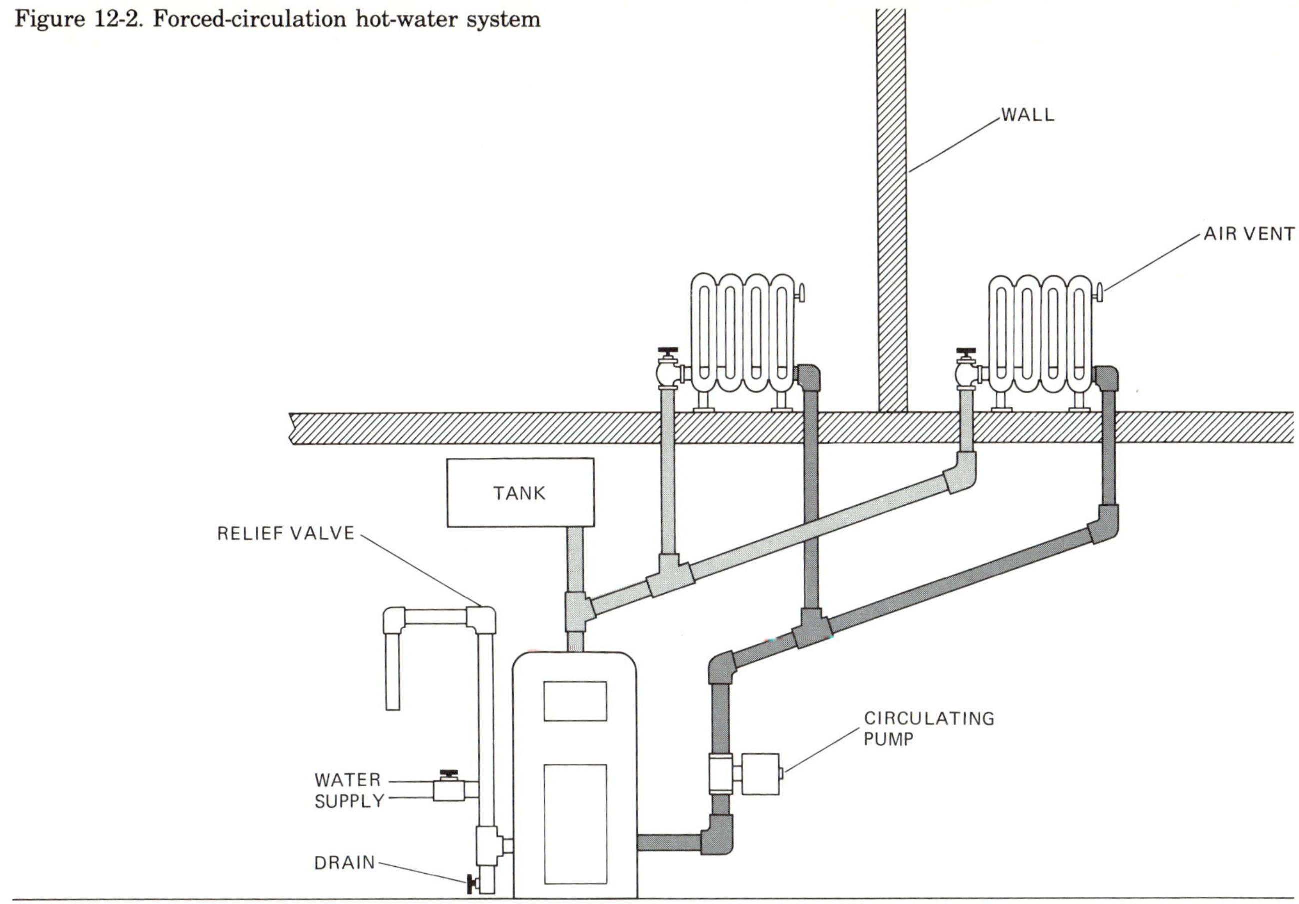

system which uses the basic principle of "warmer water rises up, cold water settles down." As the water in the boiler is heated, it rises up through the pipes and into the radiator. When the heat from the water is lost through the radiator, the water settles down through the return line and reenters the boiler. No pump is needed in this gravity-flow system.

The other type of system (the forced-circulation system) uses a pump to force the water through the pipes and radiators (Figure 12-2). This system is used when the size of the building requires a large boiler or when not all the radiators are higher than the heating source (boiler). This system requires that a pump be installed on the return line near the boiler. This pump circulates the water through the heating system and helps to maintain an even flow of hot water to each radiator.

There is one difference in these two systems besides the circulating pump. That is the location of the expansion tank. When water is heated, it expands, and the pressure built up by this expansion is enough to cause leaks, or maybe even split a pipe. One of the safety measures in a hot-water heating system is an expansion tank. This tank is put into the system to give the expanded (or overflow) water a place to go. The place where this tank is located can easily tell you if the system with which you are working is a gravity system or a circulating system. In the gravity system, the expansion tank is the highest point in the hot-water heating system. In the forced-circulation system, the expansion tank will normally be located near the boiler.

Hot-Water Heating System Boilers

The hot-water heating system boiler is made up of several individual cast-iron sections. These sections are put together by the plumber or the steamfitter using push nipples, asbestos rope or other fireproof sealant, and draw rods which secure or hold the sections of the boiler together.

There are different designs of boiler sections. Some have two equal push nipple holes, some have two unequal push nipple holes, and others have three push nipple holes. Figure 12-3 identifies three different designs of boiler sections.

Each design is made so that two or more sections can be bolted together to form a boiler large enough to provide enough hot water to heat a certain house. For a smaller house, a three-section boiler may be adequate; for a large house, the boiler may have five or six sections.

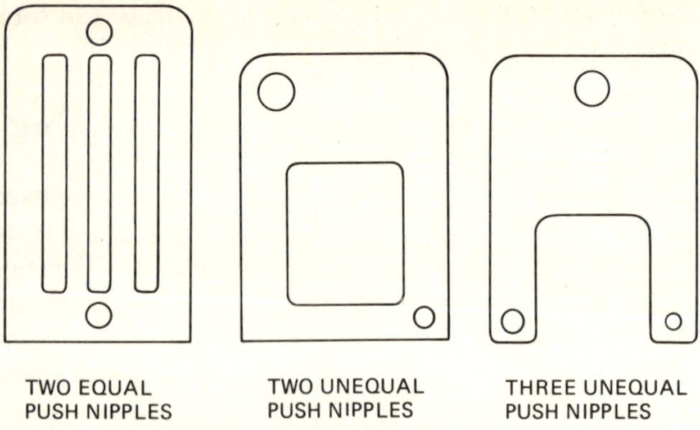

Figure 12-3. Types of heating-system boilers

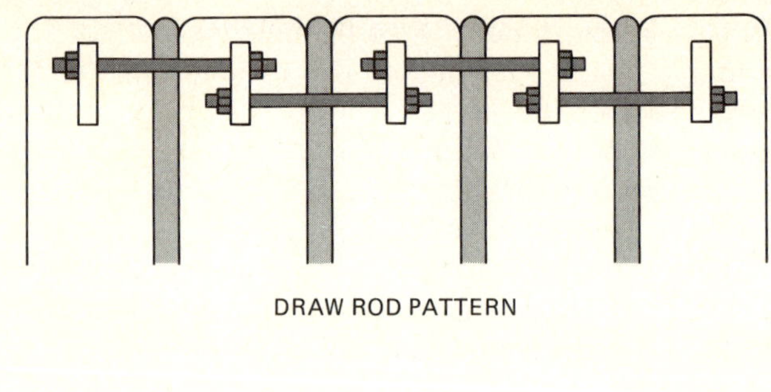

Assembling the Boiler

Each section of the boiler has push nipple holes. These are tapered holes which will seal tightly when the correct nipple is inserted and two sections are pulled together with the draw rod. Another sealing point on the boiler is the asbestos rope which is used to seal off the hot gases and the firebox from the outer shell of the boiler. Each section of the boiler is made with grooves which will hold the asbestos rope.

In assembling the boiler, the asbestos rope and the push nipples are inserted into their respective grooves or holes, and the draw rod is inserted. The nuts on the draw rods are then tightened, and the two or more sections are drawn tightly together (Figure 12-4).

Figure 12-4 also shows how hot flue gases are routed through the boiler. The heat source, burning coal, oil, or gas, is contained in the firebox, which is in the lower half of the boiler. The hot gases from the heat source heat the water in the boiler around the firebox. Then these gases are routed through the flue passages above the firebox on the way to the chimney. This routing of the flue gases also heats the water in the upper half of the boiler.

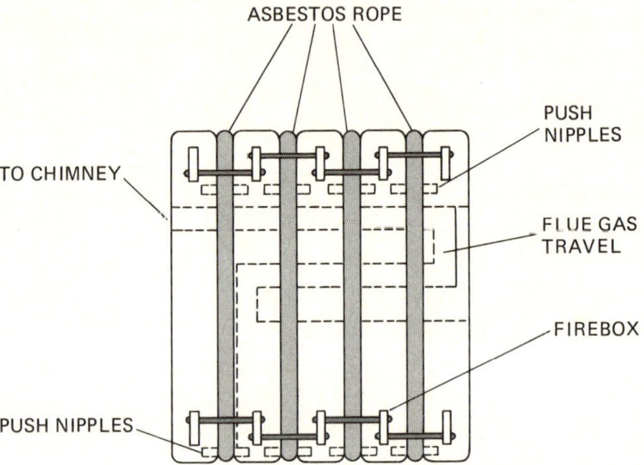

Figure 12-4. Joining heating-system boiler sections

Hot-Water Heating System Piping

Each hot-water heating system has two main lines, a hot-water flow main and a return main. The hot-water flow main carries the heated water to the radiator, and the return main moves the water from the radiator back to the boiler. These two mains are installed to have a very high pitch (steep slope). This pitch, in both cases, is to assist the water in its flow.

This pitch causes problems when the radiators are connected. The pipes which come up through the floor are affected by the slant, or pitch, of the main. To reduce this problem, the lines to the radiators from the hot-water flow main and the return main use two 90-degree ells (Figure 12-5). These two fittings allow the pitch to be removed before the line comes up through the floor.

The type of pipe normally used in a hot-water heating system is the black-painted (not galvanized) pipe. The system is always filled with water, and while there is some rusting, the black-painted pipe will normally last as long as the building. The fittings used in this system are cast iron, which also provides long service.

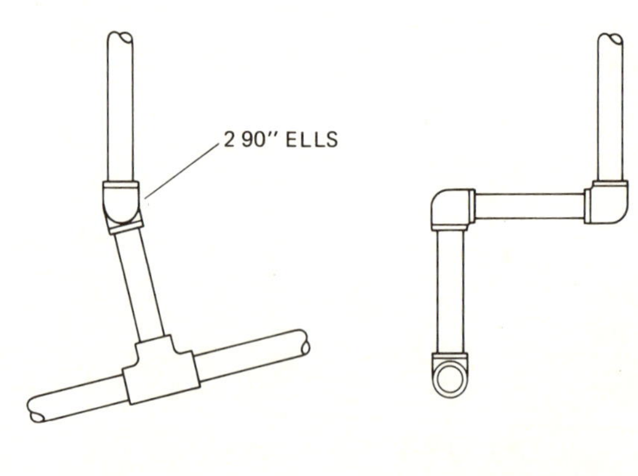

Figure 12-5. Adjusting vertical pipes for pitch

Two other items normally included in the hot-water heating system are the valves and the air vents. An on/off valve is normally placed on the inlet side of the radiator so that the hot-water flow to that radiator may be stopped if heat is not needed.

The air vents are located on each radiator so that any air in the radiator can be removed. If the radiator is half filled with air, the hot water can heat only half of the radiator. If this happens, the heating thermostat in that room will keep fuel going to the heat source (in the boiler) trying to warm the room. Air in the radiator can be an expensive waste of fuel and other energy.

• GRAVITY STEAM-HEATING SYSTEM •

The gravity steam-heating system is a more efficient system than the hot-water system. As you know, water boils at 212°F when it is in a pan on the stove. In a gravity steam-heating system, the water in the boiler is heated so that it boils, making steam. This steam is lighter and hotter than the water, so it will travel up through the heating system pipes and up to the radiator.

When the steam is in the radiator, it turns back to water, or condenses, heating the radiator and the air around it. When the steam has changed back to water, it flows through the return line down to the boiler, where it is reheated to steam.

There are two main types of gravity steam-heating systems commonly used, the one-pipe system and the two-pipe system. Figures 12-6 and 12-7 show these two systems.

The boilers used in the steam-heating system are not very different from those in a hot-water system. The main difference in the two systems is the temperature of the water in the boiler.

Gravity Steam-Heating-System Piping

You can see from Figure 12-6 that the piping used in the gravity steam-heating system is similar to that in the hot-water system. A major difference is the size of the pipes used. Since both the steam and the water (condensed steam) must use the same pipe in the one-pipe system, the diameter of the pipe must be larger.

A difference can also be noticed in the two-pipe system (Figure 12-7). There is a valve on each side of the radiator, one for the steam pipe and one for the return line. These valves must be present so that the radiator can be turned off, since there is some steam present in the return line when the system is operating.

Another difference is the type of air vent used on the radiator and in the system. The vent used in a steam system is an automatic vent; it allows air to escape but closes when steam tries to get out. These air vents serve the same purpose as the vents in a hot-water system, that is, to get the air out of the system so there is no loss of heat transfer by the steam.

Safety devices which should be used in the steam-heating system are the pressure gauge and the pressure relief valve. The pressure gauge will simply indicate the amount of pressure in the system. The steam-heating system can build up a lot of pressure if a problem occurs in the heating control unit, enough pressure to rupture or blow up the pipes or other parts of the system. The installation of a pressure relief valve at the boiler will allow the ex-

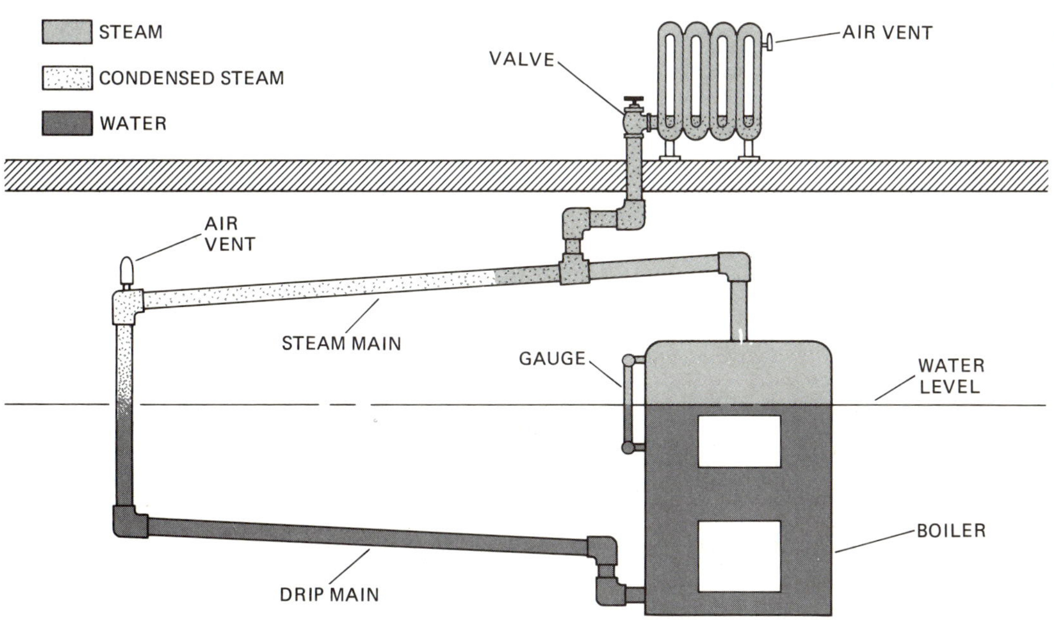

Figure 12-6. Simple one-pipe steam-heating system

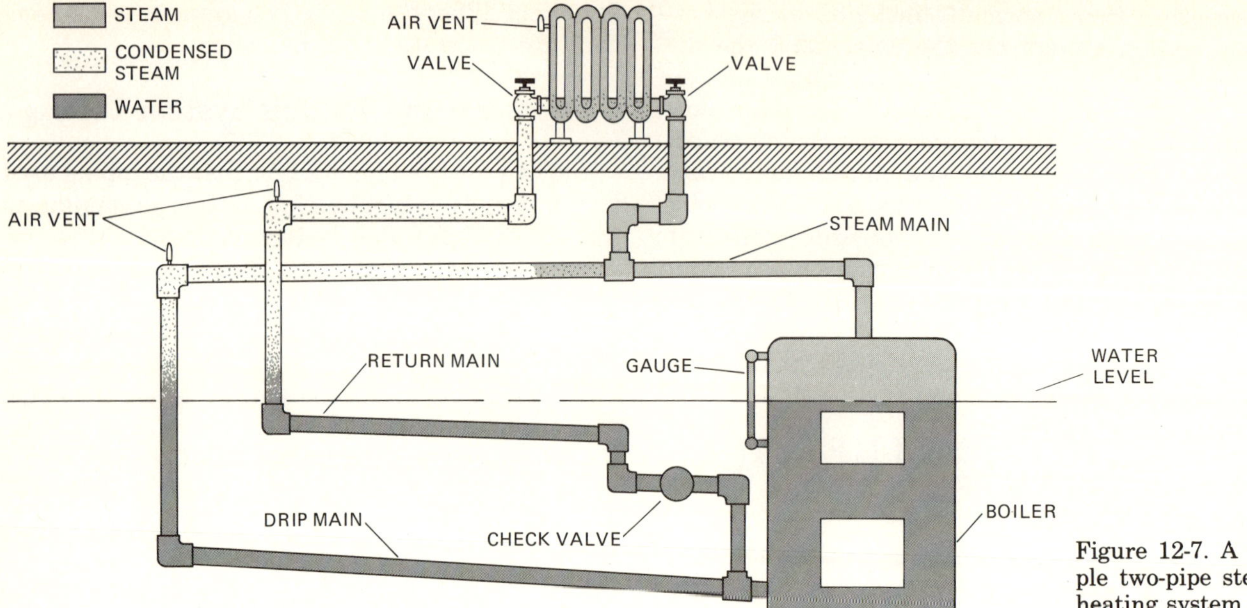

Figure 12-7. A simple two-pipe steam-heating system

cess pressure to bleed off and will also warn the people in the house of the problem.

• THE WARM-AIR HEATING SYSTEM •

The warm-air heating-system uses air to transfer the heat from the furnace to the room. The air is heated in the furnace and then blown through a series of large pipes called ducts. These heating ducts, which look like a stovepipe, are just like the pipes in the hot-water or steam-heating systems. These ducts transfer the warm air into the rooms to be heated and remove the cold air. The cold air is returned to the furnace, heated, and recirculated throughout the house.

There are two types of warm-air heating systems, the gravity warm-air heating system and the forced warm-air heating system. The difference between these two is about the same as the difference between the two hot-water heating systems; remember, the forced-circulation hot-water system used a pump. The difference here is that the forced warm-air system uses a large fan to blow the warm air through the ducts and into the rooms. We will discuss both of these warm-air systems in the following pages.

The Gravity Warm-Air System

The gravity warm-air system works on the same basic principle as the gravity hot-water system: "warm air rises, cold air settles down." Figure 12-8 shows the circulation pattern of the warm air in a gravity warm-air heating system.

You can see that the hot air is heated around the fire in the furnace and moves up through the ducts into the room. Inside the room, the warm air spreads, warming the room. As the warm air cools, it settles down toward the floor and out of the room through the cold-air duct. This cold-air duct returns the air to the furnace, where it is heated again and rises into the room.

There are some problems with the gravity heating system. One problem is that there is no filter in the system. Warm air rises slowly without much force or speed. This slow speed does not allow the use of a filter to remove dust or other dirt from the warm air.

Another problem is that the area near the outside walls may be rather cool. The warm air moves slowly, and long duct runs would allow the warm air to cool before it reached the rooms. Short warm-air ducts must be used to supply warm air to the rooms. This requires the room outlets to be near the inside walls, and by the time the warm air circulates through the room to the outside walls, it has already started to cool and may leave that part of the room slightly cooler.

The Gravity Warm-Air Furnace

The furnace used with the gravity warm-air system can use any combustible heat source, wood, coal, gas, or oil, and is made up of two parts, the firebox and the plenum. The inner part is the firebox, where the fuel is burned. This firebox has an outlet which is connected to the chimney so that the smoke and hot gases are not released into the building.

The second part of this furnace is the plenum, where the air is heated before it rises into the rooms. This plenum is just a large jacket around the firebox.

Figure 12-8. Gravity warm-air heating system

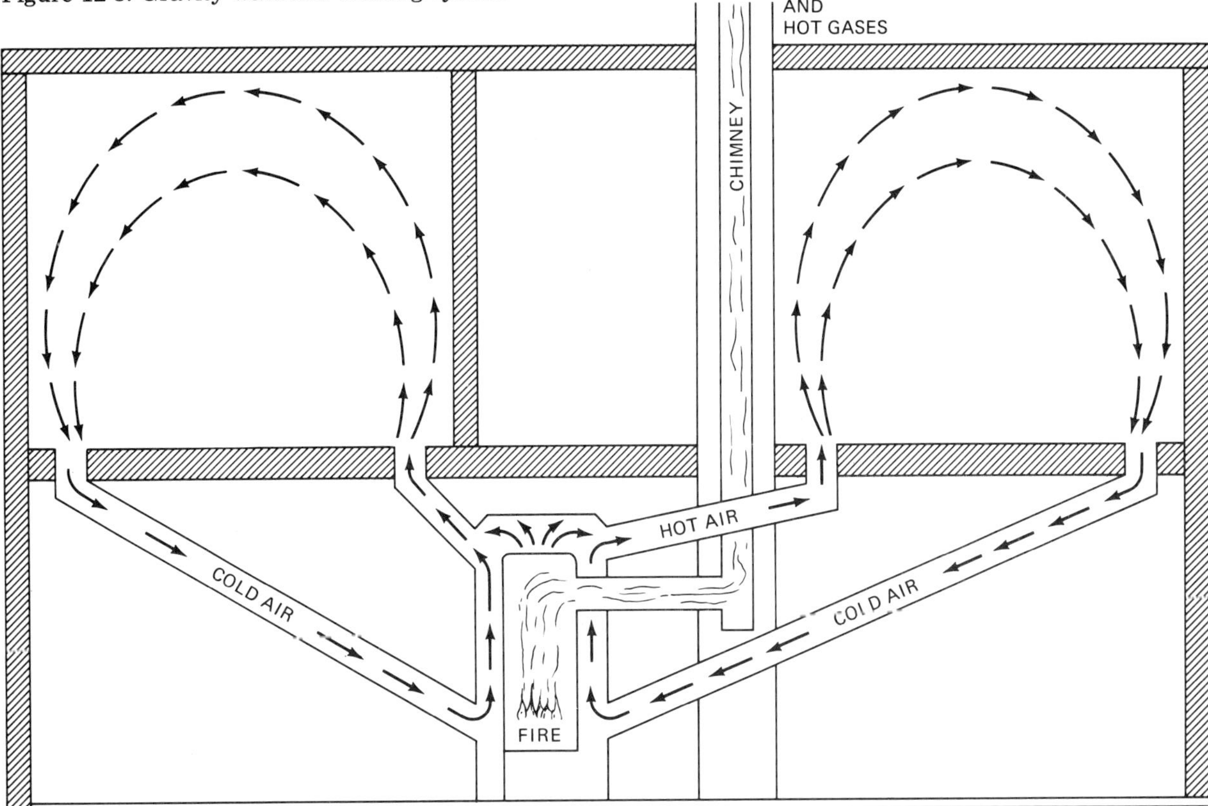

The upper part of the plenum has outlets which connect with the ducts to allow the warmed air to rise into the heated areas. The lower part of the plenum has inlets which connect to the cold-air-return ducts (Figure 12-9).

The Forced Warm-Air Heating System

The forced warm-air system is different from the gravity system because it uses a fan to force the air through the furnace and the ductwork. It is probably the most popular heating system used today. One of the reasons for its popularity is that this system can use many different heat sources. While the standard heat sources (wood, coal, gas, or oil) can be used, many new buildings have a forced warm-air system that uses such other heat sources as electric coils, heat pumps, and heat-transfer modules. In the future, some solar-heating applications are likely.

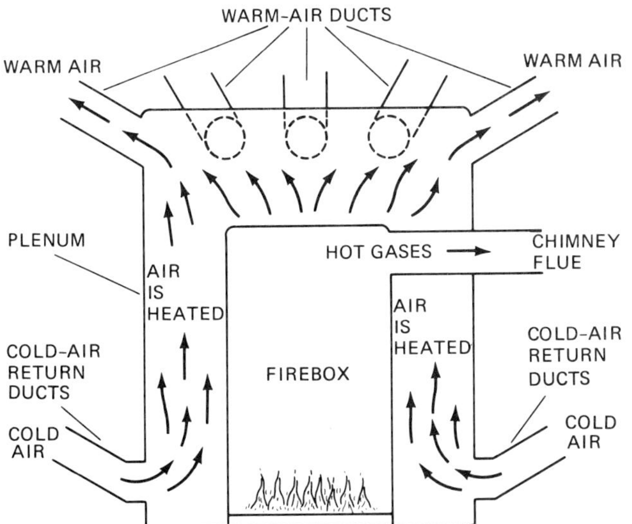

Figure 12-9. Gravity warm-air heating system furnace

The forced warm-air system uses ducts similar to those used in the gravity system. A forced warm-air heating system was used in the house which we studied in Chapter 11. That system had a warm-air supply duct and a cold-air return duct, but if you remember, the warm-air outlets were placed near the outside walls. Since the warm air is forced through the ducts and house by a fan, longer duct runs can be used. This heating system provides more even heat throughout the rooms because the room heating outlets are located near the colder outside wall and the cold-air return is located near the warmer inside walls.

Since the warm air in this system is driven by a fan, a filter and a humidifier can be used. The filter traps dirt particles and helps to keep the circulating air free of dust and other material. A humidifier puts moisture in the heated air and helps to stop furniture from drying out and the buildup of static

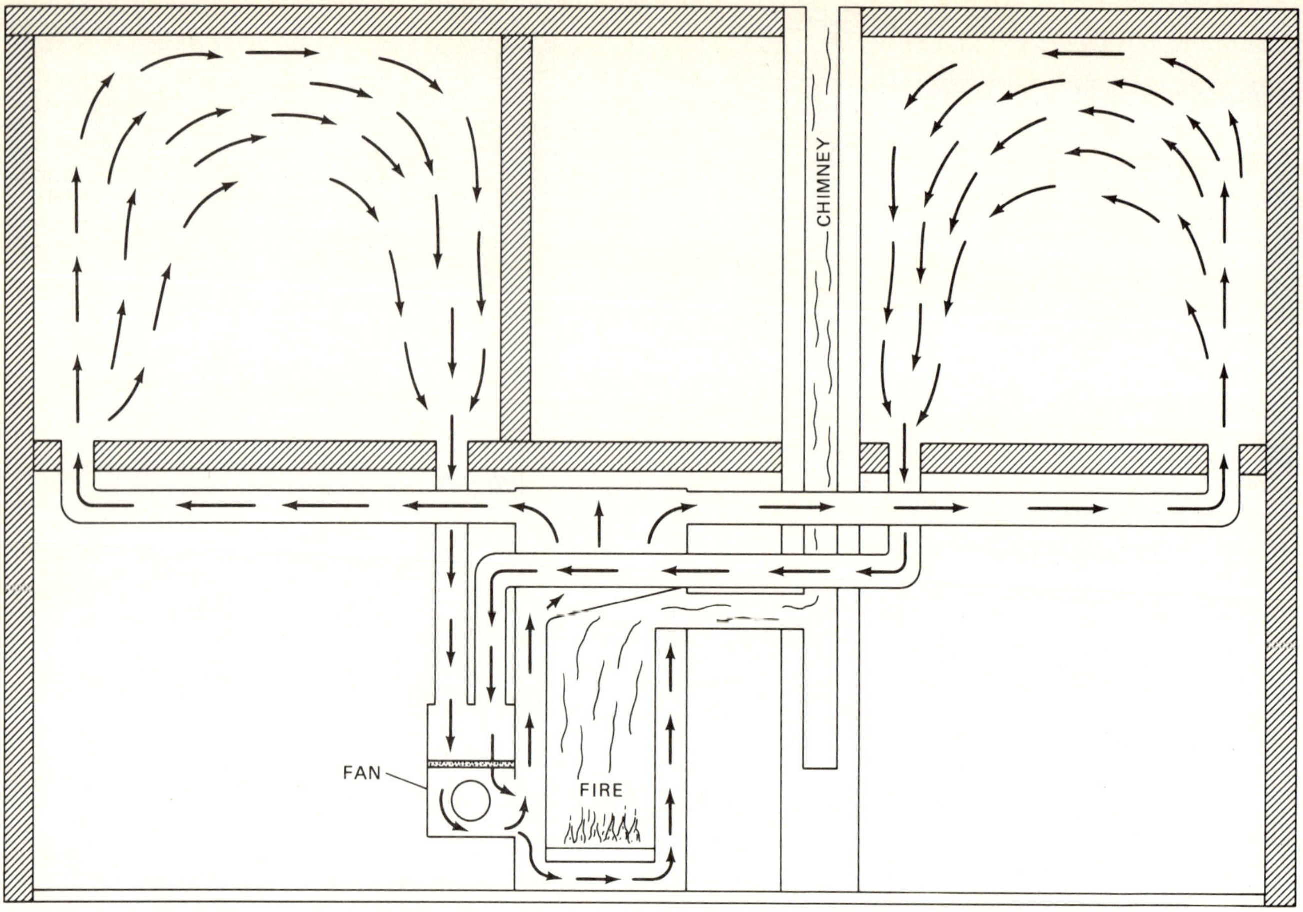

Figure 12-10. Forced warm-air heating system

electricity, two problems which are very important in cold climates. Figure 12-10 shows a simple forced warm-air heating system.

The Forced Warm-Air Furnace

The furnace used in the forced warm-air system has three parts, the fan, the heat exchanger, and a filter area. Figure 12-11 is a diagram of a forced warm-air furnace showing the three main parts.

The cold air from the return duct is pulled down through the filter by the fan and then forced up through the heat exchanger. This type of furnace has the cold-air intake on the side near the bottom, and the warm air is forced out through the top. This type is called a highboy. There are some types of furnaces which have the warm-air and cold-air outlets on top (lowboy type). Others have the incoming cold air on top and the warm-air exhaust on the bottom (counterflow type). The type of building and ductwork will identify the type of furnace to be installed.

Look back at the blueprints in Chapter 11. If the heating ductwork is installed in the crawl space below the building, a furnace with the cold-air return on top and the warm-air outlet on the bottom (counterflow type) would be the best choice. If the ductwork is installed in the attic, the best choice would be the furnace with the cold-air intake on the bottom and the warm-air outlet at the top (highboy type).

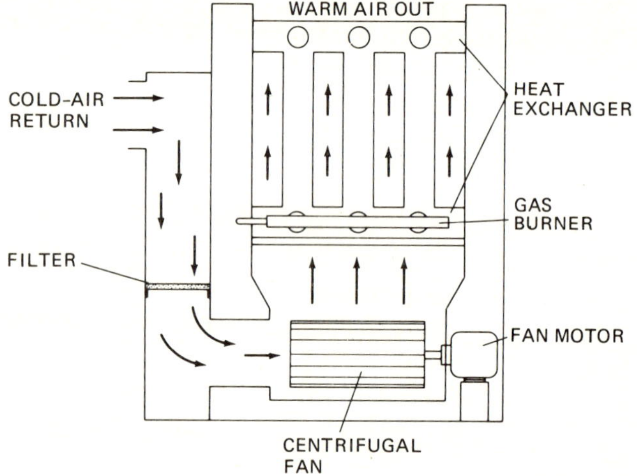

Figure 12-11. Forced warm-air heating-system furnace

The heat exchanger is basically the same for each of the types of furnaces, and they work on the same principle as the boiler used in the hot-water system. The hot gases and smoke from the burned material rise up through a series of tubes, warming them. The air being used to heat the house is forced around

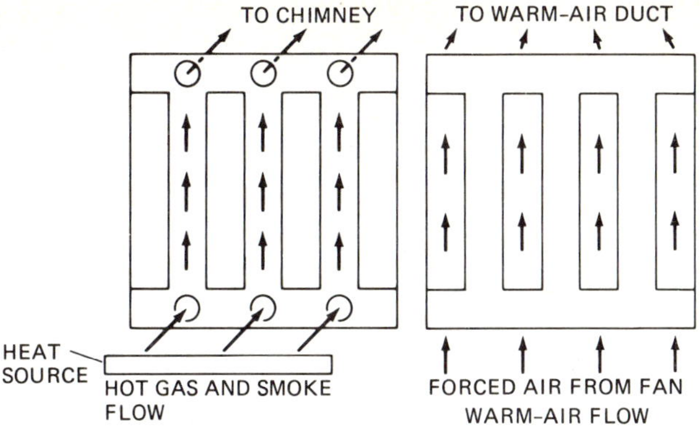

Figure 12-12. Forced warm-air heat exchanger

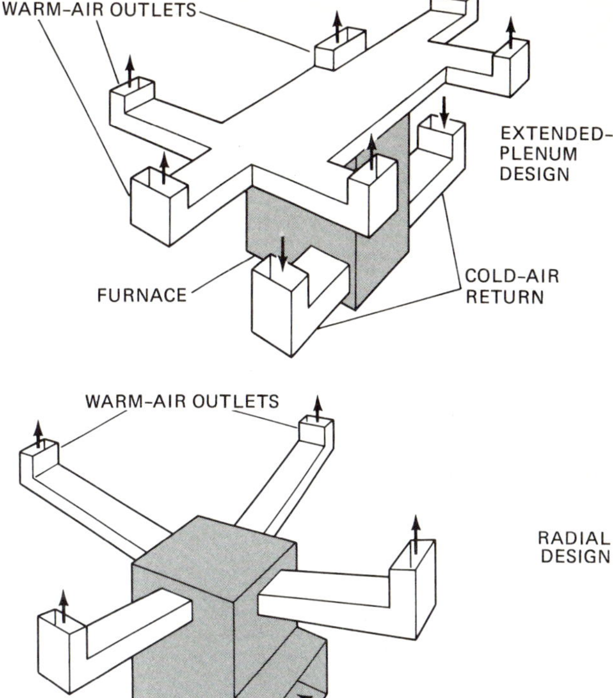

Figure 12-13. Two forced warm-air heating-system designs

these hot tubes, warmed up, and then forced out through the ductwork. Figure 12-12 shows a heat exchanger and the different airflow patterns.

Duct Systems

The duct system connects the warm-air outlets in the rooms with the furnace. This duct system can be installed in the attic, in the basement, in a crawl space below the house, or, in some cases, in the floor of a cement slab building. There are two basic designs used when planning the duct system, the extended-plenum design and the radial design. Figure 12-13 shows these two designs.

The extended-plenum design is the type which was used in the blueprints of Chapter 11. It is made up of a large, long, single duct (extended plenum) which gets smaller each time a room supply duct is connected. This reduction of the extended plenum provides an even flow of warm air to each room. This design uses both rectangular and round ducts. The cold-air return duct has two inlets in this case, but many extended-plenum designs use only one inlet.

In the radial design, each room outlet is supplied with warm air directly from the plenum on the furnace, generally using the less expensive round ducts. This also provides an even flow of warm air to each room, but it may require some long duct runs. In this radial design, a single cold-air return was used and was located at the furnace. This design is typical of the forced warm-air heating system used in a small apartment or single-level residence.

The gravity warm-air heating system and the forced warm-air heating system introduced here are just two of the warm-air heating systems in use today. However, the basic principles of heat transfer and warm-air circulation learned by studying these two systems can be applied to most of the other warm-air heating systems.

• SUMMARY •

This chapter introduced you to the three most common heating systems, the hot-water system, the steam system, and the warm-air system. You should be able to list the parts of each of these systems and identify the method of heat transfer and the circulation of the heat throughout the system and the building. While the systems described in this chapter are not the only heating systems in use today, the principles of heating and heat transfer studied here can be generally applied to most of the other systems used.

• WORDS PLUMBERS USE •

radiator	boiler	pitch	filter
function	push nipple	pressure relief valve	humidifier
gravity flow	asbestos	firebox	extended-plenum design
circulation	routed	plenum	radial design

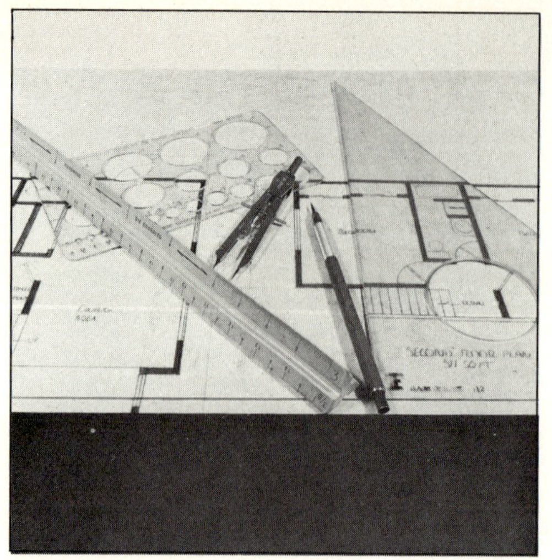

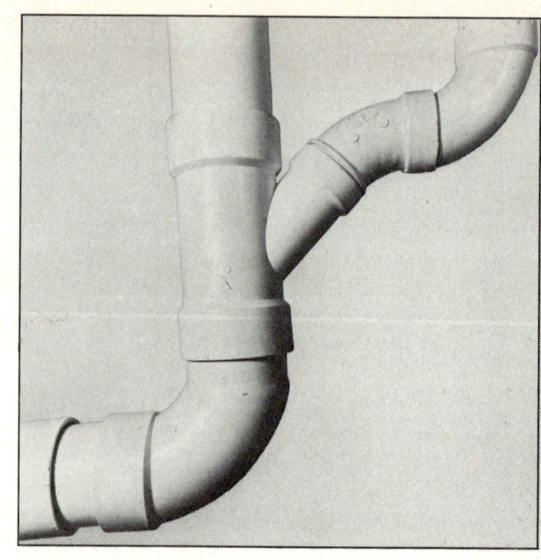

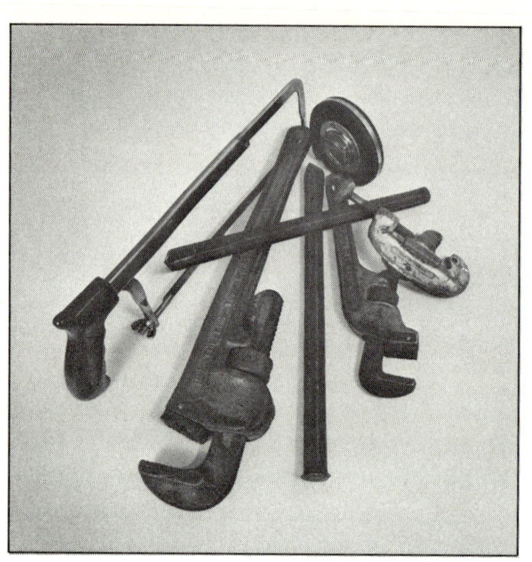

GLOSSARY

Acetylene A gas which will burn at high temperatures. It is combined with oxygen for use in oxy-acetylene welding.

Allen wrench A wrench used to tighten set screws and flush head bolts.

Apprentice plumber An inexperienced person who is learning the plumbing trade through on-the-job training with a journeyman, in addition to attending classes. The apprenticeship lasts 5 years.

Aqueduct A large pipe constructed to move water from one place to another.

Architect A person who designs and draws up the specifications for a building and supervises its construction.

Arc welding A method of welding which uses the heat of an electrical current jumping an air gap to melt and combine two pieces of metal.

Area *(A)* The number of square inches or square feet that one of the sides, bottom, or top of an object would take up. The formula is expressed as $A = R \times H$. For finding the area of a circle, the formula is $A \times \pi \times R^2$.

Asbestos A nonflammable material often used in the construction of electrical or heating systems.

Aviation snips Hard snips used to cut sheet metal or chimney flue pipe.

Bacteria Microscopic organisms that exist in a variety of places, including water. Harmful bacteria must be eliminated from municipal water supplies through proper water purification treatment.

Ball cock The water-supply valve in a commode's water tank.

Ball peen hammer A hammer used with star drills or chisels; heavier than the claw hammer, usually has a 32- to 48-oz head.

Bar hanger A fixture used to hang porcelainized cast iron sinks from the wooden backing built into the wall.

Biological process With respect to sewage treatment, this refers to the use of a trickling filter — a filter bed of gravel covered by slime, through which water flows — to remove dissolved materials.

Black pipe Steel pipe that is sprayed or dipped with black paint to prevent rust.

Blueprint A detailed, scaled drawing that shows the layout and includes the specifications for the complete construction of a building.

Boiler Vessel where fuel is burned to heat water or make steam for the heating system.

Box-end wrench A nonadjustable wrench that fits nuts or bolts; available in USCS (U.S. Customary) and metric sizes.

Bracket hangar A fixture used to hang vitreous china sinks from the wooden backing built into the wall.

Caulking iron A tool used in finishing a soil-pipe joint.

Centrifugal force The phenomenon of force on a body in spinning motion. This phenomenon is the principle behind the centrifugal pump, in which the main section spins around quickly, forcing the water or other liquid out.

Check valve A valve that allows water to flow in one direction only.

Circumference (*C*) The distance from any point on the circle all the way around the circle, stated as $C = \pi \times d$.

Claw hammer A hammer used for driving and removing nails, usually having a 16- to 20-oz head.

Cleanout An opening in a main sewer line that has a plug or cap which can be removed to allow access to a stoppage in the line.

Cold chisel Chisel used for chipping or cutting concrete, metal, or other hard materials. Has a short, blunt cutting edge.

Combination fittings Fittings that permit the joining of copper pipes to steel pipes.

Commode (water closet) The plumbing trade's term for the toilet.

Compass A V-shaped drafting tool with a metal pivot point on one leg and a pencil point on the other, used to draw circles or circular arcs.

Cross-section line Indications on a blueprint of an area that is shown in a separate cross-sectional or cutaway drawing.

Cutaway drawing A picture of an assembled object but with the side or top of the object sliced away. This gives the drawing a "cut-away" appearance so that the interior parts can be seen.

Cylinder A long, round object, which has the dimensions of radius (*r*), diameter (*d*), and height (*h*).

Decaying process The course through which waste in a pit undergoes decomposition, aided by the use of quicklime.

Detail line Similar to a cross-section line; shows the area on a blueprint that is covered in a separate detail plan.

Detail plan A subordinate drawing on a blueprint that details a particular area or construction aspect.

Diameter (*d*) The distance of a line going from one side of a circle through the center to the other line; the width of a pipe.

Discharge port The point in a pump at which liquid is released to a storage area.

Divider A V-shaped drafting tool with a metal point on each leg, used to transfer measurements from one drawing to another.

Drain and overflow type A waste connection in a lavatory that uses an additional drain near the top of sink to prevent overflow. It is connected between the plug top and the trap through a T-connection.

Drainage field A network of perforated pipes which disperse liquid waste from the septic tank.

Drainage fitting A type of fitting that must be used with steel-pipe sanitary systems which helps prevent stoppages in the line.

Drum trap A trap used to connect bathtubs to the sewer line.

Dual-control faucet A washerless faucet with two handles; one is used for cold water, and the other is used for hot water.

Elevation drawing A blueprint drawing which shows the scale of the sides, front, or rear of a building.

Elevation line A line on a blueprint that shows the height of the ground. Each line has a fixed value and follows the ground wherever it is at that level.

Engineer A person trained in the application of scientific principles to practical ends such as the construction of buildings, bridges, and so on.

Exploded drawing A drawing which shows all the different parts of an object as they would fit together along a central common axis to form the whole object.

Extended-plenum design A warm-air heating duct design that uses one long, tapering duct (the extended plenum), to which the room supply ducts are attached.

Fiber pipe Also called *Bituminous Pipe* or *Orangeburg Pipe.*

Filtration A method of removing sediment by passing the water through a series of materials that trap the mixed solids, the dissolved solids, and many germs and bacteria.

Firebox The place in the boiler where the fuel is burned.

Flaring tool The device used to spread or flare the end of the pipe.

Float ball Part connected to the ball-cock valve, which opens and closes depending on the water level in the commode's water tank.

Floor plan The drawing that identifies room, fixtures, and details of the building's layout.

Foundation plan A blueprint drawing which shows the dimensions and layout of the base of the building being constructed.

Fusible plug A plug made of a metal which has been manufactured to melt at a certain temperature. Fusible plugs are found in fire extinguisher sprinkler systems and will usually melt at 160°. The melting of the plug activates the particular sprinkler.

Galvanized pipe A pipe that has been coated with zinc, to make it rust-resistant.

Gasket joint Used instead of a caulked lead joint if allowed by the local building code. A neoprene gasket fits inside the soil-pipe hub and forms a water tight seal when connected to the spigot end.

Gate valve A valve in a building's water supply line which uses a brass gate that is raised or lowered across an opening in the valve body to control the water flow.

Glazed-tile pipe Also called *vitrified-clay pipe*: Pipe that is made of clay, which is then glazed and fired to make it hard and moisture-resistant.

Globe valve An in-line compression valve.

Grade line Also called an *elevation line.*

Gravity In physics, the pull exerted by the earth that makes an object fall. Because of the law of gravity, a liquid cannot flow uphill, and therefore, mechanical means (a pump, usually) must be used to make this possible.

Gravity-flow system Heating systems which work on the principle that warmer water (or air) rises, while cooler water (or air) settles.

Ground key valve A type of valve used for street cocks that consists of a stem with a hole through it, which fits in the valve body. When the hole through the stem is lined up with the pipe, water flows; when lined across the pipe, the flow stops.

Hanging line A main sewer line that will be suspended below the ceiling in a building's basement.

Hub end End of a soil pipe with the large bell (*the hub*).

Humidifier A device used to add moisture to heated air in forced air heating systems.

ID Abbreviation for inside diameter, the distance across the inside of a pipe.

Impeller The main section on a centrifugal pump, which spins and exerts centrifugal force on a liquid. (See Figure 3-5.)

In-line compression valve Also called a *globe valve* uses a neoprene or rubber washer which is compressed onto the valve seat to close off the supply.

Integrated drain and overflow type A waste connection for a lavatory that prevents overflow, but with the overflow hardware built into the plug top.

Isometric drawing A pictorial drawing that shows three sides of an object.

Joint runner rope Asbestos rope, usually 1 inch in diameter, that is used to seal a horizontal soil-pipe joint. It is placed around the pipe long enough for the molten lead to be poured in to fill the spaces in the joint.

Journeyman plumber A person who worked as an apprentice and passed the journeyman's test. He or she represents the average working plumber, for whom it is not necessary to enter the third stage.

Keyhole saw A specialized saw with a long, tapered blade. It is used to start a cut in a floor or wall.

Longitudinal section A blueprint drawing which shows the longest sectional dimension of a building.

Malleable steel Steel which can be curved or bent without breaking.

Master plumber A journeyman plumber who has passed an advanced, much harder test having already served 6 years as a journeyman.

Metric measuring system A measurement system in use by many countries based on the meter which is equal to 39.5 inches. Each unit in this system is either 10 times or 1/10 the size of the next.

Neoprene A soft, flexible, rubberlike material used in the neoprene gasket joint, which is sometimes used in place of the caulked lead joint.

Nipples Short pieces of pipe threaded on each end.

No-hub joint A soil-pipe joint which uses a neoprene sleeve and gasket, a stainless-steel sleeve, and two stainless-steel clamps.

Oakum A treated fiber used as a sealing agent in a soil-pipe joint.

OD Abbreviation for "outside diameter," the distance across the outside (and through the center point) of a pipe.

Offset pliers Pliers with blades that are "offset" to one side of the handles for a better hold on awkward parts.

OSHA Abbreviation for *Occupational Safety and Health Act* of 1970; a law enacted by the U.S. Congress to set up safety standards and practices for almost all the occupations in the United States.

Oxyacetylene cutting torch A torch that uses an oxyacetylene gas mixture. A cutting torch is larger and of different construction than an oxyacetylene welding torch.

Oxyacetylene welding Using oxygen gas and an acetylene gas flame to melt and combine two pieces of metal.

Perforated pipe A piece of pipe with many holes in the pipe walls.

Perspective drawing A type of drawing that uses the principle of the *vanishing point* to create a more realistic view of an object.

Pi (π) A Greek symbol meaning 3.14 which is used to find the area and circumference of a circle.

Pictorial drawing A drawing which shows the three sides of an object in one picture. Pictorial drawings can be of three types: isometric, oblique, and perspective.

Pipe cutter A tool used to trim pipe to the correct length. It makes a smooth, even cut.

Pipe fittings Apparatus which comes in many varieties, to change the direction of rigid steel pipe.

Pipe vise A vise used to hold pipe. It has V-shaped jaws with sharp teeth and has a latch or lock.

Pipe wrench An adjustable wrench with jaws that have sharp teeth. The jaws grip a pipe as turning force is applied.

Pitch The slant of heating pipes that helps the heating water flow to and from the radiators.

Plenum An area in a furnace where the circulating air is heated just before it rises into the rooms.

Plot plan A blueprint drawing also known as a site plan, which shows where a building and its associated utilities are to be located within a designated area of land.

Plug top A piece that fits into a drain hole from the top of the sink. Usually has a rubber or neoprene washer between the sink and the plug top.

Polyvinyl chloride (PVC) A type of plastic used in the construction of water-supply systems.

Porcelain Steel of cast-iron which has been coated with a glass-like liquid and baked.

Porcelainized cast iron Cast iron with baked-on porcelain (glasslike) finish. It is used for some kitchen sinks and bathtubs.

Pressure relief valve A valve in a steam heating system that keeps pressure from building up to or past the point at which pipes or the boiler would rupture or blow up.

Primary sewage treatment The first stage of the sanitation. At a primary treatment plant, the community's waste water is received and then solid particles are removed from it by means of sedimentation.

Purification A step beyond filtration, in the water treatment process, in which chemicals — chlorine and ozone, most often — are added to eliminate the remaining germs and bacteria.

Push nipple Nipple used to align and fasten boiler sections.

Quicklime The chemical calcium oxide used in sewage treatment to speed up the decaying process of waste as well as killing bacteria.

Radial design A warm-air heating duct design that connects the duct for each room outlet directly to the furnace.

Radiator A device used to transfer heat from an internal heat source to the surrounding cooler air.

Radius (*r*) The distance from the center to the edge of a circle.

Reamer A tool used to remove the lip left by a pipe cutter on the inside diameter of the pipe.

Safety hazards Dangerous conditions, tools, or working habits, which pose a threat to workers.

Scale A ruler divided into measuring units; these units will differ depending on the scale's use.

Schedules In drafting, listings of the various common parts used in constructing a building (such as doors, windows, room finishes, and so on) and their specifications.

Secondary sewage treatment In this phase of sanitation, a secondary treatment plant uses two stages of waste-water treatment, sedimentation, and a biological process to remove the dissolved material from the water.

Section drawing A drawing that shows the internal structure of a building in that it shows the foundation plan with its adjacent elevations, floor plan, and roof.

Sediment The combination of solids that are mixed but not dissolved in water, such as dirt.

Septic system An on-site sewage treatment system, which combines a large-capacity septic tank with a network of perforated pipes (drainage field).

Side cutting pliers Used for cutting wire, pipe straps, and other thin metal. Also called "electrician's pliers."

Single-control faucet A washerless faucet that uses a ball-and-socket to control water flow; has a single handle.

Siphon (siphoning) Removal of the water seal in a trap if the sewer is not vented.

Siphon-jet bowl A commode design that uses the siphon principle to help carry away the waste water.

Sludge The solid waste that is removed from raw sewage.

Soil pipe Cast-iron pipe used for sewer pipes inside a building.

Soil-pipe cutter Also called a *chain cutter* or *pipe*

cracker; used to cut soil pipe.

Solder A soft-metal mixture of tin and lead that is used to join copper pipes.

Spigot-end End of a soil pipe with a small head; fits inside the hub of a connecting soil pipe.

Stack A vertical (up and down) sewer line.

Star drill Chisel used to bore holes through concrete, brick, or tile. Has a star-shaped cutting tip.

Stop ball The device which opens and closes the commode's flushing valve.

Strap wrench A wrench used to tighten bright, plated metal pipe.

Street cock Valve that controls the flow of water from the outlet of a water main to the building site or lot.

Submersible pump A pump designed to work when it is completely covered with water and to shut off when the water level drops below a certain point.

Sump A watertight drainage pit located below the sewer level, used to collect waste water.

Sump pump A pump used to remove waste water from a sump and move it into a sewer line. Sump pumps can be of three types: the centrifugal pump, the water ejector, and the air-displacement ejector.

Swage joint A copper pipe connection made by spreading or flaring the end of the pipe.

Tailpiece Pipe fitting into the plug hole that connects the waste connection to the trap.

Template A precut guide, usually made of metal or plastic, used in drafting to reproduce certain symbols or shapes.

Terminal fixtures A category of plumbing appliances consisting of sinks, bathtubs, lavatories, showers, and so forth.

Tertiary sewage treatment This process consists of the steps used in the two previous treatments and a third, physical-chemical step, resulting in 99 percent removal of the sewage.

Thermostat A device which automatically regulates temperature by responding to changes in that temperature.

Thread joint Also called a screw-on joint, it uses threaded connections to join steel pipes.

Threading die The tool used to cut tapered threads in steel pipe.

Three-view drawing A picture that shows how an object looks from three sides, usually front, right and top views.

Transformer A deviced used to transfer electrical currents from one circuit to another.

Transverse section A blueprint drawing which shows the shortest sectional dimension of a building.

Trap (P-trap) A fixture that must be used in the waste connection of all sinks to seal off the sewer from the inside of the building. The P-trap is the most commonly used design and is shaped like a "U" to seal off toxic gases in the sewer line.

Two-view drawing A picture that shows how an object looks from two sides.

U.S. Customary Measuring System The standard system of measurement in this country, which will be converted to the metric system in the future. Units of measure commonly used by the plumber are the inch, foot, yard, and mile.

Valve Device to control the flow of water in a pipe.

Vermin Disease-carrying insects or animals, such as rats. Vermin can be blocked from entering the drainage system of a house by means of a trap.

Vertical sump pump A pump whose motor is above the sump pit, with a shaft leading down to the pump position at the bottom of the pit.

Vitreous china A material formed from clay that is fired in a kiln until it is hard, then glazed and refined. It is used for lavatories and other fixtures.

Vitrified Coated with a glasslike liquid and baked; as in vitrified-clay pipe.

Volume (*V*) The amount of space inside an object, expressed in cubic feet or cubic inches.

Volute The part of a centrifugal pump consisting of the channel, formed by the offset mounting of the impeller (see figure 3-5). Liquid flows through the volute after it has spun through the impeller.

Washdown bowl A commode design that uses the force of the water to carry away the waste water.

Weld To join pipe together using heat, pressure, or chemicals.

Welding electrodes A flux-coated steel rod or

some other solid electric conductor, used in arc-welding to conduct electricity, molten metal, and flux to the point being welded.

Welding rod Also called *filler rod*; used to add extra metal to the weld.

Wood chisel A tool with a sharp cutting edge used for chipping or cutting wood or soft metal.

Yarning iron A tool used in preparing a soil-pipe joint.

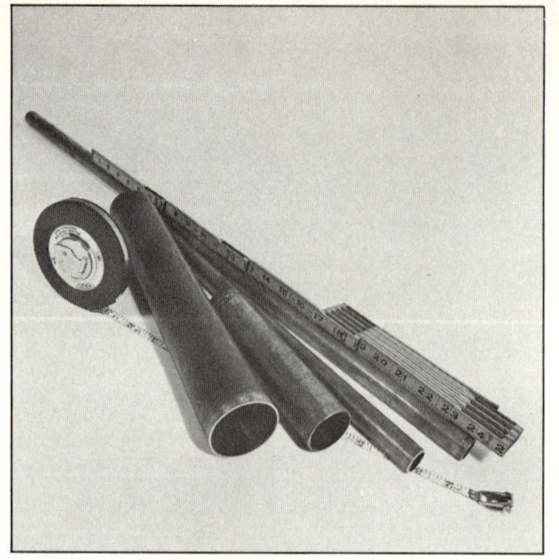

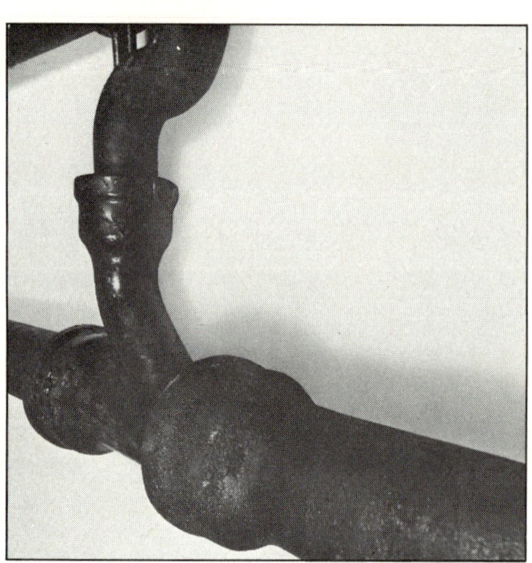
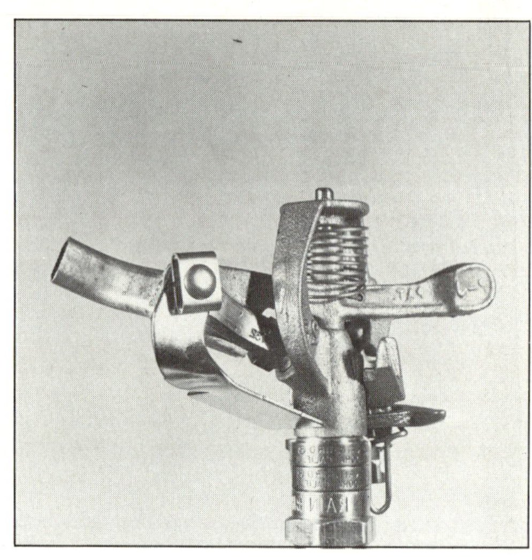

INDEX